Modeling and Modern Control of Wind Power

Modeling and Modern Control of Wind Power

Edited by

Qiuwei Wu

Technical University of Denmark, Kgs. Lyngby, Denmark

Yuanzhang Sun

Wuhan University, China

This edition first published 2018

Registered Office(s)
John Wiley & Sons, Inc., 111 River Street, Hoboken, NJ 07030, USA
John Wiley & Sons Ltd, The Atrium, Southern Gate, Chichester, West Sussex, PO19 8SQ, UK

Editorial Office
The Atrium, Southern Gate, Chichester, West Sussex, PO19 8SQ, UK

For details of our global editorial offices, customer services, and more information about Wiley products visit us at www.wiley.com.

Wiley also publishes its books in a variety of electronic formats and by print-on-demand. Some content that appears in standard print versions of this book may not be available in other formats.

Library of Congress Cataloging-in-Publication Data:

Names: Wu, Qiuwei, editor. | Sun, Yuanzhang, 1954- editor.
Title: Modeling and modern control of wind power / edited by Qiuwei Wu, Yuanzhang Sun.
Description: Hoboken, NJ : John Wiley & Sons, 2018. | Includes bibliographical references and index. |
Identifiers: LCCN 2017030336 (print) | LCCN 2017043924 (ebook) | ISBN 9781119236405 (pdf) | ISBN 9781119236399 (epub) | ISBN 9781119236269 (cloth)
Subjects: LCSH: Wind power–Mathematical models. | Wind power plants. | Wind turbines.
Classification: LCC TJ820 (ebook) | LCC TJ820 .M63 2017 (print) | DDC 621.31/2136–dc23
LC record available at https://lccn.loc.gov/2017030336

Cover Design: Wiley
Cover Image: © Stockr/Shutterstock

Set in 10/12pt Warnock by SPi Global, Chennai, India
Printed and bound in Malaysia by Vivar Printing Sdn Bhd

10 9 8 7 6 5 4 3 2 1

Contents

List of Contributors

Yongning Chi
China Electric Power Research Institute
Beijing, China

Lijun Hang
Hangzhou Dianzi University
China

Shuju Hu
Institute of Electrical Engineering
Chinese Academy of Sciences

Kim Høj Jensen
Siemens Wind Power A/S
Denmark

Guojie Li
Shanghai Jiaotong University
Shanghai, China

Yan Li
China Electric Power Research Institute
Beijing, China

Chao Liu
China Electric Power Research Institute
Beijing, China

Jacob Østergaard
Centre for Electric Power and Energy
Department of Electrical Engineering
Technical University of Denmark

Tony Wederberg Rasmussen
Centre for Electric Power and Energy
Department of Electrical Engineering
Technical University of Denmark

Ranjan Sharma
Siemens Wind Power A/S
Denmark

Lei Shi
NARI Technology Co., Ltd
Jiangsu, China

Bin Song
Institute of Electrical Engineering
Chinese Academy of Sciences

Yuanzhang Sun
Wuhan University
Wuhan, China

Haiyan Tang
China Electric Power Research Institute
Beijing, China

Xinshou Tian
China Electric Power Research Institute
Beijing, China

Ningbo Wang
Gansu Electric Power Corporation Wind
Power Technology Center
Gansu, China

Linjun Wei
China Electric Power Research Institute
Beijing, China

Qiuwei Wu
Centre for Electric Power and Energy
Department of Electrical Engineering
Technical University of Denmark

Lorenzo Zeni
DONG Energy Wind Power A/S,
Denmark

Haoran Zhao
Centre for Electric Power and Energy
Department of Electrical Engineering
Technical University of Denmark

Qiang Zhou
Gansu Electric Power Corporation Wind
Power Technology Center
Gansu, China

About the Companion Website

Don't forget to visit the companion website for this book:

www.wiley.com/go/wu/modeling

There you will find valuable material designed to enhance your learning, including:

- Matlab simulation models

  Scan this QR code to visit the companion website

1

Status of Wind Power Technologies

Haoran Zhao and Qiuwei Wu

Technical University of Denmark

1.1 Wind Power Development

Although wind power has been utilized by humans for more than 3000 years, the history of wind power for electricity production is only 120 years long.

In July 1887, Professor James Blyth (1839–1906) of Anderson's College, Glasgow built the first windmill for the production of electricity at Marykirk in Kincardineshire, Scotland [1]. The windmill was 10 m high, and was used to charge accumulators to power the lighting in the cottage. Around the same period, a wind turbine was designed and constructed in the winter of 1887-1888 by Charles F. Brush (1849–1929) in Cleveland, USA [2]. The rotor of Brush's wind turbine was 17 m in diameter and had 144 blades. The rated power was 12 kW. It was used either to charge a bank of batteries or to operate up to 100 incandescent light bulbs and various motors in Brush's laboratory.

A pioneer of modern aerodynamics, Poul la Cour (1846–1908) of Askov, Denmark, built the world's first wind tunnels for the purpose of aerodynamic tests to identify the best shape of the blades for turbines. Based on his experiments, he realized that wind turbines with fewer rotor blades were more efficient for electricity production. He designed the first four-blade wind turbine in 1891 [3].

The developments in the 20th century can be divided into two periods. From 1900–1973, the prices of wind-powered electricity were not competitive. The gradual extension of electrical networks and the availability of low-cost fossil fuels lead to the abandonment of wind turbines. Wind turbine generators (WTGs) were mainly used in rural and remote areas. Although several wind turbines in the hundred-kilowatt class were manufactured and installed for testing, due to high capital costs and reliability problems, they were not widely adopted.

The two oil crises in 1973 and 1979, with supply problems and price fluctuations for fossil fuels, spurred the adoption of non-petroleum energy sources. As an alternative to fossil fuels, wind power was once again put on the agenda. European countries and US government started to invest in research into large commercial wind turbines. The world's first multi-megawatt wind turbine was constructed in 1978, and pioneered many technologies now used in modern wind turbines. From 1975 through to the mid-1980s, NASA developed 3.2 MW and 4 MW wind turbines. Although they were

Modeling and Modern Control of Wind Power, First Edition. Edited by Qiuwei Wu and Yuanzhang Sun.

Companion website: www.wiley.com/go/wu/modeling

sold commercially, none of these were ever put into mass production. When oil prices declined, electricity generated by wind power became uneconomical and many manufacturers left the business.

At the beginning of the 21st century, although fossil fuels were still relatively cheap, concerns over energy security, global warming, and eventual fossil fuel depletion increased, and this led to an expansion of interest in renewable energy. The wind power industry has since achieved rapid development.

From the point of view of global capacity, according to statistics from the Global Wind Energy Council (GWEC), the global annual and cumulative installed wind capacities for the past ten years are as illustrated in Figures 1.1 and 1.2, respectively. In 2015, the global wind power industry installed 63.5 GW of capacity, representing annual market growth of 22%. By the end of 2015, the total installed capacity reached 432.4 GW, representing cumulative market growth of 17%. As estimated by International Energy Agency (IEA), that figure will reach 2016 GW by 2050, representing 12% of global electricity usage [5].

From the point of view of development in each country, more than 83 countries around the world were using wind power on a commercial basis by 2010. The top ten

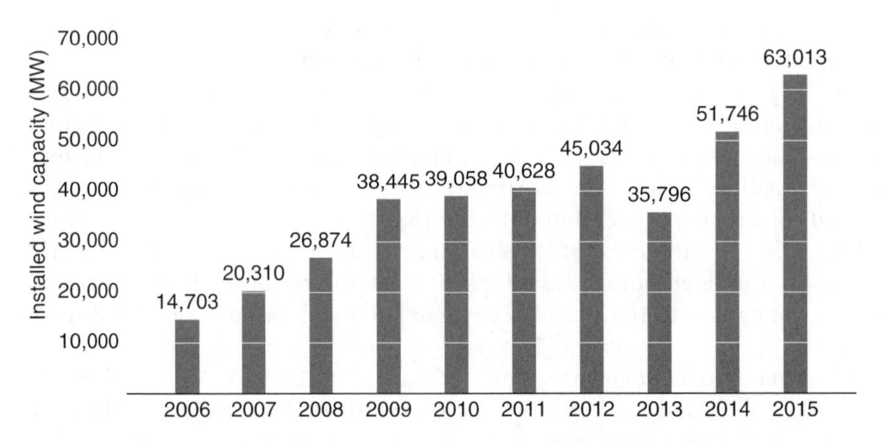

Figure 1.1 Global annual installed wind capacity 2005-2015 [4].

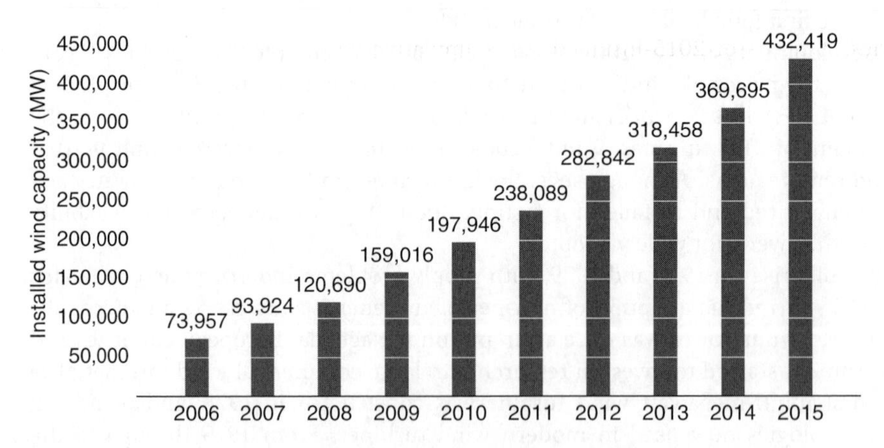

Figure 1.2 Global cumulative installed wind capacity 2005-2015 [4].

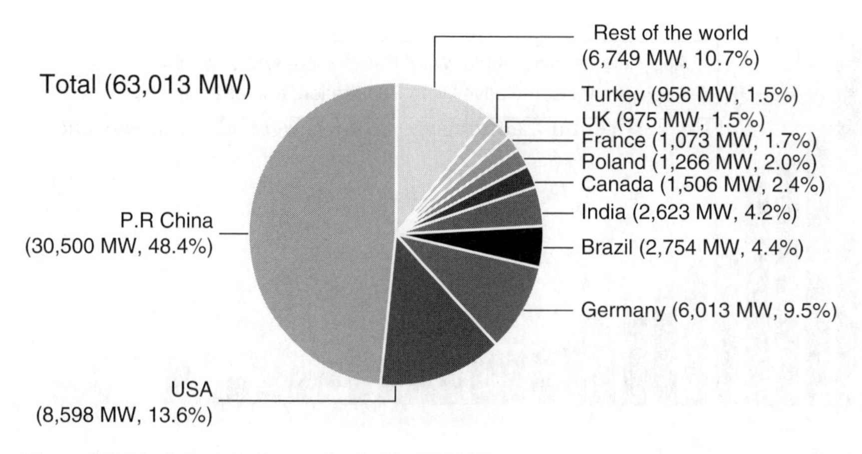

Figure 1.3 Newly installed capacity during 2015 [4].

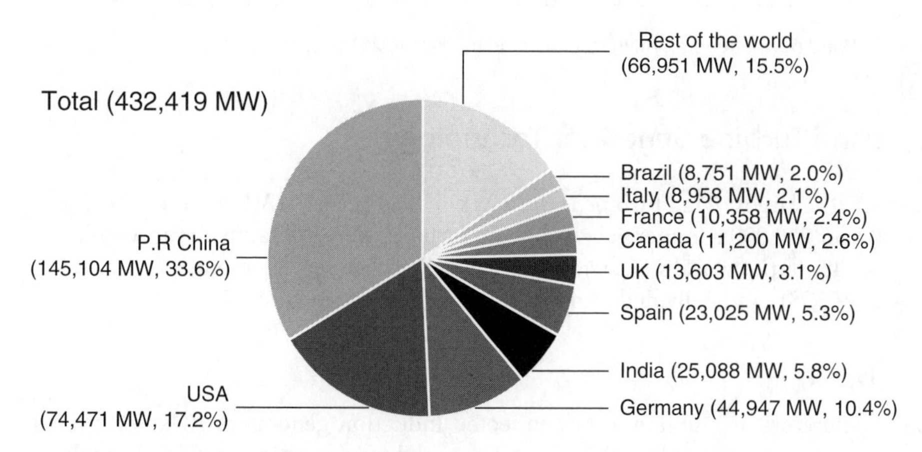

Figure 1.4 Cumulative capacities at 2015 [4].

countries in terms of 2015-installed and cumulative wind power capacities at 2015 are illustrated in Figures 1.3 and 1.4, respectively. More than half of all new installed wind power was added outside the traditional markets of Europe and North America. Asia has been the world's largest regional market for new wind power development, with capacity additions of 33.9 GW. China maintained its leadership position. China accounted for nearly half of the installations (48.4%) and its total wind power reached 145.1 GW.

In many countries, relatively high levels of wind power penetration have been achieved. Figure 1.5 presents the estimated wind power penetration in leading wind markets [6]. The installed capacity is estimated to supply around 40% of Denmark's electricity demand, and between 20% to 30% in Portugal, Ireland, and Spain, respectively. Denmark has a even more ambitious target of 50% by 2020. In the United States, 5.6% of the nation's electricity demand is estimated to be covered by the wind power. On a global basis, the contribution of wind power is estimated to be around 4.3% [6].

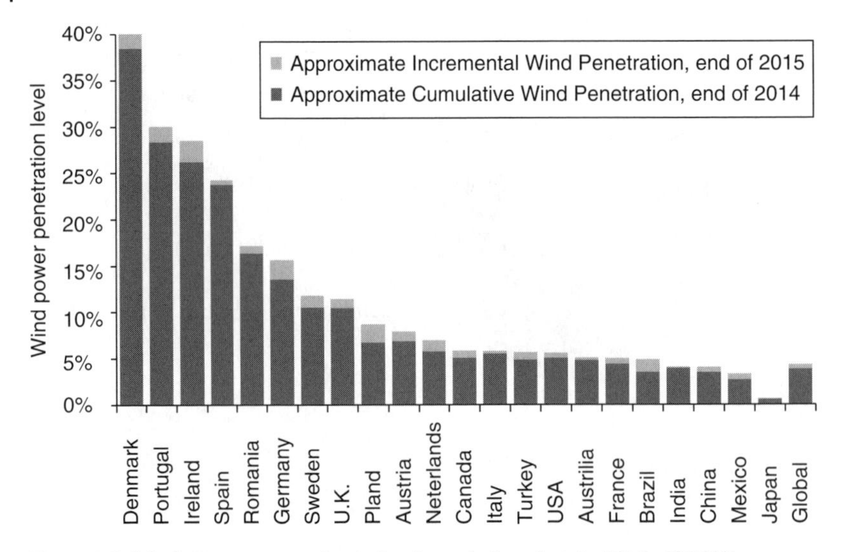

Figure 1.5 Wind power penetration in leading wind markets in 2014~2015 [6].

1.2 Wind Turbine Generator Technology

As at 2015, the largest wind turbine is the 8 MW capacity Vestas V164, for offshore use. By 2014, over 240,000 commercial-sized wind turbines were operating in the world, and these met 4% of the world's electricity demand. WTG-based wind energy conversion systems (WECS) can be divided into the following four main types [7, 8].

1.2.1 Type 1

Type 1 generators are directly grid-connected induction generators (IGs) with fixed rotor resistance. An example is the squirrel cage induction generator (SCIG). As illustrated in Figure 1.6, the wind turbine rotor (WTR) is connected to the IG via a gearbox (GB). Most Type 1 WTGs are equipped with mechanically switched capacitor (MSC) banks, which provide reactive power compensation. As the protection device, the main

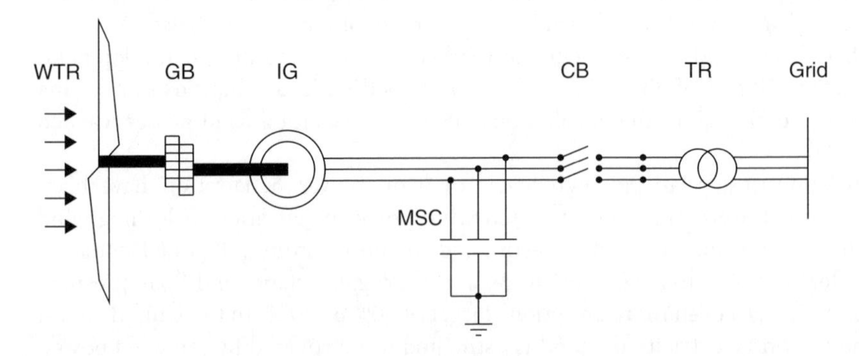

Figure 1.6 Structure of Type 1 WTG [8]. Refer to main text for explanation of acronyms.

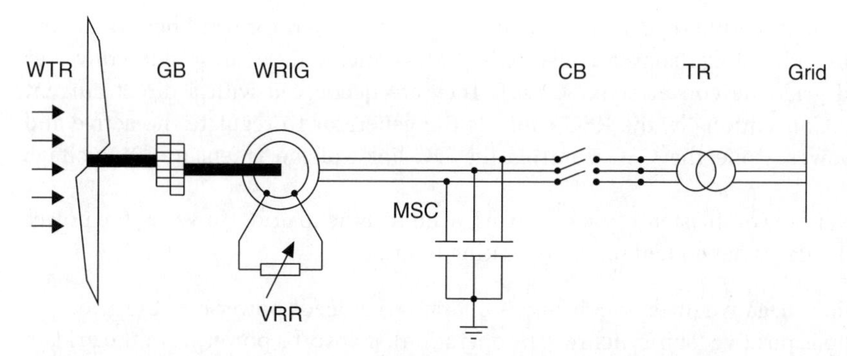

Figure 1.7 Structure of Type 2 WTG [8]. Refer to main text for explanation of acronyms.

circuit breaker (CB) disconnects the generator and capacitor from the grid in the event of a fault. Through a step-up transformer (TR), the WTG is connected to the grid.

Because of the direct connection to the grid, the IG operates at its natural mechanical characteristic, with an accentuated slope (corresponding to a small slip, normally 1–2%) from the rotor resistance [7]. The rotational speed of the IG is close to the synchronous speed imposed by grid frequency, and is not affected significantly by wind variation.

1.2.2 Type 2

Type 2 generators are directly grid-connected IGs with variable rotor resistance (VRR).

Figure 1.7 illustrates the general structure of a Type 2 WTG. As an evolution of Type 1 WTGs, using regulation through power electronics, the total (internal plus external) rotor resistance is adjustable. In this way, the slip of the generator can be controlled, which affects the slope of the mechanical characteristic. The range of dynamic speed variation is decided by the additional resistance. Usually, the control range is up to 10% over the synchronous speed.

1.2.3 Type 3

Type 3 generators are double-fed induction generators (DFIGs). As illustrated in Figure 1.8, the DFIG is an induction generator with the stator windings connected

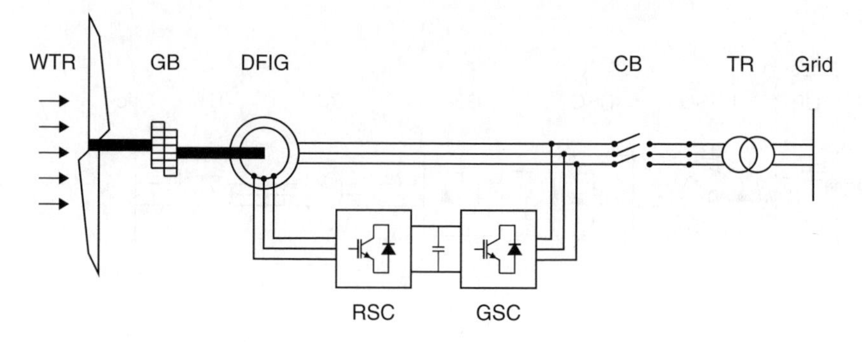

Figure 1.8 Structure of Type 3 WTG [8]. Refer to main text for explanation of acronyms.

directly to the three-phase, constant-frequency grid and the rotor windings connected to back-to-back voltage source converters (VSCs), including a rotor-side converter (RSC) and grid-side converter (GSC) [9]. They are decoupled with a direct current (DC) link. Conventionally, the RSC controls the generator to regulate the active and reactive power, while the GSC controls the DC-link voltage to ensure DC voltage stability.

The power flow of the stator is always from wind turbine to grid. However, the power flow of the rotor is dependent on the operating point:

- If the slip is negative (over-synchronous operation), it feeds power into the grid.
- If the slip is positive (sub-synchronous operation), it absorbs power from the grid.

In both cases, the power flow in the rotor is approximately proportional to the slip. By regulation of the generator behaviour through the GSC controller, the rotation speed is allowed to operate over a larger, but still restricted range (normally 40%).

1.2.4 Type 4

Type 4 WTGs have the wind turbine connected fully through a power converter. Figure 1.9 shows the general structure of Type 4 WTG. The generator type can be either an induction generator or a synchronous generator. Furthermore, the synchronous generator can be either a wound-rotor synchronous generator (WRSG) or a permanent-magnet synchronous generator (PMSG). Currently, the latter is widely used by the wind turbine industry. The back-to-back VSC configuration is used. The RSC ensures the rotational speed is adjusted within a large range, whereas the GSC transfers the active power to the grid and attempts to cancel the reactive power consumption [7].

The PMSG configuration is considered a promising option. Due to its self-excitation property, its gives high power factors and efficiency. As it is supplied by permanent magnets, a PMSG does not require an energy supply for excitation. Moreover, since the salient pole of a PMSG operates at low speeds, the gearbox (Figure 1.9) can be removed. This is a big advantage of PMSG-based WECS, as the gearbox is a sensitive device in wind power systems. The same thing can be achieved using direct driven multipole PMSGs (DD-PMSGs) with large diameters.

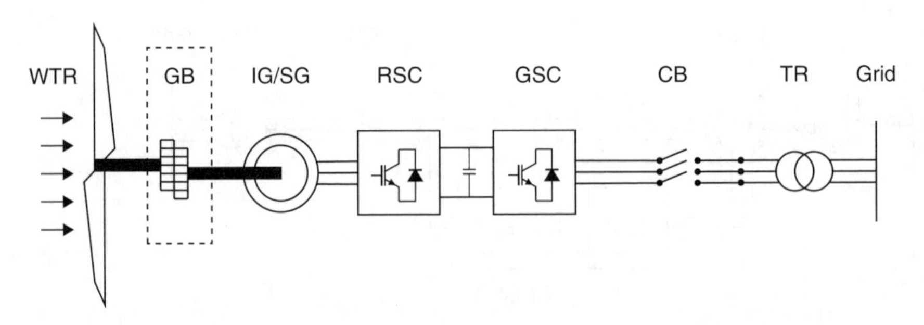

Figure 1.9 Structure of Type 4 WTG [8]. Refer to main text for explanation of acronyms.

1.2.5 Comparison

These four WTG types can also be classified into two categories according to the rotor speed control criterion: fixed-speed wind turbines (FSWTs), including Type 1, and variable-speed wind turbines (VSWTs), including Types 2–4 [10].

FSWTs have the advantage of being simple, robust and reliable, with simple and inexpensive electric systems. They are well-proven in operation. Moreover, they can naturally provide the inertial response. However, as FSWTs have limited controllability of rotational speed, the captured aerodynamic efficiency is restricted. Due to the fixed-speed operation, mechanical stress is important. All fluctuations in wind speed are transmitted into mechanical torque and then, as electrical fluctuations, into the grid.

Due to the regulation of rotor speed within a larger range, VSWTs, especially Types 3 and 4, are highly controllable, allowing maximum power extraction over a large range of wind speeds. In addition, the active and reactive power control can be fully decoupled and implemented separately, and they are therefore they are more flexible. VSWTs dominate the marketplace, especially in the megawatt class. Due to the electrical decoupling between the generator and the grid, they cannot contribute to the power system apparent inertial as conventional synchronous generators [11]. However, the inertial response can be emulated by an additional power or torque loop [12–15].

1.2.6 Challenges with Wind Power Integration

Due to the inherent variability and uncertainty of the wind, the integration of wind power into the grid has brought challenges in several different areas, including power quality, system reliability, stability, and planning. The impact of each is largely dependent on the level of wind power penetration in the grid [16].

Impact on Power Quality

Power quality is evaluated as a deviation from the normal sinusoidal voltage and current waveforms in power system network. Power quality distortions of a power system include flickers and harmonic distortions.

Flickers are periodic voltage and frequency variations, typically of between 0.5 and 25 Hz. The oscillatory output power produced by WTGs can cause flickers in a power system. Fluctuations due to the tower shadow and turbulence effects in the wind may cause flickers too. The IEC 61400-21 standard furnishes a measurement procedure to calculate the flicker impact of wind turbines.

Harmonics can be injected on both the generation and consumer sides. On the consumer side, harmonics are caused by non-linear loads. On the generation side, sources of harmonics include flexible alternating current transmission systems (FACTS), such as reactive power compensators and power electronics devices. The power electronic converters used by VSWTs are considered sources of harmonics.

Impact on System Reliability

The uncertainty of wind generation will increase the requirement for operating reserve, which will in turn increase generation costs. When the wind penetration level is low, the wind power fluctuation is comparable to existing load fluctuations. Committed conventional generators, such as thermal or hydro units, have sufficient load-tracking capability, so no additional operating reserve is required. However, load balancing becomes challenging at high wing-power penetration levels. An extra reserve of 3–6% of the rated capacity of the wind plant is required at 10% wind penetration and 4–8% for 20% wind penetration.

Impact on System Stability

Frequency stability Conventional synchronous generators can provide inertia response, which plays a significant role in stabilizing system frequency during a transient scenario. The inertia value dictates the frequency deviations due to a sudden change in the generation and load power balance. It affects the eigenvalues and vectors that determine the stability and mode shape of the transient response [17].

The contributions to the system inertia of WTGs are dependent on the WTG type. Due to the direct connection of the power system, fixed-speed induction generators can provide inertia response. Modern VSWTs, whose rotation speed is normally decoupled from grid frequency by a power electronic converter, may decrease the system inertia [18]. With high wind power penetration, this decrease aggravates the grid frequency instability.

Voltage stability Many power system faults are cleared by the relay protection of the transmission system, either by disconnection or by disconnection and fast reclosure. There is a short period with a voltage drop beyond a specified threshold, followed by a period when the voltage returns. Previously, when the voltage dip occurred, the wind turbine was simply disconnected from the grid. When the fault was cleared and the voltage returned to normal, the wind turbine was reconnected. When the wind power penetration level is low, the impact on system stability is limited. However, with high levels of wind penetration, if the entire wind farm is suddenly disconnected while at full generation, the system will lose further production capability [17]. This can lead to a further large frequency and voltage drop and possibly complete loss of power. It is very important to maintain the connection of WTGs when there are disturbances in the network. Therefore, modern WTGs are required to have the fault ride-through (FRT) capability by grid codes.

Since large wind farms are mainly located in areas far from load centres, the short-circuit ratio (SCR) is small [19], and the grid at the connection point is weak. Voltage fluctuations caused by the intermittent power of the wind farms are large.

Impact on System Planning

As wind resources are often located far from load centres, it is critical to develop sufficient transmission to transport wind power to load centres. Old transmission lines must be updated. On the one hand, transmission planning processes are highly dependent on regional politics. The generation capacity, transmission location and load size are different from one place to another. These disparities make the development of transmission for wind power contentious and complex. On the other hand, in order to transfer variable and unpredictable wind power, new requirements for transmission technology arise [20].

Microgrids are considered as an alternative vision of the future grid, with energy generated and consumed locally. They can significantly reduce the long-distance energy transmission requirement and transmission losses. The electricity grid could be conceptualized as a collection of independent microgrids [20].

1.3 Conclusion

There is potential for wind energy to play an important role in future energy supply. With the development of wind turbine technology, wind power will become more controllable and grid-friendly. It is desirable to make wind farms operate as conventional power plants. To achieve this objective, more advanced control strategies for both wind turbines and wind farms are required.

References

1 Price, T.J. (2005) James Blyth – Britain's first modern wind power pioneer. *Wind Engineering*, **29** (3), 191–200.

2 Anonymous (1890) Mr Brush's windmill dynamo. *Scientific American*, **63** (25), 389.

3 Vestergaard, J., Brandstrup, L., and Goddard, R.D. (2004) A brief history of the wind turbine industries in Denmark and the United States, in *Academy of International Business (Southeast USA Chapter) Conference Proceedings*, pp. 322–327.

4 GWEC (2015) Global wind statistics 2015, *Tech. rep.*, Global Wind Energy Council.

5 IEA (2009) Wind energy roadmap, *Tech. rep.*, International Energy Agency.

6 USDEA (2015) 2015 Wind technologies market report, *Tech. rep.*, US Department of Energy.

7 Munteanu, I., Bratcu, A.I., Cutululis, N.A., and Ceanga, E. (2008) *Optimal Control of Wind Energy Systems: Towards a global approach*, Springer Science & Business Media.

8 Zhao, H., Wu, Q., Rasmussen, C., and Xu, H. (2014) *Coordinated control of wind power and energy storage*, PhD thesis, Technical University of Denmark, Department of Electrical Engineering.

9 Akhmatov, V. (2003) *Analysis of dynamic behaviour of electric power systems with large amount of wind power*, PhD thesis, Technical University, Denmark.

10 Lalor, G., Mullane, A., and O'Malley, M. (2005) Frequency control and wind turbine technologies. *IEEE Transactions on Power Systems*, **20** (4), 1905–1913.

11 Conroy, J.F. and Watson, R. (2008) Frequency response capability of full converter wind turbine generators in comparison to conventional generation. *IEEE Transactions on Power Systems*, **23** (2), 649–656.

12 Morren, J., De Haan, S.W., Kling, W.L., and Ferreira, J. (2006) Wind turbines emulating inertia and supporting primary frequency control. *IEEE Transactions on Power Systems*, **21** (1), 433–434.

13 Keung, P.K., Li, P., Banakar, H., and Ooi, B.T. (2009) Kinetic energy of wind-turbine generators for system frequency support. *IEEE Transactions on Power Systems*, **1** (24), 279–287.

14 Ullah, N.R., Thiringer, T., and Karlsson, D. (2008) Temporary primary frequency control support by variable speed wind turbines – potential and applications. *IEEE Transactions on Power Systems*, **23** (2), 601–612.

15 Mauricio, J.M., Marano, A., Gómez-Expósito, A., and Ramos, J.L.M. (2009) Frequency regulation contribution through variable-speed wind energy conversion systems. *IEEE Transactions on Power Systems*, **24** (1), 173–180.

16 Wang, P., Gao, Z., and Bertling, L. (2012) Operational adequacy studies of power systems with wind farms and energy storages. *IEEE Transactions on Power Systems*, **27** (4), 2377–2384.
17 Abo-Khalil, A.G. (2013) *Impacts of Wind Farms on Power System Stability*, Intech.
18 Sun, Y.Z., Zhang, Z.S., Li, G.J., and Lin, J. (2010) Review on frequency control of power systems with wind power penetration, in *Power System Technology (POWERCON), 2010 International Conference on*, IEEE, pp. 1–8.
19 Neumann, T., Feltes, C., and Erlich, I. (2011) Response of DFG-based wind farms operating on weak grids to voltage sags, in *2011 IEEE Power and Energy Society General Meeting*, IEEE, pp. 1–6.
20 IEC (2012) Grid integration of large-capacity renewable energy sources and use of large-capacity electrical energy storage, *Tech. rep.*, International Electrotechnical Commission.

2

Grid Code Requirements for Wind Power Integration

Qiuwei Wu

Centre for Electric Power and Energy, Department of Electrical Engineering, Technical University of Denmark

2.1 Introduction

With rapid growth of installed capacity, wind power will have a big impact on the electricity system steady state and its dynamic operation. In the past, wind power plants (WPPs) were allowed to be disconnected under system disturbances. However, with high penetration of wind power in electric power systems, the disconnection of WPPs will make it difficult for the system to withstand system disturbances. Therefore, transmission system operators (TSOs) around the world have specified requirements for WPPs under steady-state and dynamic conditions in their grid codes. WPPs must have voltage- and frequency-regulation capabilities under steady-state conditions. Under specified voltage dip conditions caused by the faults within the grid, WPPs have to stay connected and provide reactive power for the duration of the voltage dip, and to fulfill recovery requirements.

An analysis of large-scale integration of wind power in Europe has been carried out [1], with reference to power and energy balancing, grid connection and system stability, grid infrastructure extension and reinforcement, power system adequacy, market design, demand-side management and storage. Recommendations were made for power system operation when there is a large amount of wind power. The main problems with connecting wind farms to the grid were reviewed by de Alegria et al. [2], who made suggestions for the amendment of the grid codes to help integrate wind power into power systems without affecting the quality and stability of the system. An analysis of grid codes from Canada, Denmark, Ireland, Scotland, Germany and UK was conducted by Christiansen at al. [3], with regard to the technical requirements for integrating wind power into grids.

In this chapter, the requirements on WPPs from the latest grid codes are reviewed and discussed. The grid codes reviewed are from the UK, Ireland, Germany, Denmark, Spain, Sweden, the USA, and Canada.

Modeling and Modern Control of Wind Power, First Edition. Edited by Qiuwei Wu and Yuanzhang Sun.

Companion website: www.wiley.com/go/wu/modeling

2.2 Steady-state Operational Requirements

Steady-state operational requirements concern the power factor requirement, voltage operating range, frequency operating range and voltage quality. Usually, the steady-state operational requirements for WPPs are specified at the point of connection (POC).

2.2.1 Reactive Power and Power Factor Requirements

The reactive power and power factor requirements concern the reactive power capability of the WPP within the specified voltage range at the POC under the steady-state conditions. The requirements for reactive power and power factor are similar in different grid codes and are listed in Table 2.1.

The reactive power requirement in the UK is quite complex, and is shown in Figure 2.1. WPPs are requested to have reactive power capability as follows:

- from 100% to 50% active power production, from −32.87% to +32.87% of the rated active power
- from 50% to 20% active power production from −32.87% to −12% for the leading part up to +32.87% of the rated active power
- from 20% to 0% active power production, −5% to +5% of the rated active power.

Table 2.1 Reactive power/power factor requirement.

Country or region	PF requirements
UK	0.95 lagging to 0.95 leading [4]
Ireland	0.95 lagging to 0.95 leading at full production; the same reactive power capability as at 100% production, when the active power output decreases from 100% to 12% [5]
Germany	0.95 lagging to 0.925 leading; depending on the voltage at the POC, the requirement changes [6]
Denmark (WPP 1.5–25 MW)	Reactive capability equal to 0.975 inductive and capacitive at 1 pu active power production from 1 pu to 0.2 pu active power production, straight line from 0.975 inductive and capacitive according to the rated active power to 0 from 0.2 pu to 0 active power production [7]
Denmark (WPP >25 MW)	0.975 inductive and capacitive at 1 pu active power production; straight line to 0.95 inductive and capacitive according to the rated active power from 1–0.8 pu active power production; 0.95 inductive and capacitive according to the rated active power from 0.8–0.2 pu active power production; straight line from 0.95 inductive and capacitive according to the rated active power to 0 from 0.2 pu to 0 active power production [7]
Spain	0.98 leading to 0.98 lagging without any penalty [8]
Sweden	Reactive exchange can be regulated to zero [9]
US	0.95 leading to 0.95 lagging [10]
Quebec	0.95 leading to 0.95 lagging of the rated capacity over the entire active power generation range [11]
Alberta	0.90 lagging to 0.95 leading [12]

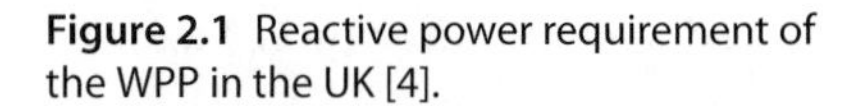

Figure 2.1 Reactive power requirement of the WPP in the UK [4].

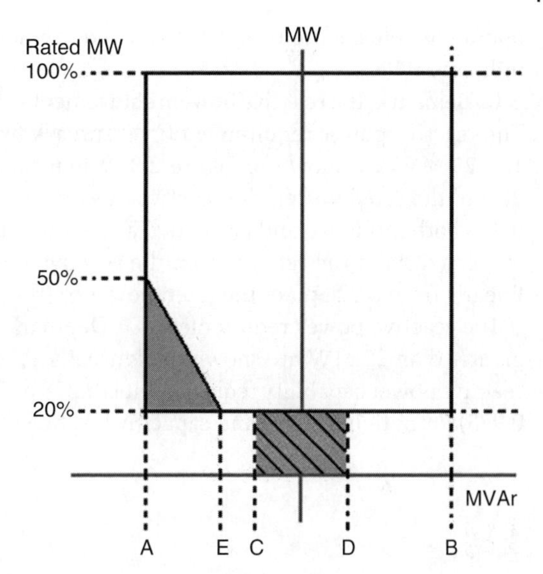

Figure 2.2 Reactive power requirement of WPPs in Ireland [5].

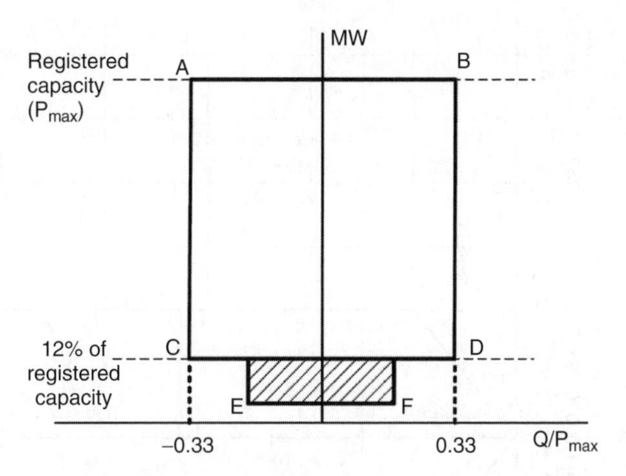

The reactive power requirement in Ireland is shown in Figure 2.2. The WPP operating in power factor control mode, voltage control mode or constant reactive power mode should be at least capable of operating at any point within the P-Q capability ranges shown. From 100% to 12% active power output of the installed capacity, the WPP is required to have reactive power capability from −33% to +33% of the rated active power. Points E and F represent the minimum M_{var} absorption and production capability at the cut-in speed of individual wind turbines.

For WPPs in Germany, the reactive power requirement is voltage dependent. When the voltage at the POC increases from the minimum operating level to the voltage with full power factor range, the power factor range changes from capacitive 0.925 to capacitive 0.95 to the full power factor range of 0.95 inductive to 0.925 capacitive, and the middle part is the voltage range with the full power factor requirement. When the voltage at the POC is above the maximum voltage with the full power factor requirement, the power factor requirement is moving to the left. At the maximum operating voltage, the reactive power requirement is reduced to unity power factor to inductive 0.95.

In Denmark, the reactive power requirements are different for WPPs of different sizes. The reactive power requirements in Denmark for WPPs with a power output range of 1.5–25 MW are shown in Figure 2.3. When the active power output is from 100% to 20% of the rated power, the reactive power capability should be equal to power factor 0.975 both inductive and capacitive at the rated power. When the active power output is below 20% of the rated power, the reactive power capability requirement decreases linearly from 0.228 pu of the rated power to 0.

The reactive power requirements in Denmark for WPPs with a power output range greater than 25 MW are shown in Figure 2.4. At the full rated active power output, the reactive power capability requirement is 0.228 pu of the rated active power (power factor 0.975) for both inductive and capacitive. When the active power output decreases from

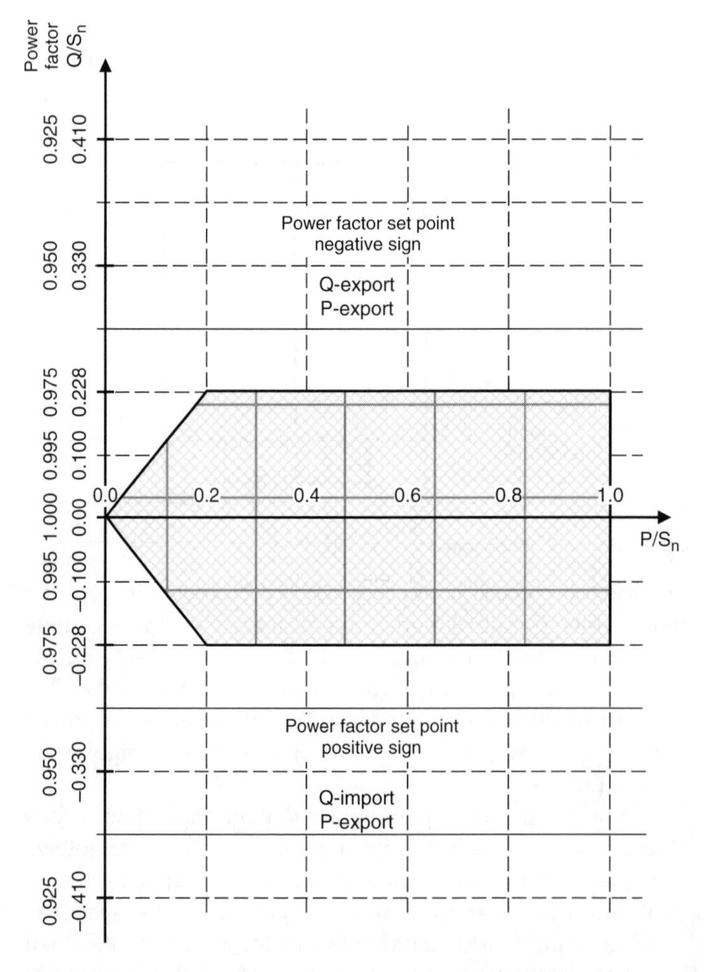

Figure 2.3 Reactive power requirement in Denmark for WPPs with a power output range of 1.5–25 MW [7].

1 pu to 0.8 pu, the reactive power capability requirement linearly increases from 0.228 pu to 0.33 pu of the rated active power for both inductive and capacitive (power factor 0.95). When the active power output is 0.8 pu to 0.2 pu of the rated active power, the reactive power capability is the same as at the 0.8 pu active power output. When the active power output decreases from 0.2 pu to 0 of the rated active power, the reactive power capability requirement linearly decreases from 0.33 pu to 0 of the rated active power.

The grid code in Denmark also specifies requirements for the voltage control range, as shown in Figure 2.5. The WPP should be able to operate at any point in the hatched area in Figure 2.5.

The reactive power requirements versus voltage and active power for WPPs in Quebec are shown in Figure 2.6 and Figure 2.7. The requirements are specified on the HV side of the WPP main transformer. Reactive power must be available for the entire range of normal operating voltages (0.9–1.1 pu). At a voltage of 0.9 pu, the WPP is not required to absorb reactive power corresponding to a lagging power factor 0.95. However, the WPP still must be able to supply reactive power corresponding to a leading power factor of 0.95. Similarly, at a voltage of 1.1 pu, the WPP is not required to supply reactive power

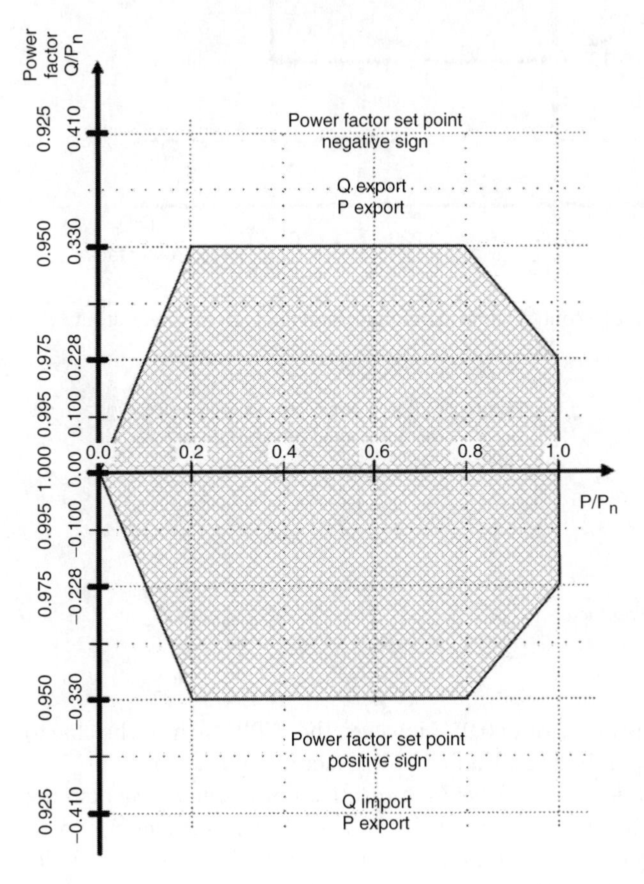

Figure 2.4 Reactive power requirement in Denmark for WPPs with a power output greater than 25 MW [7].

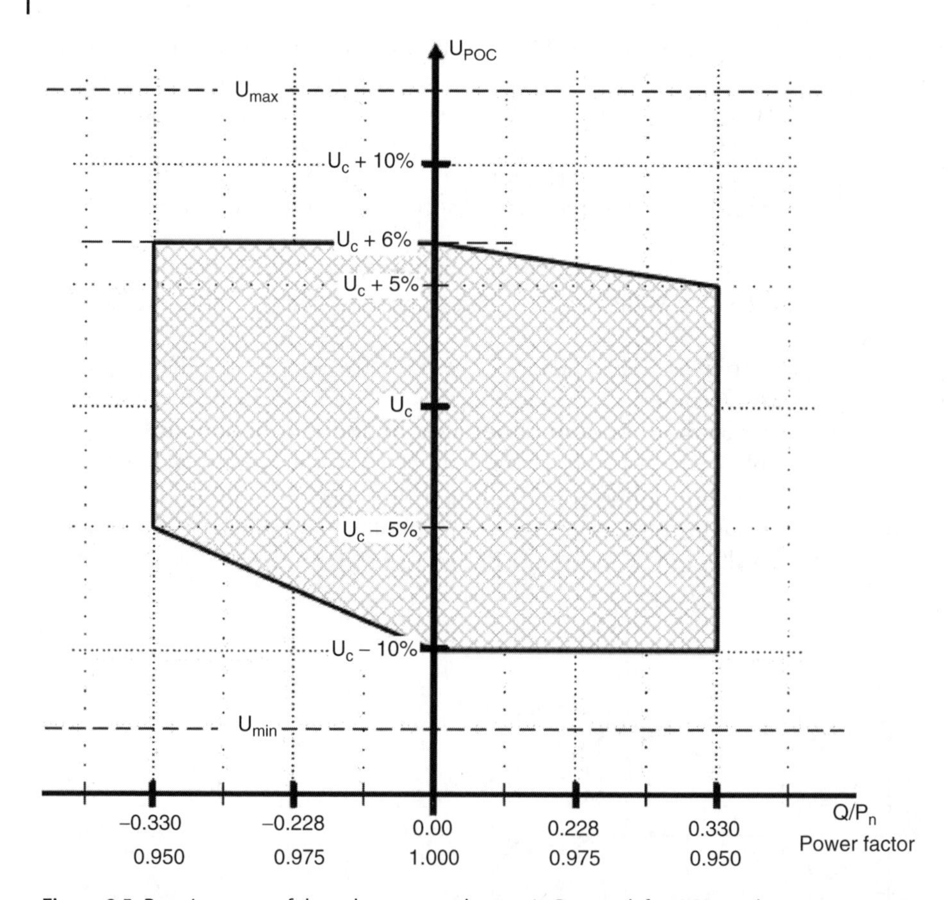

Figure 2.5 Requirements of the voltage control range in Denmark for WPPs with a power output greater than 25 MW [7].

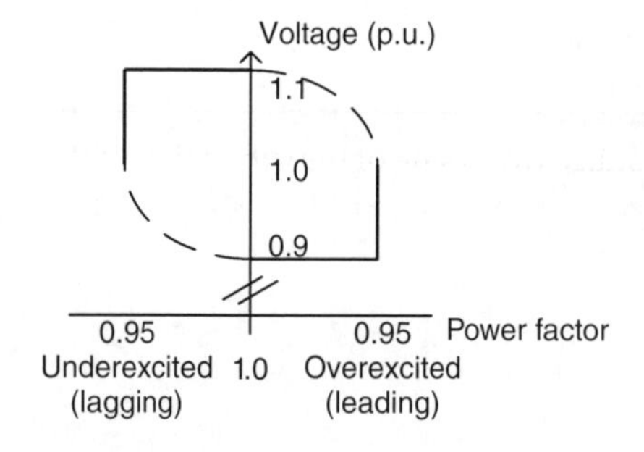

Figure 2.6 Reactive power requirement vs. voltage on the HV side of the main transformer of a WPP in Quebec [11].

corresponding to a lagging power factor of 0.95. However, the WPP still must be able to absorb reactive power corresponding to a leading power factor of 0.95.

In Alberta, there are dynamic and non-dynamic reactive power requirements. The minimum continuous reactive power supply capability of a WPP must meet or exceed a +0.9 power factor, and the minimum continuous reactive power absorption capability of a WPP must meet or exceed a −0.95 power factor, based on the gross real power upto and including the maximum authorized real power of the WPP. The minimum dynamic

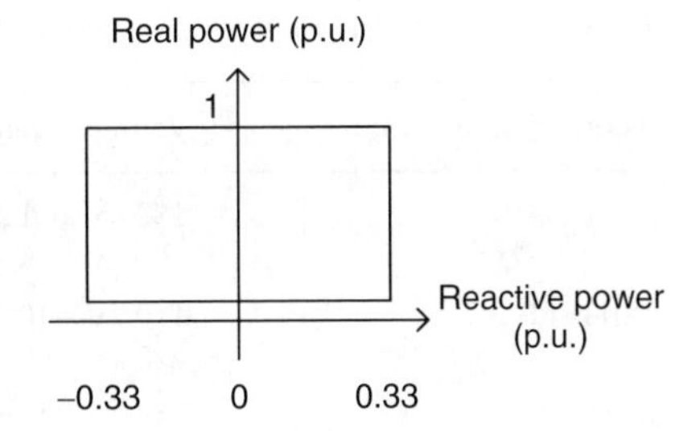

Figure 2.7 Reactive power requirement vs. active power on the HV side of the main transformer of a WPP in Quebec [11].

reactive power supply capability of a WPP must meet or exceed a +0.95 power factor, and the minimum dynamic reactive power absorption capability of a WPP must meet or exceed a −0.985) power factor, based on the gross real power up to and including the maximum authorized real power of the WPP.

In order to fulfill the reactive power and power factor requirements at the POC, it is recommended to use the reactive power capability of wind turbine generators (WTGs) as much as possible.

The reactive power capability of the doubly-fed induction generator (DFIG) wind turbines (WTs) and full-scale converter (FSC) WTs increases when the active power output decreases from 100% active power output to a specified active power output. After the specified active power level, the reactive power capability of wind turbines will keep the maximum reactive power capability within a range of the active power output level. When the active power output of the wind turbine is lower than the minimum active power level with the maximum reactive power, the reactive power capability of the wind turbines is kept in the range specified by an inductive power factor and a capacitive power factor.

Therefore, to ensure that WPPs can fulfill the reactive power requirements throughout the active power output range of WTGs, a power flow study at 100% active power output is needed. In most cases, it is good enough to specify an inductive power factor and a capacitive power factor. If the reactive power requirement range at low active power output is quite big, it is advised to carry out a power flow study at low active power output to check the power factor requirement compliance.

If the WPPs cannot fulfill the reactive power requirements, shunt reactive power compensation equipment is needed at the medium voltage (MV) side of the substation of the WPPs. If the POC is far from the substation of the WPPs, it is advised to install reactive power compensation equipment at the busbar of the POC too. The size of the reactive power compensation equipment can be determined through a power flow study, with the specified voltage operating range and the power factor range at the POC.

2.2.2 Continuous Voltage Operating Range

The continuous voltage operating requirements are illustrated in Table 2.2.

The WPPs are required to be able to operate under the condition when the voltage at the POC is within the continuous operating voltage range.

The reactive power requirement must be coordinated with the voltage control. It must be ensured that, when the voltage at the POC is within the specified continuous operating voltage range, the voltages inside the WPPs are within the operating voltage range of

Table 2.2 Continuous voltage operating range.

Country or region	Voltage operating requirement
UK	132 kV and 275 kV ±10% 400 kV ±5% [4]
Ireland	110 kV −10% to +12% 220 kV −9% to +12% 400 kV −13% to +5% [5]
Germany	110 kV −12.7% to +11.8% 220 kV −12.3% to +11.4% 380 kV −7.9% to +10.5% [6]
Denmark	132 kV −10% to +10% 150 kV −10% to +13% 400 kV −10% to +5% [7]
Spain	132 kV ±10% [8]
Sweden	−10% to +5% [9]
US	±10% [10]
Quebec	44 kV and 49 kV ±6% 69 kV and higher voltages except 735 kV ±10% 735 kV −5% – +4% [11]
Alberta	±5% [12]

the electrical equipment. Normally, the operating voltage range of the electrical equipment inside WPPs is 0.9–1.1 pu.

Sometimes, it is difficult to maintain both the voltage operating requirement and the reactive power requirement due to the topology of the WPPs and the grid conditions. It is advised for the WPP owner to liaise with the TSO for a reduced reactive power requirement or a voltage-dependent reactive power requirement.

2.2.3 Frequency Operating Range and Frequency Response

Frequency Operating Range

The frequency operating ranges for WPPs specified in the grid codes reviewed are listed in Table 2.3.

Frequency Response

Regarding the frequency response, WPPs in the UK are required not to change the active power output when the system frequency is 40.5–50.5 Hz. When the system frequency changes within the range of 47 Hz and 49.5 Hz, WPPs should maintain their active power output at a level not lower than the one determined by the linear relationship in Figure 2.8. When the system frequency decreases to 47 Hz, the active power change of WPPs should not be bigger than 5%. WPPs are also required to be capable of contributing to frequency control by continuous modulation of active power to the National Electricity Transmission System or the user system in which it is embedded.

Table 2.3 Frequency operating range.

Country or region	Frequency operating range
UK	51.5–52 Hz: Operation for a period of at least 15 min each time frequency above 51.5 Hz 51–51.5 Hz: Operation for a period of at least 90 min each time frequency above 51Hz 49.0–51 Hz: Continuous operation required 47.5–49.0 Hz: Operation for a period of at least 90 min each time frequency is below 49.0 Hz 47.0–47.5 Hz: Operation for a period of at least 20 s each time frequency is below 47.5Hz [4]
Ireland	Continuous operating range 49.5–50.5 Hz 60 min: 50.5–52.0 Hz and 49.5–47.5 Hz 20 s: <47.5 Hz Remain connected if rate of frequency change up to 0.5 Hz/s No additional WTGs connected when the system frequency is above 50.2 Hz [5]
Germany	Continuous operating range 47.5–51.5 Hz [6]
Denmark	Continuous operating range 49.5–50.2 Hz Minimum 5 h 49.0–49.5 Hz Minimum 30 min 48.0–49.0 Hz Minimum 15 min 50.2–52.0 Hz Minimum 3 min 47.5 – 48 Hz Minimum 20 s 47.0–47.5 Hz [7]
Spain	Continuous operating range 49.5–50.5 Hz [8]
Sweden	Continuous operating range 49.0–51.0 Hz >30 min 51.0–52.0 Hz and 47.5–49.0 Hz [9]
USa	Not specified [10]
Quebec	Continuous operating range 59.4–60.6 Hz 11 min 58.5–59.4 Hz and 60.6–61.5 Hz 1.5 min 57.5–58.5 Hz and 61.5–61.7 Hz 10 s 57.0–57.5 Hz 2 s 56.5–57.0 Hz 0.35 s 55.5–56.5 Hz Instantaneous <55.5 Hz or >=61.7 Hz[11]
Alberta	Continuous operating range 59.4–60.6 Hz 3 min 60.6–61.6 Hz and >58.4–59.4 Hz 30 s 61.6 Hz–61.7 Hz and >57.8–58.4 Hz 7.5 s >57.3–57.8 Hz 45 cycles >57.0–57.3 Hz 0 s >61.7 Hz and ≤57.0 Hz [12]

For primary frequency control, the droop setting should be equivalent of a fixed setting of 3–5% applied to each wind turbine. The deadband should not be greater than ±0.015 Hz.

In the Ireland, in wind-following mode, the WPP should have the capabilities in the power–frequency response curve in Figure 2.9, where the power and frequency ranges required for points A, B, C, D, and E are as defined in Tables 2.4 and 2.5.

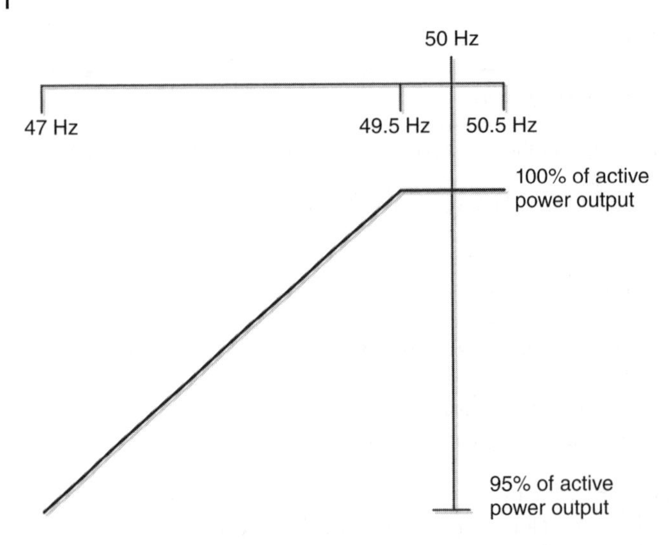

Figure 2.8 Active power output according to the system frequency in the UK [4].

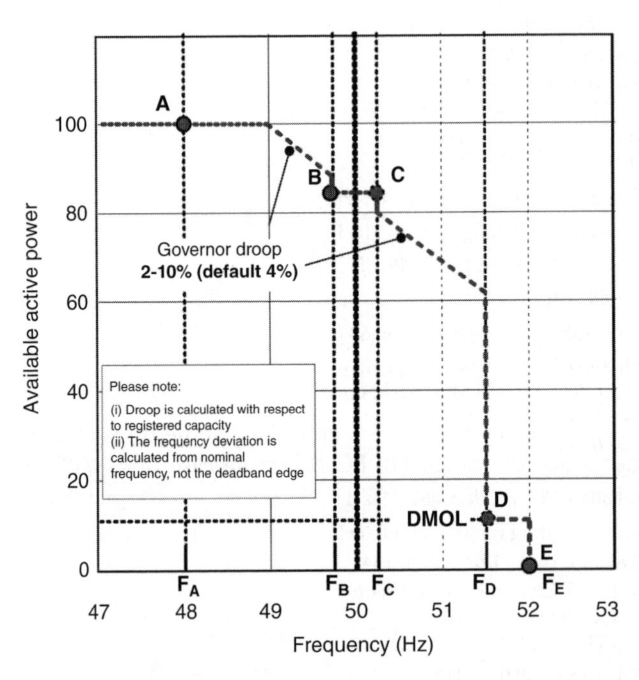

Figure 2.9 Power-frequency response curve for wind-following mode in Ireland [5].

The frequency response system should adjust the active power output of the WPP according to a governor droop, settable by the TSO in a range from 2% to 10% and defaulting to 4%, when operating in the ranges outside the deadband range FB–FC in the power–frequency response curve. The WPP frequency response and governor droop should be calculated with respect to registered capacity. When in active power

Table 2.4 Transmission system frequency and available active power settings for points A–E in Figure 2.9.

Point	Transmission system frequency (Hz)	WPP active power output (% of available active power)
A	F_A	P_A
B	F_B	Minimum of P_B or active power set-point (converted to % of available active power)
C	F_C	Minimum of P_C or active power set-point (converted to % of available active power)
D	F_D	Minimum of P_D or active power set-point (converted to % of available active power)
E	F_E	$P_E = 0\%$

Table 2.5 Transmission system frequency and active power ranges appropriate to Figure 2.9.

	Transmission system frequency (Hz)		WPP active power output (% of available active power)
			Registered capacity ≥5 MW
F_A	47.0–49.5	P_A	50–100
F_B	49.5–50	P_B	15–100
F_C	50–50.5	P_C	
F_D	50.5–52.0	P_D	15–100 but not less than DMOL
F_E		P_E	0

control mode, the WPP should always operate in frequency sensitive mode, with a governor droop as set out above and with a deadband of ±15 mHz.

When acting to control transmission system frequency, the WPP should provide at least 60% of its expected additional active power response within 5 s, and 100% of its expected additional active power response within 15 s of the start of the transmission system frequency's excursion outside the range FB–FC or, in the case of a WPP in active power control mode, when the transmission system frequency goes outside the deadband.

When the transmission system frequency is in the range F_C–F_D, the WPP should ensure that its active power output does not increase beyond the level when the transmission system frequency first exceeded F_C, due to an increase in available active power in that period.

If the frequency drops below F_A, then the frequency response system should act to maximize the active power output of the WPP, irrespective of the governor droop setting. If the frequency rises above F_D, then the frequency response system should act to reduce the active power output of the WPP to its design minimum operating level (DMOL) value. If the frequency rises above F_E, then the frequency response system should act to reduce the active power output of the WPP to zero. Any WTG that has disconnected should be brought back on load as fast as technically feasible, provided the transmission system frequency has fallen below 50.2 Hz.

In the area controlled by Tennet TSO GmbH in Germany, WPPs with a rated capacity of ≥100 MW must be capable of supplying primary control power. The following requirements must be met for the primary frequency control:

- the primary control band must be at least ±2% of the rated power
- the frequency power droop characteristic must be adjustable
- given a quasi-stationary frequency deviation of ±200 mHz, it must be possible to activate the total primary control power range required by the generating plant evenly in 30 s and to supply it for at least 15 min
- the primary control power must be again available 15 min after activation, provided that the setpoint frequency has been reached again
- in the event of smaller frequency deviations, the same rate of power change applies until the required power is reached
- the insensitivity range must be less than ±10mHz.

In Denmark, when there are frequency deviations in the public electricity supply network, the WPP must be able to provide frequency control to stabilize the grid frequency (50.00 Hz). The metering accuracy for the grid frequency must be ±10 mHz or better. It must be possible to set the frequency control function for all frequency points shown in Figures 2.10 and 2.11. It must be possible to set the frequencies f_{min}, f_{max}, as well as f1–f7 to any value in the range 50.00 Hz ± 3.00 Hz with an accuracy of 10 mHz.

The purpose of frequency points f1–f4 is to form a deadband and a control band for primary control. The purpose of frequency points f5–f6 is to supply critical power/frequency control. The droop required to perform control between the various frequency points is shown in Figures 2.10 and 2.11. P_{Delta} is the setpoint to which the available active power is reduced in order to provide frequency stabilization (upward regulation) in the case of falling grid frequency. Different P_{Delta} values with the same

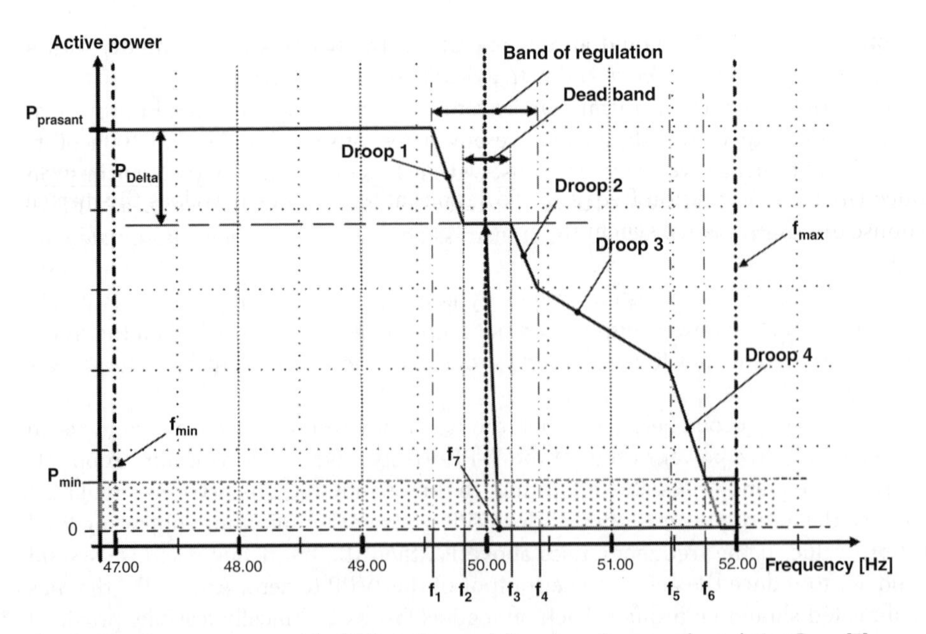

Figure 2.10 Frequency control for WPPs in Denmark for minor downward regulation P_{Delta} [7].

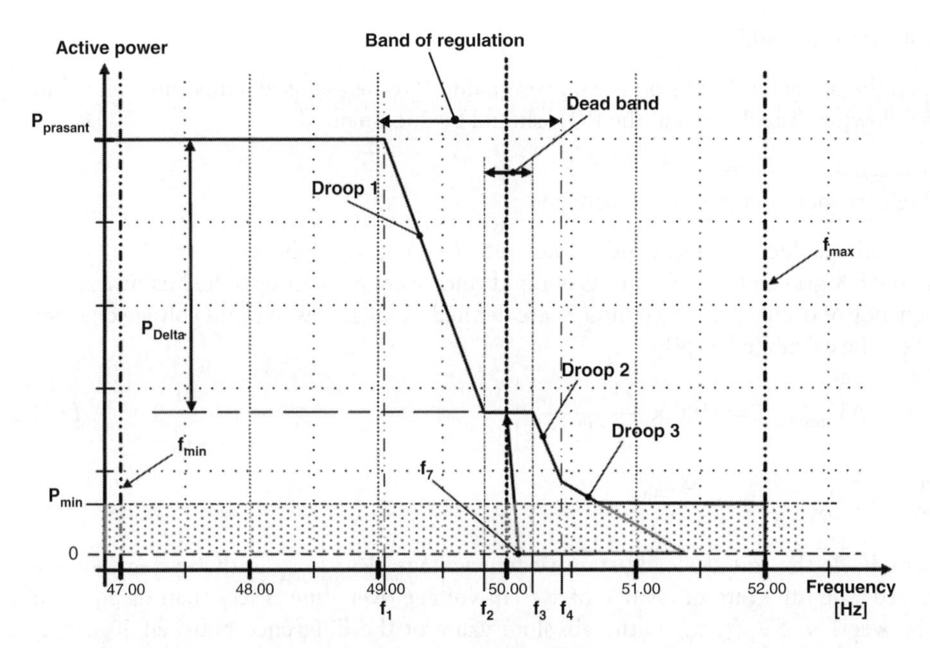

Figure 2.11 Frequency control for WPPs in Denmark for major downward regulation P_{Delta} [7].

droop (droops 1, 2, 3 and 4) are shown in Figures 2.10 and 2.11. It must be possible to activate the frequency control function in the interval from f_{min} to f_{max}.

If the frequency-control setpoint is to be changed, such a change must be commenced within 2 s and completed not later than 10 s after receipt of an order to change the setpoint. The accuracy of the control performed and of the setpoint must not deviate by more than ±2% of the setpoint value or by ±0.5% of the rated power, depending on which yields the highest tolerance.

In Quebec, WPPs with a rated output greater than 10 MW must be equipped with a frequency control system for inertial response. The system must be continuously in service but only act during major frequency deviations. To achieve this, the system must reduce large, short duration frequency deviations at least as much as does the inertial response of a conventional synchronous generator whose inertia equals 3.5 s.

In Alberta, the WPP must have an over-frequency control system that continuously monitors the frequency of the transmission system at a sample rate of at least 30 per second and a resolution of at least 0.004 Hz, and the over-frequency control system must automatically control the gross real power output of the WPP at all times. The over-frequency control system may have an intentional deadband of up to 0.036 Hz. It must be designed and calibrated to reduce the gross real power output at the collector bus based on the capability of all on-line WTs producing real power during an over-frequency excursion, and such reductions must be:

- proportional to the frequency increase by a factor of 33% per Hz of the gross real power output
- at a rate of 5% of the gross real power output per second.

In addition, no intentional time delay should be added to the control system.

2.2.4 Power Quality

When the impact of WPPs on the power quality is to be assessed, emissions related to the following disturbances at the POC should be documented.

- voltage fluctuations
- high-frequency currents and voltages.

The voltage fluctuations include rapid voltage changes and flickers.

In the UK grid code [4], the limits of rapid voltage changes are specified using a stated frequency of occurrence. Two indices are defined as measures of rapid voltage changes and can be calculated as [4]:

$$\%\Delta V_{\text{steadystate}} = \left| 100 \times \frac{\Delta V_{\text{steadystate}}}{V_0} \right| \tag{2.1}$$

$$\%\Delta V_{\max} = 100 \times \frac{\Delta V_{\max}}{V_0} \tag{2.2}$$

where V_0 is the initial steady-state system voltage, $V_{\text{steadystate}}$ is the system voltage reached when the rate of change of system voltage over time is less than or equal to 0.5% over 1 s, $\Delta V_{\text{steadystate}}$ is the absolute value of the difference between $V_{\text{steadystate}}$ and V_0, and $V_{\max}$ is the absolute value of the maximum change in the system voltage relative to the initial steady-state system voltage of V_0. All voltages are the root mean square of the voltage measured over one cycle refreshed every half a cycle, as per IEC 61000-4-30. The limits for the stated frequency of occurrence are listed in Table 2.6. Voltage changes in Category 3 do not exceed the limits depicted in the time-dependent characteristic shown in Figure 2.12.

Regarding flicker, the limits are flicker severity (short-term) of 0.8 units and a flicker severity (long-term) of 0.6 units for voltages above 132 kV, and flicker severity (short-term) of 1.0 unit and a flicker severity (long-term) of 0.8 units for voltages of 132 kV and below.

The UK grid code [4] refers to the standard ER G5/4, "Planning levels for harmonic voltage distortion and the connection of non-linear loads to the transmission

Table 2.6 Limits of rapid voltage changes in the UK grid code [4].

Category	Maximum number of occurrences	$\%\Delta V_{\text{steadystate}}$ and $\%\Delta V_{\max}$
1	No limit	$\lvert\%\Delta V_{\max}\rvert \le 1\%$ and $\lvert\%\Delta V_{\text{steadystate}}\rvert \le 1\%$
2	$\frac{3600}{\sqrt[0.304]{2.5 \times \%\Delta V_{\max}}}$ occurrences per hour with events evenly distributed	$1\% \le \lvert\%\Delta V_{\max}\rvert \le 3\%$ and $\lvert\%\Delta V_{\text{steadystate}}\rvert \le 3\%$
3	No more than 4 per day for commissioning, maintenance and fault restoration	For decrease in voltage: $\%\Delta V_{\max} \le 12\%$ and $\%\Delta V_{\text{steadystate}} \le 3\%$ For increase in voltage: $\%\Delta V_{\max} \le 5\%$ and $\%\Delta V_{\text{steadystate}} \le 3\%$

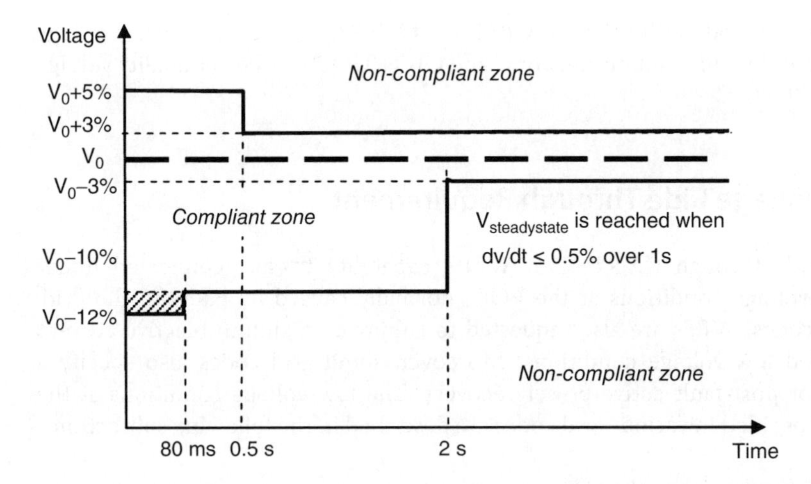

Figure 2.12 Time and magnitude limits for Category 3 rapid voltage changes in the UK [4].

systems and public electricity supply systems in the United Kingdom". IEC 61000-3-6 "Electromagnetic compatibility (EMC) – Part 3-6: Limits – Assessment of emission limits for the connection of distorting installations to MV, HV and EHV power systems" and IEC 61000-3-7 "Electromagnetic compatibility (EMC) – Part 3–7: Limits – Assessment of emission limits for the connection of fluctuating installations to MV, HV and EHV power systems" are referred to in the Irish grid codes, regarding harmonics, voltage fluctuation and flicker.

In the Danish grid code, the specific requirements for voltage quality are defined. The detailed requirements are presented below.

Rapid voltage changes [7]
- If $U_n > 35$ kV, the limit is 3.0%
- If $U_n \leq 35$ kV, the limit is 4.0%

Voltage variations and flicker
- If $U_n \leq 35$ kV, $P_{lt} < 0.50$, no requirement on P_{st}
- If 35 kV $< U_n \leq 100$ kV, $P_{lt} < 0.35$, no requirement on P_{st}
- If $U_n > 100$ kV, $P_{lt} < 0.20$, $P_{st} < 0.30$

Harmonic voltages
- If $U_n > 35$ kV, total harmonic distortion (THD) <3.0%
- If $U_n \leq 35$ kV, THD < 6.5%

In Alberta, with regard to voltage flicker at the POC, the legal owner of any WPP must comply with the specifications set out in the IEC 61000-3-7 standard, "Electromagnetic compatibility (EMC)", and in particular parts 3-7: "Limits – Assessment of emission limits for the connection of fluctuating installations to MV, HV and EHV power systems". The version used should be the one in effect as of the date of the first ISO approved revision of the functional specification for the project. With regard to harmonics at the POC, the legal owner of any WPP must comply with Section 11 of IEEE Standard 519, "Recommended practices and requirements for harmonic control in electrical power systems".

The Chinese grid code [13] refers to GB/T 14549-1993, "Quality of electric energy supply harmonics in public supply network" and GB 12326-2000 "Power quality voltage fluctuation and flicker".

2.3 Low-voltage Ride Through Requirement

Low-voltage ride through (LVRT) is a WPP's capability to stay connected under specified low-voltage conditions at the POC, normally caused by faults in the grid. In some grid codes, WPPs are also requested to supply a maximum reactive current during specified low-voltage conditions. Moreover, some grid codes also specify a requirement for post-fault active power recovery. The low-voltage conditions at the POC that are specified in various grid codes are described in the following subsections.

2.3.1 LVRT Requirement in the UK

The UK grid code specifies LVRT requirements for both onshore and offshore WPPs.

LVRT Requirements for Onshore WPPs

For short-circuit faults on the transmission system of up to 140 ms duration, resulting in zero voltage on the faulted phase(s) at the point of fault, all wind turbines should remain transiently stable and connected to the system without tripping. Two examples of short-circuit faults on the transmission system cleared within 140 ms are shown in Figures 2.13 and 2.14. Following fault clearance, recovery of the voltage on the transmission system to 90% may take longer than 140 ms. During the fault, wind turbines in WPPs should generate maximum reactive current without exceeding the transient rating limit.

When the grid voltage dips are longer than 140 ms (up to 3 min), the LVRT requirements are as shown in Figure 2.15. This profile is not a voltage–time response curve

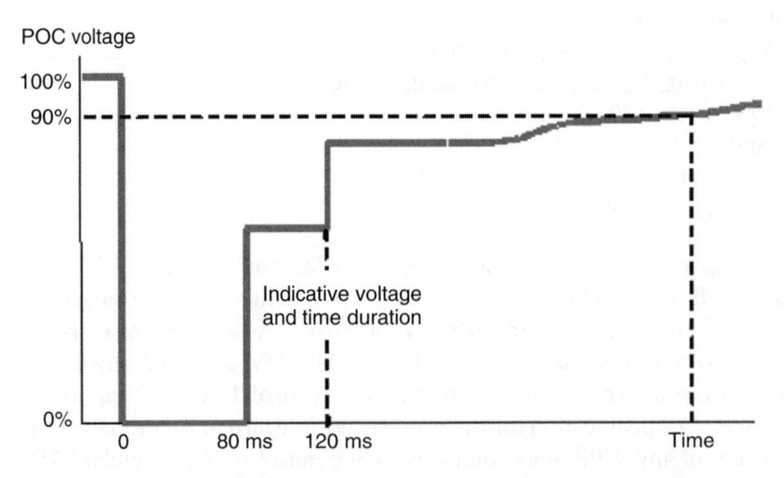

Figure 2.13 Voltage recovery example of short-circuit faults cleared within 140 ms by two circuit breakers, as specified in the UK grid code for onshore WPPs [4].

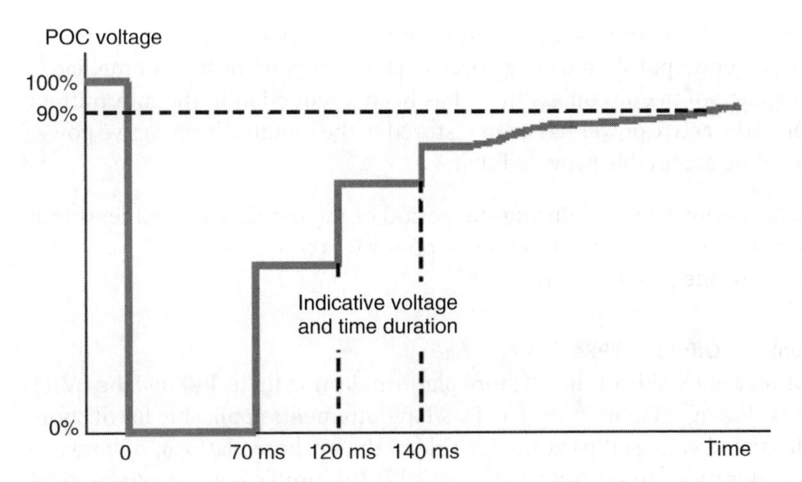

Figure 2.14 Voltage recovery example of short-circuit faults cleared within 140 ms by three circuit breakers, as specified in the UK grid code for onshore WPPs [4].

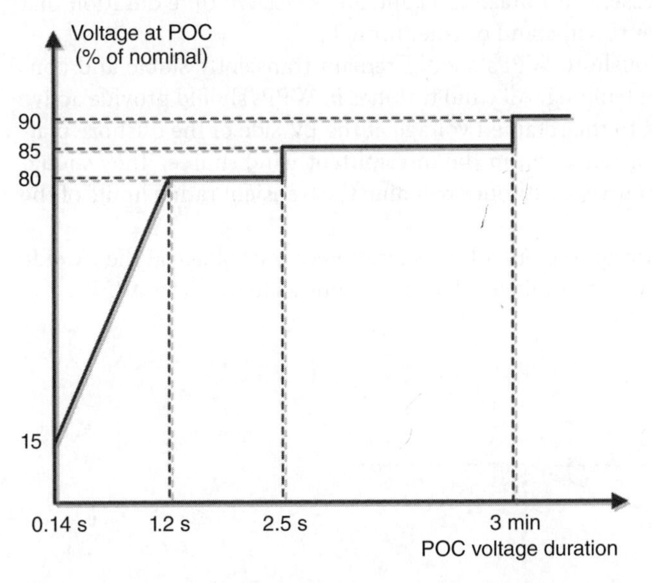

Figure 2.15 LVRT requirements with voltage dips greater than 140 ms and up to 3 min for onshore WPPs in the UK [4].

that would be obtained by plotting the transient voltage response at a point on the transmission system to a disturbance. Rather, each point on the profile represents a voltage level and an associated time duration that connected WPPs must withstand or ride through. For the voltage level with the associated time duration according to the profile, all wind turbines within WPPs should remain transiently stable and connected without tripping. Active power output should be at least proportional to the retained voltage on the transmission system unless there has been a reduction in the intermittent wind source.

Within 1 s of restoration of the voltage on the transmission system to 90% of the nominal value, active power output should be restored to at least 90% of the level immediately before the occurrence of the dip unless there has been a reduction in the intermittent wind source. Once the active power has been restored to the required level, active power oscillations should be acceptable provided that:

- the total active energy delivered during the period of the oscillations is at least that which would have been delivered if the active power was constant
- the oscillations are adequately damped.

LVRT Requirements for Offshore WPPs

For voltage dips on the LV side of the offshore platform lasting up to 140 ms, the LVRT requirement is as shown in Figure 2.16. The LVRT requirements applicable for offshore WPPs during balanced voltage dips at the LV side of the offshore platform of between 140 ms and 3 min duration are as shown in Figure 2.17. This profile is not a voltage–time response curve that would be obtained by plotting the transient voltage response at the LV side of the offshore platform to a disturbance. Rather, each point on the profile (that is, the heavy black line) represents a voltage level and an associated time duration that connected offshore WPPs must withstand or ride through.

All wind turbines in the offshore WPPs should remain transiently stable and connected to the system without tripping. All wind turbines in WPPs should provide active power at least in proportion to the retained voltage at the LV side of the offshore platform unless there has been a reduction in the intermittent wind source. They should generate maximum reactive current without exceeding the transient rating limits of the wind turbines.

The active power recovery requirement after restoration of the voltage at the LV side of the offshore platform to 90% of nominal value is the same as for onshore WPPs.

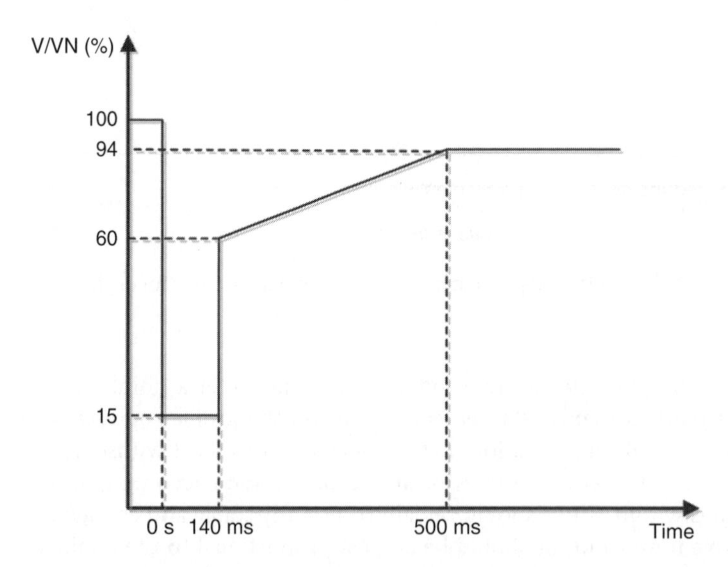

Figure 2.16 UK FRT requirements for voltage dips on the LV side of the offshore platform of up to 140 ms [4].

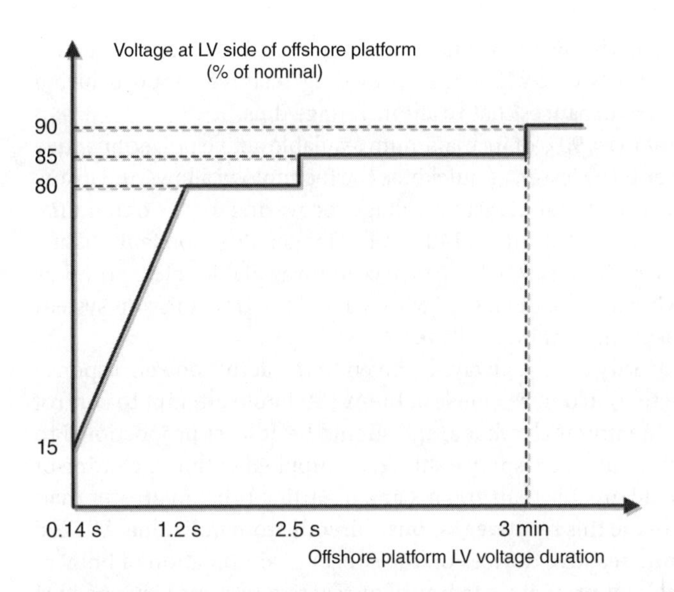

Figure 2.17 UK FRT requirements for balanced voltage dips greater than 140 ms and up to 3 min [4].

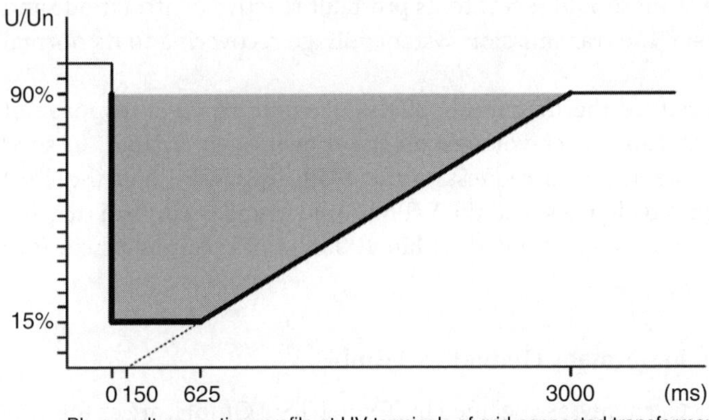

Figure 2.18 FRT capability requirements for controllable WPPs in Ireland [5].

2.3.2 LVRT Requirement in Ireland

In Ireland, a controllable WPP should remain connected to the transmission system during transmission system voltage dips on any or all phases, and should remain stable, with the transmission system phase voltage measured at the HV terminals of the grid-connected transformer remaining above the heavy black line in Figure 2.18.

In addition to remaining connected to the transmission system, the WPP should have the technical capability to provide the following functions:

- During transmission system voltage dips, the WPP should provide active power in proportion to retained voltage and provide reactive current to the transmission system. The provision of reactive current should continue until the transmission system voltage recovers to within the normal operational range, or for at least 500 ms, whichever is the sooner. The WPP may use all or any available reactive sources, including installed STATCOMs or SVCs, when providing reactive support during transmission system fault disturbances that result in voltage dips.

- The WPP should provide at least 90% of its maximum available active power or active power set-point, whichever is the lesser, as quickly as the technology allows and in any event within 500 ms of the transmission system voltage recovering to 90% of nominal voltage for fault disturbances cleared within 140 ms. For longer-duration fault disturbances, the WPP should provide at least 90% of its maximum available active power or active power set-point, whichever is the lesser, within 1 s of the transmission system voltage recovering to 90% of the nominal voltage.
- During and after faults, priority should always be given to the active power response as defined above. The reactive current response of the WPP should attempt to control the voltage back towards the nominal voltage, and should be at least proportional to the voltage dip. The reactive current response should be supplied within the rating of the WPP, with a rise time no greater than 100 ms and a settling time no greater than 300 ms. The WPP may provide this reactive response directly from individual WTGs, or other additional dynamic reactive devices on the site, or a combination of both.
- The WPP should be capable of providing its transient reactive response irrespective of the reactive control mode in which it was operating at the time of the transmission system voltage dip. The WPP should revert to its pre-fault reactive control mode and set-point within 500 ms of the transmission system voltage recovering to its normal operating range.
- The TSO may seek to reduce the magnitude of the dynamic reactive response of the WPP if it is found to cause over-voltages on the transmission system. In such a case, the TSO will make a formal request to the WPP. The WPP and the TSO should agree on the required changes, and the WPP should formally confirm that any requested changes have been implemented within 120 days of receipt of the TSO's formal request.

2.3.3 LVRT Requirement in Germany (Tennet TSO GmbH)

The LVRT requirements in Germany, as specified by Tennet TSO GmbH, are as follows. Three-phase short circuits or fault-related symmetrical voltage dips at the POC above a limit line must not lead to disconnection of the WPP. The pair of values (voltage, time) are also required for asymmetrical faults with reference to the positive sequence system.

Within a different set of limits, the WPP should stay connected and the active power output must be continued immediately after fault clearance and increased to the original value with a gradient between 10% and 20% of the rated power per second.

The WPP must support the grid voltage with additional reactive current during a voltage dip. In the event of a voltage dip of more than 10% of the effective value of the POC voltage, voltage control must be activated. Voltage control must take place within 20 ms of fault recognition, providing a reactive current on the LV side of the WPP transformer amounting to at least 2% of the rated current for each percentage of the voltage dip. A reactive power output of at least 100% of the rated current must be possible if required.

After the voltage returns to the deadband, the voltage support must be maintained for a further 500 ms in accordance with the specified characteristic. The transient balancing procedures following voltage return must be complete after 300 ms.

2.3.4 LVRT Requirement in Denmark

In Denmark, WPPs with active power output greater than 1.5 MW must meet the LVRT requirements in Figure 2.19. The requirements must be complied with in the event of symmetrical as well as asymmetrical faults; that is, the requirements apply in case of faults in one, two or three phases. For Area A, the WPP must stay connected to the network and maintain normal production; for Area B, the WPP must stay connected to the network and provide maximum voltage support by supplying a controlled amount of reactive power so as to ensure that the WPP helps to stabilize the voltage within the design framework offered by the current technology, as shown in Figure 2.20; for Area C, disconnecting the WPP is allowed. If the voltage U reverts to area A during a fault sequence, subsequent voltage drops will be regarded as a new fault situation. If several successive fault sequences occur within area B and evolve into area C, disconnection is allowed. During LVRT in Area B, reactive power has the first priority.

The WPP is also required to ride through recurring fault conditions. The details are listed in Table 2.7.

2.3.5 LVRT Requirement in Spain

The LVRT requirements of the Spanish TSO are shown in Figure 2.21. When the voltage at the POC is above the thick black curve, the WPP must stay connected to the grid. Moreover, the WPP should provide reactive power to support the voltage at the POC. The reactive current from the WPP is dependent on the residue voltage at the POC, as shown in Figure 2.22. When the residue voltage at the POC is below 0.85, the WPP should supply reactive current and the reactive current linearly increases from 0 to 0.9

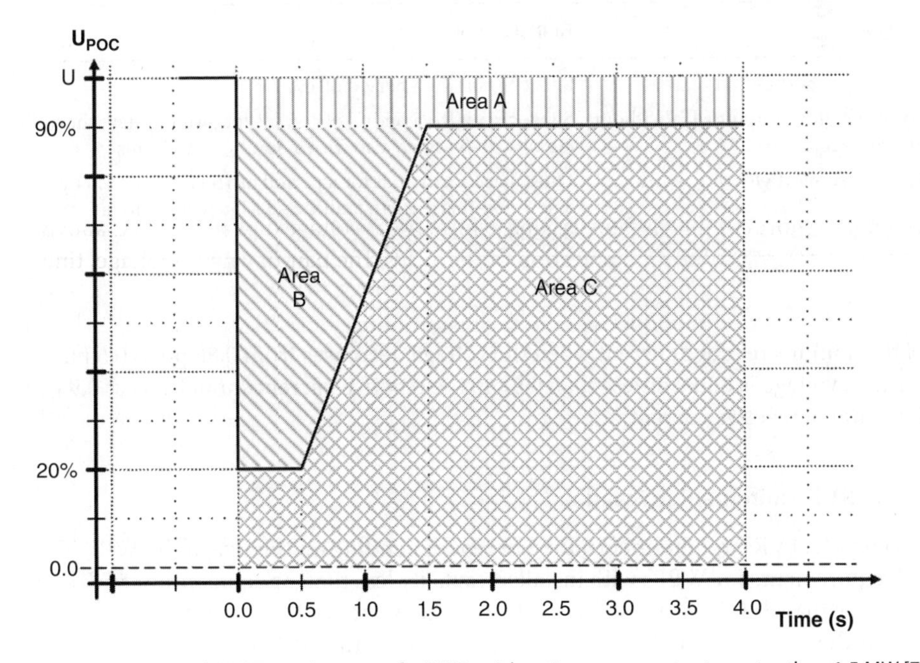

Figure 2.19 Danish LVRT requirements for WPPs with active power output greater than 1.5 MW [7].

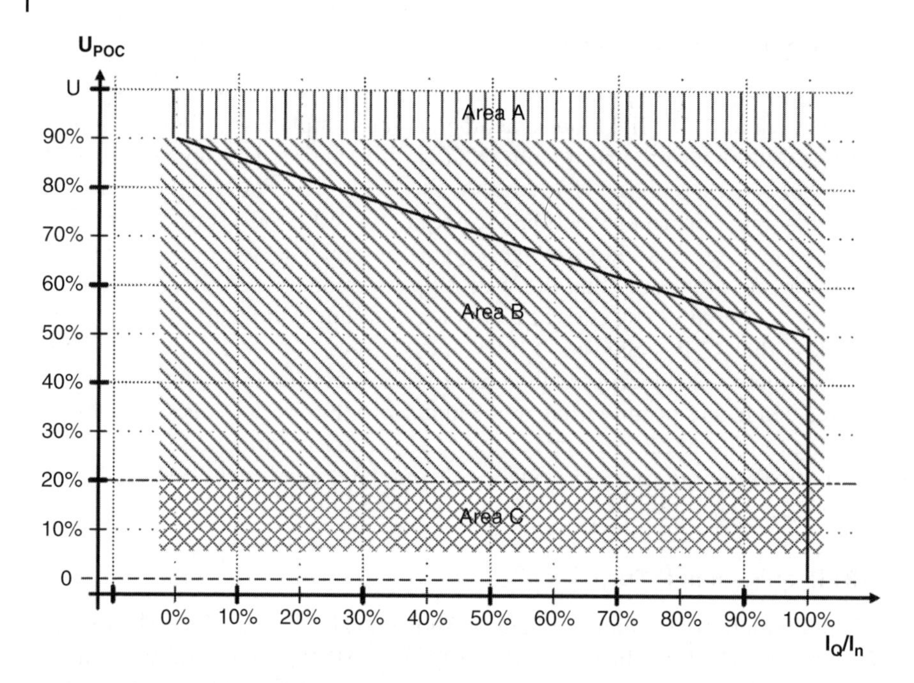

Figure 2.20 Danish requirement for reactive power supply during LVRT for WPPs with active power output greater than 1.5 MW [7].

Table 2.7 Requirements of recurring fault conditions in Denmark.

Fault type	Fault duration
Three-phase short circuit	Short circuit for a period of 150 ms
Two-phase short -circuit with/without earth contact	Short circuit for a period of 150 ms followed by new short circuit 0.5–3 s later, also with a duration of 150 ms
Single-phase short circuit to earth	One-phase earth fault for a period of 150 ms followed by a new one-phase earth fault 0.5–3 s later, also with a duration of 150 ms

pu of the nominal reactive current when the voltage decreases from 0.85 pu to 0.5 pu. If the residue voltage is equal or less than 0.5 pu, the reactive current should stay at 0.9 pu of the nominal reactive current.

2.3.6 LVRT Requirement in Sweden

In Sweden, the LVRT requirements are different depending on the size of the WPP. The LVRT requirements for WPPs with installed capacity less than or equal to 100 MW and those larger than 100 MW are shown in Figures 2.23 and 2.24, respectively. When the voltage at the POC is at or above the thick black curve, the WPP must stay connected to the grid.

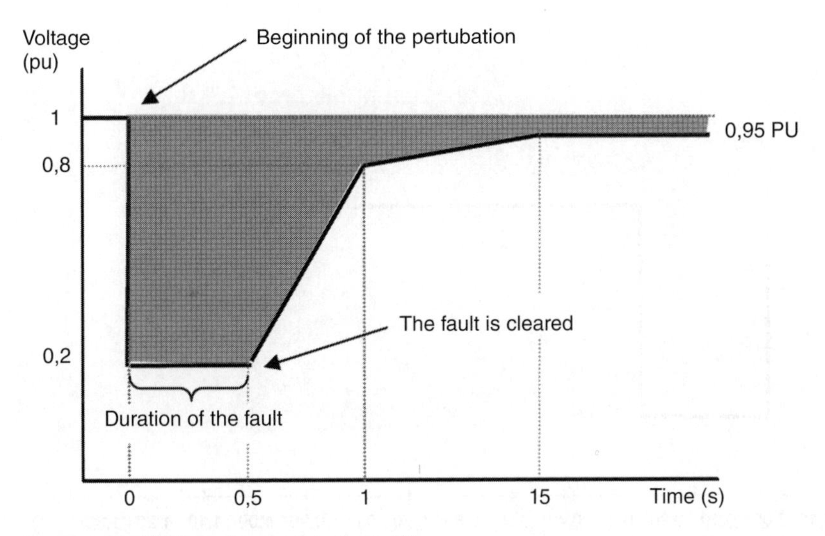

Figure 2.21 LVRT requirements in Spain [8].

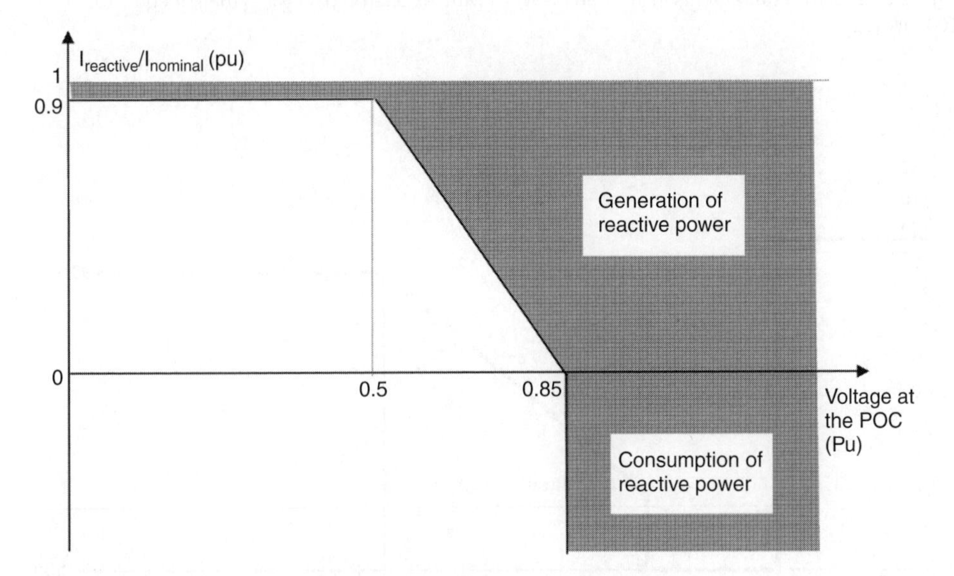

Figure 2.22 Requirement for reactive power supply during LVRT in Spain [8].

2.3.7 LVRT Requirement in the USA

In the USA, a WPP should be able to remain online during voltage disturbances of the durations and associated voltage levels shown in Figure 2.25. Before time 0.0, the voltage at the transformer is the nominal voltage. At time 0.0, the voltage drops. If the voltage remains at a level greater than 15% of the nominal voltage for a period that does not exceed 0.625 s, the plant must stay online. Further, if the voltage returns to 90% of the nominal voltage within 3 s of the beginning of the voltage drop (with the voltage at any given time never falling below the minimum voltage indicated by the solid line in Figure 2.25), the WPP must stay online.

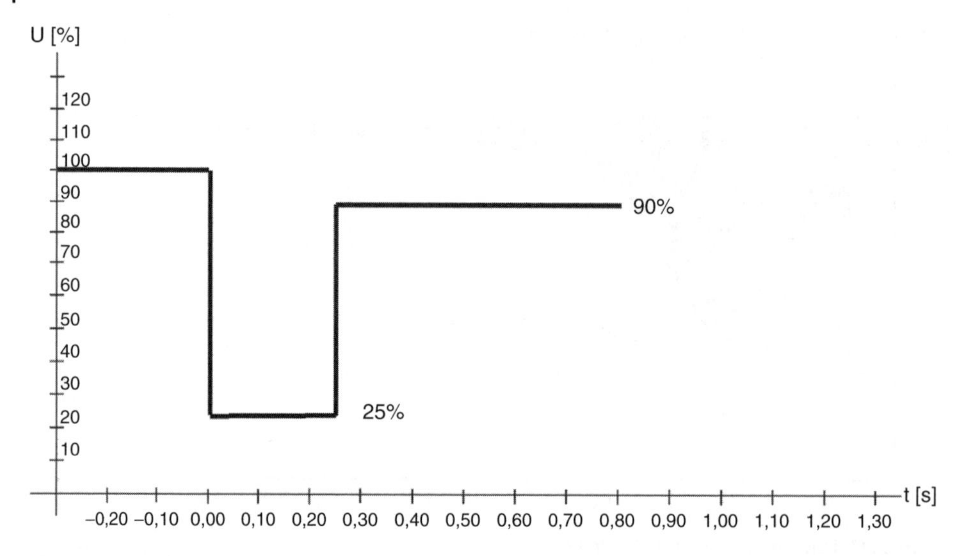

Figure 2.23 LVRT requirements in Sweden for WPPs with installed capacity less than or equal to 100 MW [9].

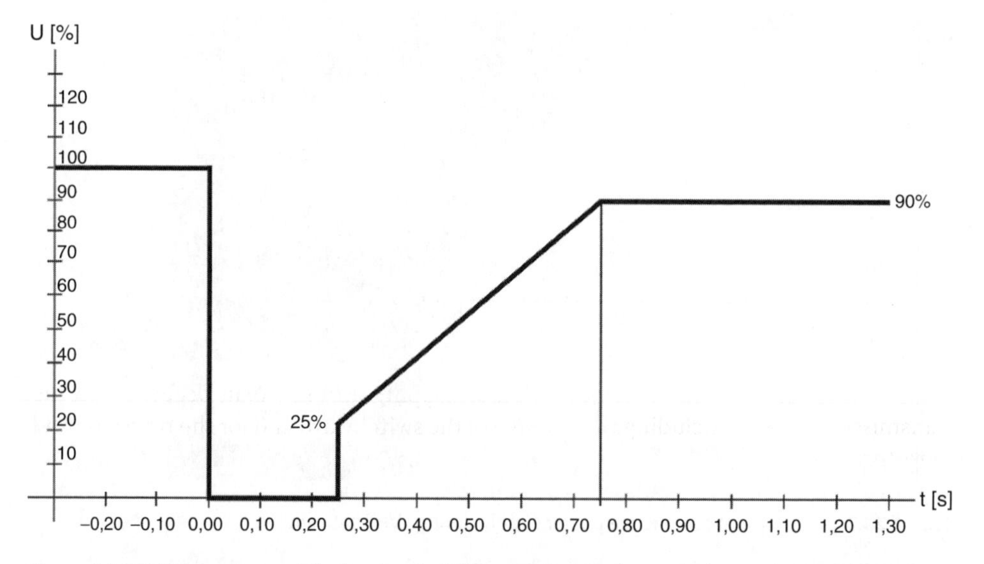

Figure 2.24 LVRT requirements in Sweden for WPPs with installed capacity larger than 100 MW [9].

2.3.8 LVRT Requirement in Quebec and Alberta

The LVRT requirements in Quebec are shown in Figure 2.26. WPPs must remain in service without tripping when, following a disturbance, the positive sequence voltage on the HV side of the switchyard is:

- less than 0.9 pu, but greater than or equal to 0.85 pu, for less than 30 s
- less than 0.85 pu but greater than or equal to 0.75 pu for less than 2 s.

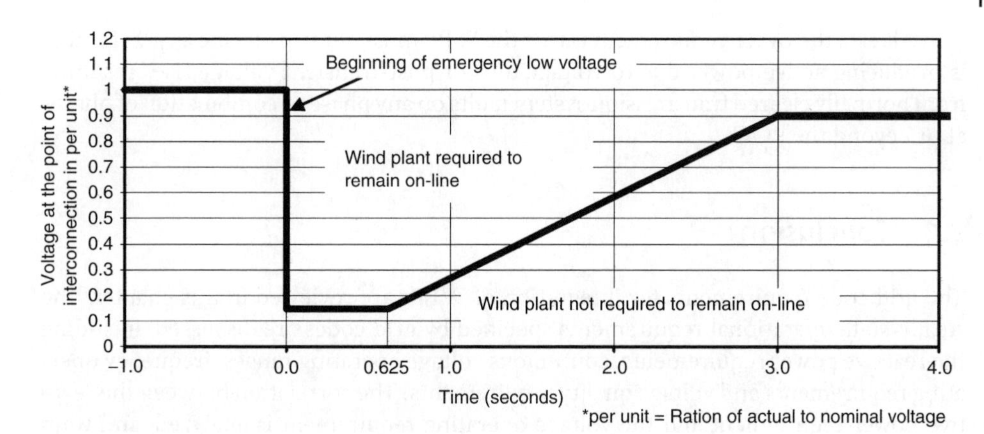

Figure 2.25 US LVRT requirements for WPPs [10].

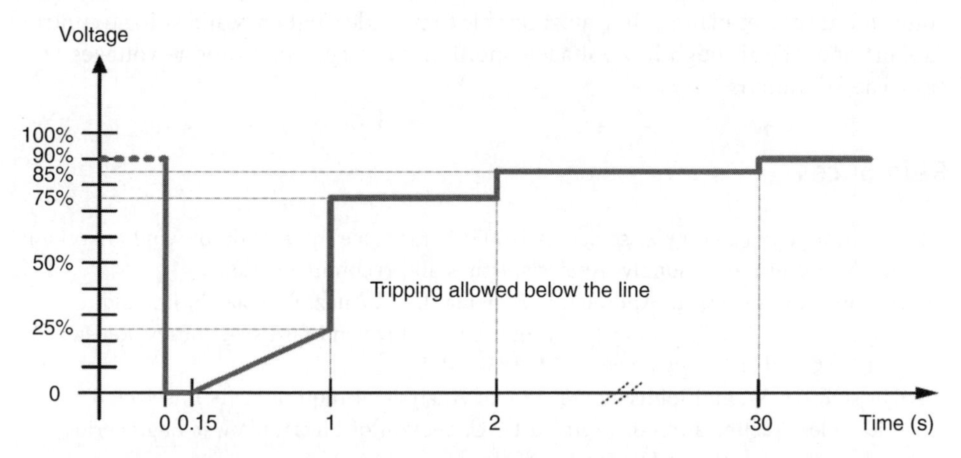

Figure 2.26 LVRT requirements for WPPs in Quebec [11].

WPPs must also remain in service without tripping during a fault occurring on the transmission system (including the HV side of the switchyard) and for the time required to restore voltage after the fault is cleared. The fault can be a three-phase fault cleared in 0.15 s or a two-phase-to-ground fault or phase-to-phase fault cleared in 0.15 s or a single-phase-to-ground fault cleared in 0.30 s. WPPs must also remain in service without tripping during a remote fault cleared by a slow protective device and for the time required to restore voltage after the fault is cleared. The remote fault can be:

- a three-phase fault with the positive sequence voltage at the HV side of the switchyard not below 0.25 pu
- a two-phase-to-ground fault with the positive sequence voltage at the HV side of the switchyard not below 0.5 pu
- a phase-to-phase fault with the positive sequence voltage at the HV side of the switchyard not below 0.6 pu
- a single-phase-to-ground fault with the positive sequence voltage at the HV side of the switchyard not below 0.7 pu.

In Alberta the LVRT requirement is that the WPP must not trip off-line any WTG that is producing active power due to voltage dips or post-transient voltage rises resulting from normally cleared transmission system faults on any phase or combination of phases at or beyond the POC.

2.4 Conclusion

The grid code requirements for integration of WPPs are reviewed in this chapter. The steady-state operational requirements specified by grid codes are discussed, including the reactive power requirements, continuous voltage operating ranges, frequency operating requirements and voltage quality requirements. The correlation between the reactive power requirement and the voltage operating requirement is analyzed, and ways to meet both requirements are presented. The dynamic LVRT requirement is also discussed. With increasing penetration of wind power, WPPs are required to operate like conventional power plants. They must be able to provide frequency and voltage control support and ride through low voltage conditions with specified residue voltages and associated durations.

References

1. European Wind Energy Association (2005) Large scale integration of wind energy in the European power supply: Analysis, issues and recommendations.
2. Martinez de Alegrıa, I., Andreua, J., Martina, J.L., Ibanez, P., Villateb, J.L. and Camblong, H. (2007) Connection requirements for wind farms: A survey, *Renewable and Sustainable Energy Reviews,* **11**, 1858–1872.
3. Christiansen W. and Johnsen, D.T. (2006) Analysis of requirements in selected grid codes, Technical report, Ørsted DTU, Section of Electric Power Engineering, Technical University of Denmark (DTU).
4. National Grid (2010) The grid code. http://www2.nationalgrid.com/UK/Industry-information/Electricity-codes/Grid-code/The-Grid-code/.
5. EirGrid (2009) EirGrid grid code V6. http://www.eirgridgroup.com/site-files/library/EirGrid/GridCodeVersion6.pdf.
6. Tennet TSO GmbH (2015) Onshore grid code 2015. http://www.tennet.eu/electricity-market/german-customers/grid-customers/grid-connection-regulations/.
7. Energine.dk (2010) Technical regulation 3.2.5 for wind power plants with a power output greater than 11 kW.
8. RED Electrica de Espana (2005) PO 12.3. Requisitos de respuesta frente a huecos de tension de las instalaciones de produccion de regimen especial.
9. Svenska Kraftna (2005) SvKFS 2005:2 Driftsäkerhetsteknisk utformning av produktionsanläggningar.
10. Federal Energy Regulatory Commission (2005) Interconnection for wind energy.
11. Hydro Quebec Transenergie (2009) Transmission provider technical requirements for the connection of power plants to the Hydro-Quebec transmission system.
12. Alberta Electric System Operator (2015) Wind power facility technical requirements, Rev. 0.

3

Control of Doubly-fed Induction Generators for Wind Turbines

Guojie Li[1] and Lijun Hang[2]

[1] *Department of Electrical Engineering, Shanghai Jiaotong University, China*
[2] *Hangzhou Dianzi University, Hangzhou, China*

3.1 Introduction

The doubly-fed introduction generator (DFIG) has a growing market. The term "doubly-fed" refers to the fact that the voltage on the stator is applied from the grid and the voltage on the rotor is induced by the power converter. It also means there is both stator- and rotor-fed power to the grid. This system allows a variable-speed operation over a large but restricted range (±30% of synchronous speed). The converter compensates for the difference between the mechanical and electrical frequency by injecting a rotor current with a variable frequency. Both during normal operation and faults, the behavior of the generator is thus governed by the power converter and its controllers [1, 2].

This chapter is arranged as follows. The principles of DFIGs are introduced in Section 3.2. PQ control and direct torque control of DFIGs is presented in Sections 3.3 and 3.4, respectively. Low-voltage ride through of the DFIG is described in Section 3.5. In Section 3.6, conclusions are drawn.

3.2 Principles of Doubly-fed Induction Generator

Doubly-fed induction generator (DFIG) technology is widely used in wind turbine applications. This is mainly because the power electronic converter only has to handle a fraction (20–30%) of the total power and therefore is cheaper. Figure 3.1 shows a schematic diagram of a DFIG-based wind power generation system [1–3].

As shown in Figure 3.1, the DFIG consists of a wound rotor induction generator (WRIG) with the stator windings directly connected to the constant-frequency three-phase grid and with the rotor windings mounted to a bidirectional back-to-back voltage source power converter. The power converter consists of two parts: the rotor-side converter and grid-side converter.

One task of the grid-side converter in Figure 3.1 is to keep the DC-link voltage constant and convey the slip power from the rotor to the grid. Its secondary role is to provide reactive power to assist in grid voltage support. The main control strategy of the DFIG

Modeling and Modern Control of Wind Power, First Edition. Edited by Qiuwei Wu and Yuanzhang Sun.

Companion website: www.wiley.com/go/wu/modeling

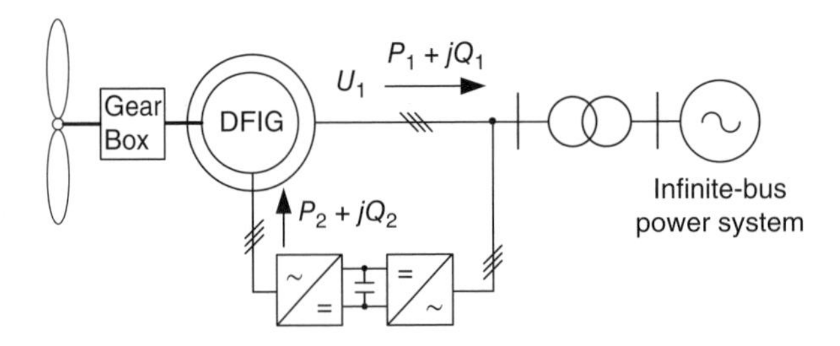

Figure 3.1 Schematic of a DFIG-based wind power generation system.

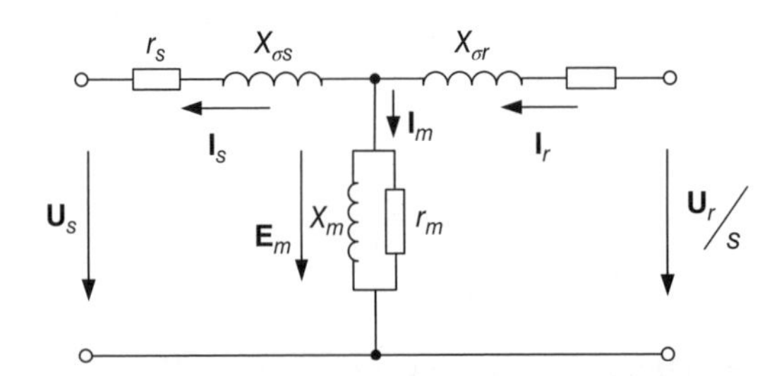

Figure 3.2 Equivalent circuit of DFIG. U_s, stator voltage vector; I_s, stator current vector; r_s, stator resistance; $X_{\sigma s}$, stator leakage reactance; U_r, rotor voltage vector; I_r, rotor current vector; r_r, rotor resistance; $X_{\sigma r}$, rotor leakage reactance; E_m, magnetizing voltage vector; I_m, magnetizing current vector; X_m, magnetizing reactance; r_m, magnetizing resistance.

is implemented on the rotor-side converter, regulating the active and reactive power to the grid [1–3].

The equivalent circuit of the DFIG, including the magnetizing losses, can be seen in Figure 3.2. It is similar to that of the synchronous generator except the excitation voltage is different, because it is controlled from a more complex rotor equivalent circuit.

The slip, s, equals

$$s = \frac{w_s - n_p w_m}{w_s} = \frac{w_r}{w_s} \tag{3.1}$$

where ω_s is the angular speed of the stator current flux, ω_r is the angular speed of the rotor current flux, ω_m is the angular speed of the generator (mechanical), and n_p is the number of pole pairs of the machine.

Alternatively,

$$f_s = n_p\ f_m \pm f_r$$

where f_s, f_m, and f_r are the stator current frequency, the rotor mechanical frequency, and the rotor current frequency, respectively. In a sub-synchronous mode, "+" is adopted for f_r, while "−" is adopted in an over-synchronous mode.

Neglecting the magnetizing losses, the following equations can be obtained:

$$\begin{cases} \mathbf{U}_s = -r_s\mathbf{I}_s - X_{\sigma s}\mathbf{I}_s + \mathbf{E}_m \\ \dfrac{\mathbf{U}_r}{s} = \dfrac{r_r}{s}\mathbf{I}_r + X_{\sigma s}\mathbf{I}_r + \mathbf{E}_m \end{cases} \tag{3.2}$$

The electromagnetic power of the DFIG presents the mechanical power arriving at the stator side through the air gap. It can be expressed as:

$$P_{em} = \text{Re}(\mathbf{E}_m \mathbf{I}_s^*) = \text{Re}(X_m \mathbf{I}_m \mathbf{I}_s^*) \tag{3.3}$$

where, $\mathbf{I}_s^*$ is the conjugate of $\mathbf{I}_s$. Substituting $\mathbf{I}_m = -\mathbf{I}_s + \mathbf{I}_r$ into the above equation, we obtain:

$$P_{em} = \text{Re}(X_m \mathbf{I}_r \mathbf{I}_s^*) \tag{3.4}$$

The stator power can be determined as

$$P_s = \text{Re}(\mathbf{U}_s \mathbf{I}_s^*) \tag{3.5}$$

Substituting the first equation of (3.2) into the above, we get

$$P_s = -r_s I_s^2 + \text{Re}(\mathbf{E}_m \mathbf{I}_s^*) = -P_{cus} + P_{em} \tag{3.6}$$

The rotor power can be determined as

$$P_r = \text{Re}(\mathbf{U}_r \mathbf{I}_r^*) \tag{3.7}$$

Multiplying by $\mathbf{I}_r^*$ the second equation of (3.2), we have:

$$P_r = r_r I_r^2 + s\,\text{Re}(\mathbf{E}_m \mathbf{I}_r^*) = P_{cur} + s\,\text{Re}(\mathbf{E}_m \mathbf{I}_r^*) \tag{3.8}$$

Through (3.3) and (3.8), we get the rotor power as:

$$P_r = P_{cur} + sP_{em} \tag{3.9}$$

P_r represents the active power injected to the rotor side.

Because

$$P_m = P_{em} - sP_{em} = (1 - s)P_{em} \tag{3.10}$$

combining (3.6), (3.9), and (3.10), we obtain:

$$P_m = P_s + P_{cus} - P_r + P_{cur} \tag{3.11}$$

Thus the power flow of the DFIG according to the above equations is as shown in Figure 3.3. Depending on the operating conditions of the drive, power is fed into or out of the rotor:

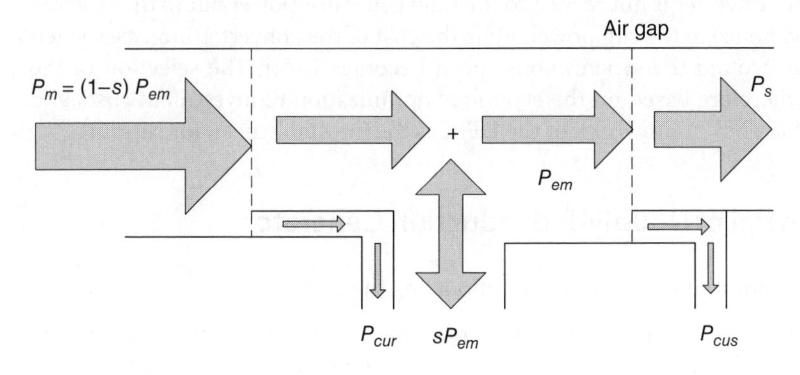

Figure 3.3 Power flow of a DFIG system.

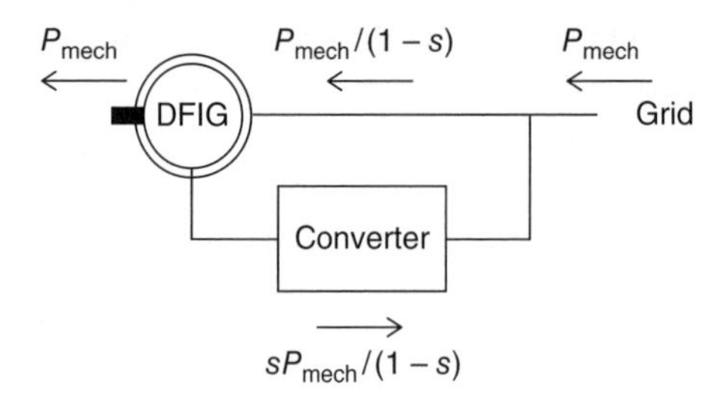

Figure 3.4 Power flow of a DFIG system neglecting losses.

- In sub-synchronous mode, $s > 0$, power is fed into the rotor via the converter.
- In over-synchronous mode, $s < 0$, it flows from the rotor via the converter to the grid.

In both cases – sub-synchronous and over-synchronous – the stator feeds energy into the grid. In addition, if $s = 0$, the converter only provides reactive power to the rotor.

Neglecting the resistive and magnetizing losses, the stator power can be expressed as:

$$P_s \approx P_{\text{mech}}/(1 - s) \tag{3.12}$$

and the rotor power as:

$$P_r \approx -sP_{\text{mech}}/(1 - s) \tag{3.13}$$

Therefore, the rotor power is approximately minus the stator power times the slip:

$$P_r \approx -sP_s \tag{3.14}$$

Thus, it can be seen that the rotor power is a fraction of the stator power. Therefore, the power through the converter is much less than the rating of the DFIG.

Finally, the power flow of the DFIG is as shown in Figure 3.4.

The DFIG has several advantages. It has the ability to control reactive power and to decouple active and reactive power control by independently controlling the rotor excitation current. It does not necessarily have to be magnetized from the power grid; it can be magnetized from the rotor circuit, too. It is capable of generating reactive power that can be delivered to the stator by the grid-side converter. However, the grid-side converter normally operates at unity power factor and is not involved in the reactive power exchange between the turbine and the grid. In the case of a weak grid, where the voltage may fluctuate, the DFIG may be ordered to produce or absorb an amount of reactive power to or from the grid, with the purpose of voltage control.

The size of the converter is not related to the total generator power but to the selected speed range and hence to the slip power. Thus the cost of the converter increases when the speed range around the synchronous speed becomes wider. The selection of the speed range is therefore based on the economic optimization of investment costs and on increased efficiency. A drawback of the DFIG is the inevitable need for slip rings.

3.3 PQ Control of Doubly-fed Induction Generator

The following assumptions are adopted for modeling the DFIG:

- It is a symmetrical three-phase induction machine, the magnetic potential of each phase current in the air-gap space is sinusoidal in distribution, and the cogging effect is ignored.

- The influence of temperature and frequency change on the DFIG parameters are not considered.
- The saturation of the magnetic circuit and the loss of the core are not considered.

The DFIG model is obtained under *dq* synchronous coordinates. The voltage equations and flux equations are derived as follows [2–4]:

$$\begin{cases} u_{ds} = -r_s i_{ds} + p\psi_{ds} - \psi_{qs}\omega_s \\ u_{qs} = -r_s i_{qs} + p\psi_{qs} + \psi_{ds}\omega_s \\ u_{dr} = r_r i_{dr} + p\psi_{dr} - \psi_{qr}\omega_r \\ u_{qr} = r_r i_{qr} + p\psi_{qr} + \psi_{dr}\omega_r \end{cases} \tag{3.15}$$

$$\begin{cases} \psi_{ds} = -L_s i_{ds} + L_m i_{dr} \\ \psi_{qs} = -L_s i_{qs} + L_m i_{qr} \\ \psi_{dr} = L_r i_{dr} - L_m i_{ds} \\ \psi_{qr} = L_r i_{qr} - L_m i_{qs} \end{cases} \tag{3.16}$$

where:

- p is the differential operator
- subscripts d and q represent d, q components, respectively
- u_s, ψ_s, i_s, L_s, and r_s are the stator voltage, flux, current, inductance, and resistance, respectively
- u_r, ψ_r, i_r, L_r, and r_r are the rotor voltage, flux, current, inductance, and resistance, respectively.

3.3.1 Grid-side Converter

The grid-side converter (GSC) is a pulse-width modulation (PWM) voltage source converter (VSC). The structure of the GSC is shown in Figure 3.5. The GSC connects with the grid via equivalent inductance and resistance.

According to the figure, the equations of the ac side of the GSC under the *dq* synchronous reference frame are obtained as:

$$\begin{cases} pi_{dg} = -\dfrac{r_g}{L_g} i_{dg} + \omega_s i_{qg} + \dfrac{1}{L_g}(u_{ds} - u_{dg}) \\ pi_{qg} = -\dfrac{r_g}{L_g} i_{qg} - \omega_s i_{dg} + \dfrac{1}{L_g}(u_{qs} - u_{qg}) \end{cases} \tag{3.17}$$

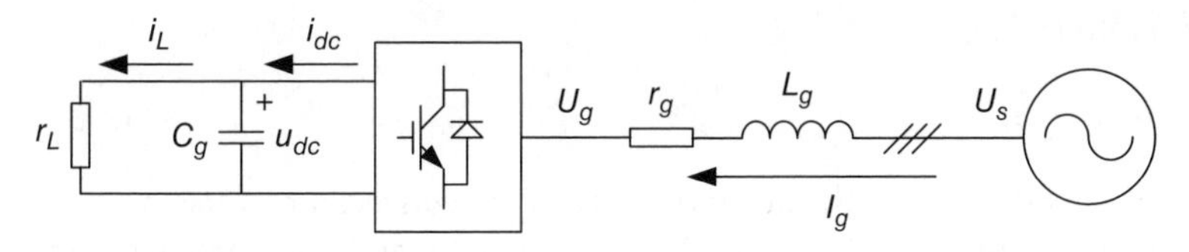

Figure 3.5 Grid-side converter structure.

where

- u_{dg} and i_{dg} are d components of the grid voltage and current, respectively
- u_{qg} and i_{qg}, are q components of the grid voltage and current, respectively
- L_g is the equivalent inductance from the converter to the grid.

The d axis of the dq reference frame will be aligned with the grid voltage. So,

$$\begin{cases} u_{ds} = u_s \\ u_{qs} = 0 \end{cases} \tag{3.18}$$

Substituting (3.18) into (3.17), we get:

$$\begin{cases} pi_{dg} = -\dfrac{r_g}{L_g} i_{dg} + \omega_s i_{qg} + \dfrac{1}{L_g}(u_s - u_{dg}) \\ pi_{qg} = -\dfrac{r_g}{L_g} i_{qg} - \omega_s i_{dg} - \dfrac{1}{L_g} u_{qg} \end{cases} \tag{3.19}$$

Equation 3.19 can be rewritten as:

$$\begin{cases} u_{dg} = -(r_g + L_g p) i_{dg} + \omega_s L_g i_{qg} + u_s \\ u_{qg} = -(r_g + L_g p) i_{qg} - \omega_s L_g i_{dg} \end{cases} \tag{3.20}$$

The dc-side equations can be expressed as:

$$C_g \frac{d}{dt} u_{dc} = i_{dc} - i_L \tag{3.21}$$

For constant power ac/dq transformation, the active and reactive power can be expressed as:

$$\begin{cases} P_g = u_{ds} i_{dg} + u_{qs} i_{qg} \\ Q_g = u_{qs} i_{dg} - u_{ds} i_{qg} \end{cases} \tag{3.22}$$

Substituting (3.18) into (3.22), we have:

$$\begin{cases} P_g = u_s i_{dg} \\ Q_g = -u_s i_{qg} \end{cases} \tag{3.23}$$

The above equations show that the active power is proportional to i_{dg} when the grid voltage u_s remains constant, while the reactive power is proportional to i_{qg}. That means the active and reactive power can be controlled by i_{dg} and i_{qg} independently.

Neglecting converter losses, the following is obtained:

$$P_g = u_s i_{dg} = u_{dc} i_{dc} \tag{3.24}$$

Substituting (3.24) into (3.21),

$$C_g u_{dc} \frac{d}{dt} u_{dc} + u_{dc} i_L = u_s i_{dg} \tag{3.25}$$

From the above, it can be seen that the dc bus voltage of the converter can also be controlled by i_{dg}. According to (3.22), (3.23), and (3.25), the decoupled PQ control strategy can be shown as in Figure 3.6.

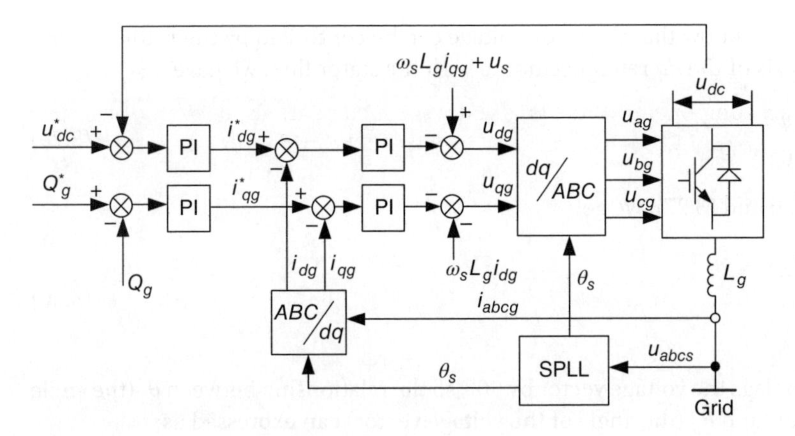

Figure 3.6 PQ decoupled control strategy of the DFIG.

From the above control strategy, we can see that the main objective of the grid-side converter is to control the DC-link voltage. The control of the GSC consists of a fast inner current control loop, which controls the current through the grid filter, and an outer slower control loop that controls the DC-link voltage. The reference frame of the inner current control loop will be aligned with the grid voltage. This means that the *d* component of the grid-filter current will control the active power delivered from the converter and the *q* component of the filter current will, accordingly, control the reactive power. This implies that the outer DC-link voltage control loop has to act on the *d* component of the grid-filter current.

3.3.2 Rotor-side converter

DFIG Control Strategy Before Connecting to the Grid

Before connecting to the grid, the voltage of the DFIG needs to be set up to follow the grid. Afterwards the DFIG can connect to the grid smoothly. The following shows how to establish the stator voltage by the rotor-side converter (RSC).

The stator current is zero, then

$$\begin{cases} i_{ds} = 0 \\ i_{qs} = 0 \end{cases} \tag{3.26}$$

Substituting (3.26) into (3.16), we get

$$\begin{cases} \psi_{ds} = L_m i_{dr} \\ \psi_{qs} = L_m i_{qr} \\ \psi_{dr} = L_r i_{dr} \\ \psi_{qr} = L_r i_{qr} \end{cases} \tag{3.27}$$

Then, the last two equations in (3.15) can be simplified as:

$$\begin{cases} u_{dr} = (r_r + L_r p) i_{dr} - \omega_r L_r i_{qr} \\ u_{qr} = (r_r + L_r p) i_{qr} + \omega_r L_r i_{dr} \end{cases} \tag{3.28}$$

From the above, we know that the rotor voltage can be controlled by the rotor current. Aligning the d axis of the dq reference frame with the stator flux, we have

$$\begin{cases} \psi_{ds} = \psi_s \\ \psi_{qs} = 0 \end{cases} \tag{3.29}$$

Combining (3.29) and (3.27), we get:

$$\begin{cases} i_{dr} = \dfrac{\psi_s}{L_m} \\ i_{qr} = 0 \end{cases} \tag{3.30}$$

The flux vector lags the voltage vector by 90°, so the relationship between θ_s (the angle of the flux vector) and θ_u (the angle of the voltage vector) can expressed as:

$$\theta_s = \theta_u - \frac{\pi}{2} \tag{3.31}$$

The rotor voltage and current frequency is the slip frequency. Therefore, the coordinate transformation between abc and dq needs a differential rotation angle θ_{sr}. This can be calculated as:

$$\theta_{sr} = \theta_s - \theta_r \tag{3.32}$$

where θ_r is the rotor position angle. It can normally be obtained through the encoder.

Thus, before connecting to the grid, we can obtain the DFIG control strategy as in Figure 3.7.

Figure 3.7 shows that the magnitude, frequency, and phase angle of the stator voltage can be regulated through the rotor current. Similar to the grid connection for the synchronous machine, the DFIG will be connected to the grid when there is no difference

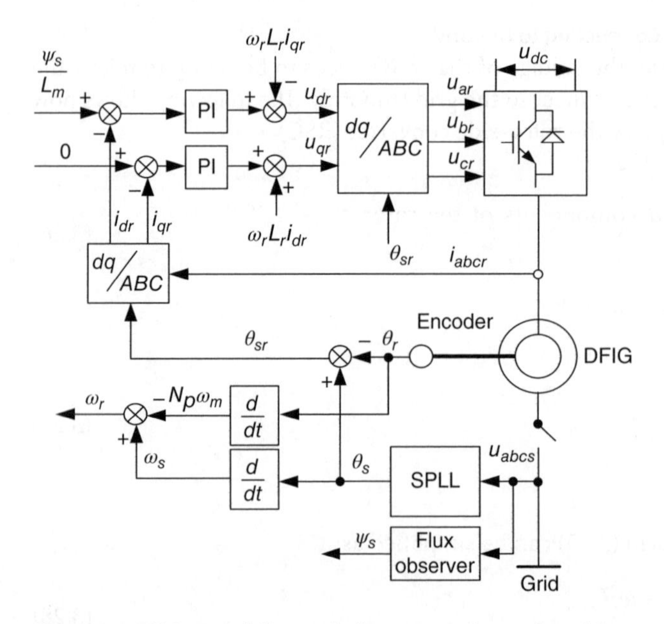

Figure 3.7 DFIG control diagram before connecting to the grid.

between the stator and grid voltages. The RSC can control the rotor current rapidly and accurately. Therefore, the grid connection for the DFIG will be carried without inrush current.

DFIG Control Strategy After Connecting to the Grid

After connecting to the grid, the DFIG is to control the active power and reactive power of the stator. The active power and reactive power of the stator are expressed as [3–7]:

$$\begin{cases} P_s = u_{ds}i_{ds} + u_{qs}i_{qs} \\ Q_s = u_{qs}i_{ds} - u_{ds}i_{qs} \end{cases} \tag{3.33}$$

Neglecting flux transient and stator resistance, the following can be found through (3.15) and (3.29):

$$\begin{cases} u_{ds} = 0 \\ u_{qs} = \psi_s\omega_s = u_s \end{cases} \tag{3.34}$$

Substituting (3.34) into (3.33), we obtain:

$$\begin{cases} P_s = \psi_s\omega_s i_{qs} \\ Q_s = \psi_s\omega_s i_{ds} \end{cases} \tag{3.35}$$

The stator current can be found from (3.16) and (3.29):

$$\begin{cases} i_{ds} = \dfrac{L_m i_{dr} - \psi_s}{L_s} \\ i_{qs} = \dfrac{L_m i_{qr}}{L_s} \end{cases} \tag{3.36}$$

Thus the active and reactive power of the stator can be derived as:

$$\begin{cases} P_s = \dfrac{L_m\psi_s\omega_s}{L_s} i_{qr} \\ Q_s = \dfrac{L_m\psi_s\omega_s}{L_s} i_{dr} - \dfrac{\psi_s^2\omega_s}{L_s} \end{cases} \tag{3.37}$$

From the above, it can be seen that the active and reactive power of the stator are proportional to the q and d components of the rotor current if the stator flux and frequency remain constant.

According to (3.16) and (3.36), we have:

$$\begin{cases} \psi_{dr} = \sigma L_r i_{dr} + \dfrac{L_m}{L_s}\psi_s \\ \psi_{qr} = \sigma L_r i_{qr} \end{cases} \tag{3.38}$$

where $\sigma = 1 - \frac{L_m^2}{L_s L_r}$ represents the leakage coefficient of the DFIG.

The rotor voltage equations can be expressed through (3.15) and (3.38) as:

$$\begin{cases} u_{dr} = (r_r + \sigma L_r p)i_{dr} - \sigma\omega_r L_r i_{qr} \\ u_{qr} = (r_r + \sigma L_r p)i_{qr} + \sigma\omega_r L_r i_{dr} + \dfrac{L_m}{L_s}\psi_s\omega_r \end{cases} \tag{3.39}$$

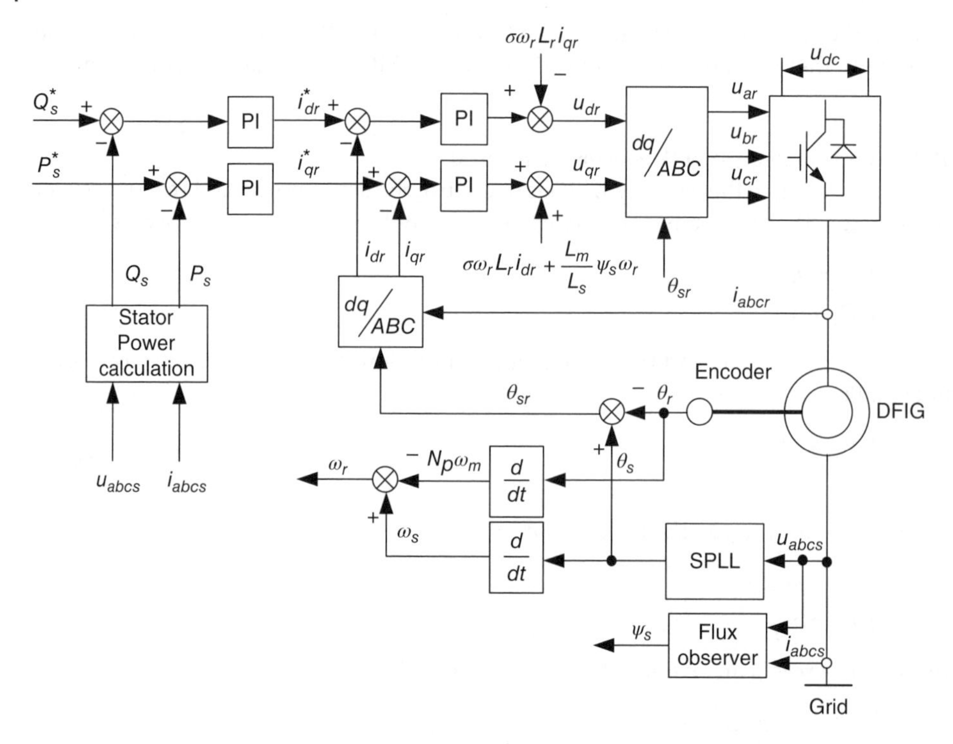

Figure 3.8 DFIG control diagram after connecting to the grid.

Then the DFIG control strategy after connecting to the grid is as shown in Figure 3.8.

Figure 3.8 and equations derived in this section show that the DFIG can be controlled in terms of power, flux, speed, current control, and so on. The rotor position angle and speed can be obtained by the encoder, and some sensorless controls have also been studied [8].

Here, only inner and outer closed-loop current control are described for the grid-side and rotor-side converters, and they are widely used in practice. Inner current control uses DC control to ensure good dynamic responses of the DFIG. There are, however, some other control methods, such as proportion and resonant control, hysteresis control, and sliding mode control. The most popular is the proportion and integration (PI)-based space vector control [2, 5–7]. Decoupled control of active and reactive power is accomplished by aligning the d-axis either to the stator flux linkage [5–6] or the stator voltage [7].

3.4 Direct Torque Control of Doubly-fed Induction Generators

The direct torque control (DTC) technique was first proposed by I. Takahashi and T. Noguchi in 1986. Since then, DTC has gained popularity and several commercial drives are based on it. DTC originates in the fact that torque and flux are directly controlled by an instantaneous space voltage vector, unlike field oriented control

(FOC). The configuration of DTC is much simpler than FOC. In fact, it consists only of torque and stator flux estimators, two hysteresis comparators, a switching table and a voltage source inverter [9–11].

3.4.1 Features of Direct Torque Control

DTC, as the name indicates, is the direct control of the torque and flux of an electrical motor by the selection, through a look-up table, of the power converter voltage space vectors. The main advantage of DTC is its structural simplicity, since no coordinate transformations, current controllers, and modulations are needed. Moreover, the controller does not depend on the motor parameters. DTC is considered to be a simple and robust control scheme that achieves quick and precise torque control response.

However, the torque and flux modulus values and the sector of the flux are needed and, more importantly, short sampling times for the torque and flux control loops are needed in order to keep electromagnetic torque ripple within acceptable limits [12].

The basic principle of DTC is the selection of an appropriate voltage vector from a switching table. This is used to restrict both torque and flux errors in their respective hysteresis bands. Compared to prior stator flux/voltage orientation control, DTC is characterized by a quick dynamic response, a simple structure and low parameter dependency, so it has become the focus of research throughout the world. Despite the merits above, DTC has some notable drawbacks, including variable switching frequency and irregular torque and flux ripples in the steady state, especially at low speeds.

The basic idea of DTC is control of the flux of the machine by control of the input voltage of the windings. The key point is to establish the relationship between the flux and the switching mode of the inverter. The rotor windings of the inverter machine are connected with a two-level VSC, which can be seen as a universal bridge block, as shown in Figure 3.9.

The DTC strategy is based on space vector pulse width modulation (SVPWM), which is a widely used modulation in motor control. In the theory of SVPWM, there are eight voltage space vectors and each voltage vector represents a switching state for the six switching devices (Q1–Q6). When one state of switching mode is applied to these devices, a respective voltage is applied on the rotor windings of the DFIG. These eight states of the switching devices can be regarded as voltage space vectors V0–V7, as shown in Figure 3.10. For example, the space vector V1(100) means that the devices Q1,Q6, and Q2 are turned on, and Q4, Q3, and Q5 are turned off. V4(011) means Q4, Q3, and Q5 are turned on and Q1, Q6, and Q2 are turned off.

As for the principle of the machine, when the flux ψ is located in the position shown in Figure 3.10, the voltage space vector V3 is applied. According to the relationship between

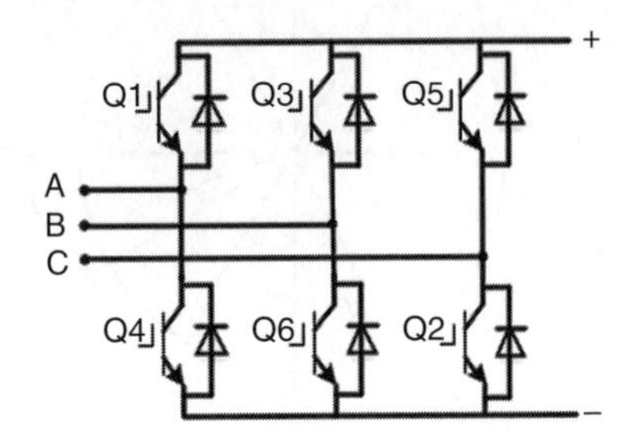

Figure 3.9 Universal bridge block.

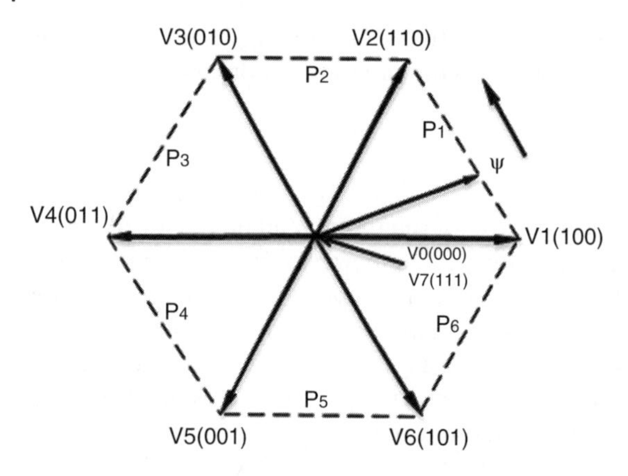

Figure 3.10 Voltage space vectors.

the flux space vector and the voltage space vector:

$$\psi(t) = \int V(t)dt \tag{3.40}$$

As shown in Figure 3.10, the position of the flux ψ will move towards the direction of V3, following the path P1. If the voltage vectors are applied at the right time in proper order, the trajectory of the flux will follow the path P1–P2–P3–P4–P5–P6 and form a hexagon to control the machine. However, the hexagonal trajectory of the flux will cause a large torque ripple and can only be used in certain fields.

In order to reduce the torque ripple, a round flux trajectory is widely used. In order to accurately locate the position of the flux at a particular time, the space is divided into six sectors S1–S6, as shown in Figure 3.11

For example, when the flux position is in sector S1, as shown in Figure 3.12, the voltage vectors V1, V2, and V6 will increase the amplitude of the flux and V3, V4, V5 will reduce the amplitude of the flux; V2 and V3 will make the flux move anticlockwise and V5 and V6 will make the flux move clockwise. When the voltage vectors are applied in the order V3 > V2 > V3, as shown in Figure 3.12, the flux will move towards an approximately arc-shaped trajectory. Therefore, when the proper voltage vectors are applied, the flux will move towards an approximate circle, indicated by the dotted line in Figure 3.11.

The basic idea of DTC is control the flux of the machine by controlling the input voltage of the windings, the input voltage being determined by the voltage vectors that are

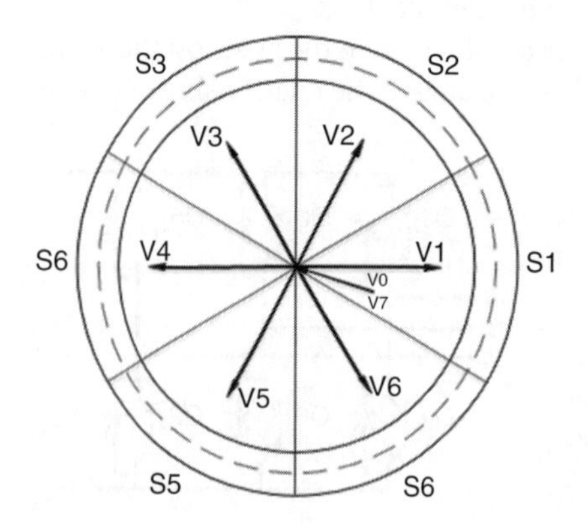

Figure 3.11 Division of space.

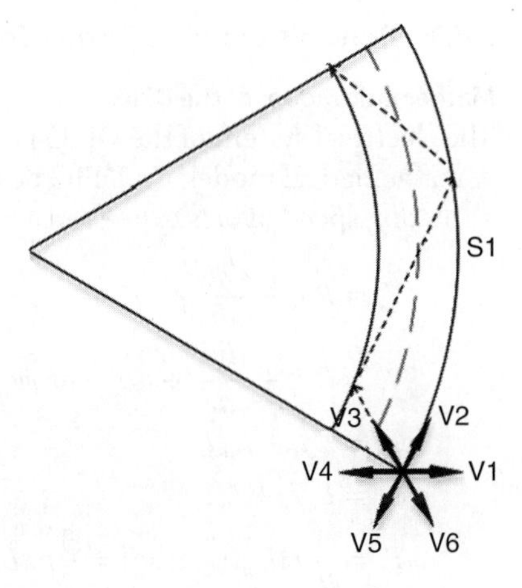

Figure 3.12 Forming a circular flux trajectory.

applied to the VSC. More details of the realization of DTC in DFIGs will be introduced in the next section.

3.4.2 Application of Direct Torque Control in DFIGs

Due to the improvements in power electronics and their high reliability, DFIGs are clearly superior in wind power applications. DTC has been widely applied in DFIGs because of its good performance, simple structure, and low parameter dependency. In a DTC system, a suitable voltage vector is selected from a switching table, which is determined by three factors: the torque hysteresis, the flux hysteresis, and the position of the rotor flux.

As shown in Figure 3.13, the stator windings of the DFIG are directly connected to the grid and the rotor windings are fed through back-to-back variable frequency converters [13].

The focus of this chapter is on the rotor-side converter, in which the DTC strategy is applied.

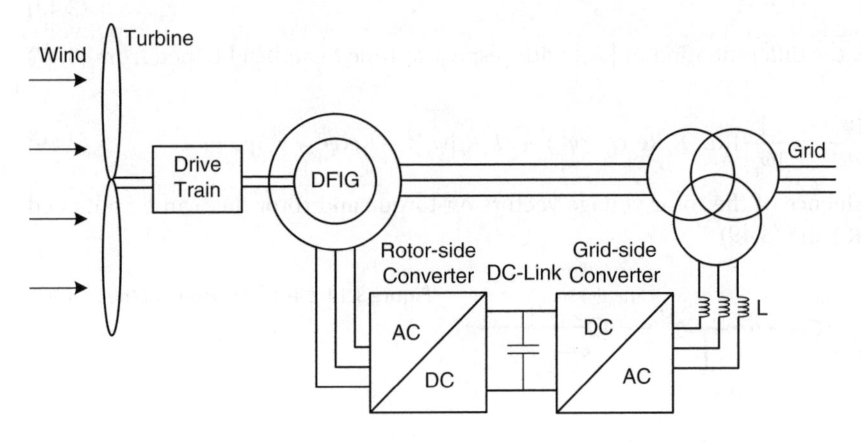

Figure 3.13 Wind energy conversion system of a DFIG.

3.4.3 Principle of Direct Torque Control in DFIG

Mathematic Model of the DFIG

The electrical system of the DFIG is shown in Figure 3.14:

A mathematical model of a DFIG described by space vectors in an arbitrary frame with a rotating speed of ω is expressed as:

$$V_s = R_s i_s + \frac{d\psi_s}{dt} + j\omega_s \tag{3.41}$$

$$V_r = R_r i_r + \frac{d\psi_r}{dt} + j(\omega - \omega_r)\psi_r \tag{3.42}$$

$$\psi_s = L_s i_s + L_m i_r \tag{3.43}$$

$$\psi_r = L_m i_s + L_r i_r \tag{3.44}$$

$$T_e = \frac{3}{2} p\lambda L_m(\psi_r \times \psi_s) = \frac{3}{2} p\lambda L_m(|\psi_r||\psi_s| \sin\theta) \tag{3.45}$$

where V_s, i_s, V_r, and i_r are the stator voltage vector, stator current vector, rotor voltage vector, and rotor current vector, respectively; ψ_s and ψ_r are the stator flux linkage vector and rotor flux linkage vector and θ is the phase angle difference between them; R_s, R_r, L_s, L_r, and L_m are the stator resistance, rotor resistance, stator inductance, rotor inductance and mutual inductance, respectively; ω_r and p are the electrical rotor speed and pole pairs and $\lambda = 1/(L_s L_r - L_m^2)$ [14].

From (3.41) to (3.44), the DFIG model can be expressed in a stationary frame with ψ_s and ψ_r as state variables:

$$\frac{d\psi_s}{dt} = -\lambda L_r R_s \psi_s + \lambda L_m R_s \psi_r + V_s \tag{3.46}$$

$$\frac{d\psi_r}{dt} = \lambda L_m R_r \psi_s - (\lambda L_s R_r - j\omega_r)\psi_r + V_r \tag{3.47}$$

The differentiation of T_e with respect to time t can be obtained from (3.45)–(3.47) as:

$$\frac{dT_e}{dt} = \frac{3}{2} p\lambda L_m[-\lambda(L_s R_r + L_r R_s)\mathrm{Im}(\psi_r * \psi_s) - \omega_r \mathrm{Re}(\psi_r * \psi_s) + \mathrm{Im}(V_r * \psi_s) + \mathrm{Im}(\psi_r * V_s)] \tag{3.48}$$

Similarly, the differentiation of $|\psi_r|$ with respect to time t can be obtained from (3.47) as

$$\frac{d|\psi_r|}{dt} = \frac{1}{|\psi_r|}[\lambda L_m R_r \mathrm{Re}(\psi_r * \psi_s) - \lambda L_s R_r |\psi_r|^2 + \mathrm{Re}(\psi_r * V_r)] \tag{3.49}$$

The influence of the rotor voltage vectors on torque and rotor flux can be obtained from (3.48) and (3.49).

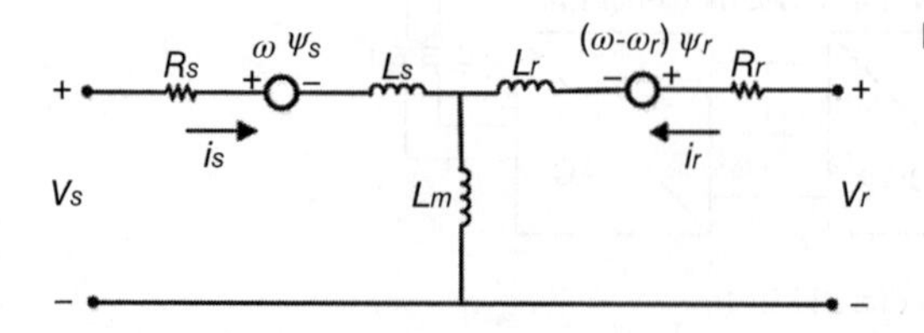

Figure 3.14 Electrical system of the DFIG.

As the stator winding of DFIG is connected to the power grid, the magnitude of the stator flux $|\Psi_s|$ can be considered to be constant when the voltage drop on the stator winding resistance is neglected and the supply voltage does not fluctuate [12]. Therefore, the torque of DFIG T_e is controlled by the magnitude of the rotor flux $|\Psi_r|$ and the phase angle difference θ.

Realization of Direct Torque Control

The principle of DTC strategy based on DFIG system is to directly select proper rotor voltage vectors according to the rotor flux, torque errors, and the rotor flux sector [14]. As shown in Figure 3.15, the torque T_e^* and the rotor flux Ψ_r^* references are compared with the corresponding estimated values. The T_e^* reference comes from the rotor speed ω^* through a PI regulator and the rotor speed ω^* comes from the maximum power point tracking (MPPT) strategy. The torque and flux hysteresis H_T and $H\psi$ are discretized from the torque and rotor error E_T and $E\psi$ by using hysteresis band comparators.

In the torque hysteresis band comparator, the input is E_T (error of T_e^* and T_e) and the output is H_T. When E_T drops down to the lower limit $-\varepsilon$ of the comparator's tolerance, H_T is set to 0 and when E_T rises up to the upper limit $+\varepsilon$ of the comparator's tolerance, H_T is set to 1. When E_T is between $+\varepsilon$ and $-\varepsilon$, the value of H_T remains unchanged. The flux hysteresis band comparator has the same function.

In the DTC method, as shown in Figure 3.16, there are six rotor flux position sectors, S1–S6, and each sector occupies 60°. There are, in total, eight switching combinations, and each one has one equivalent basic vector, as mentioned before. Six are active vectors V1–V6 and the other two are zero vectors, V0 and V7. As an example, when $\Psi_r(t)$ lies

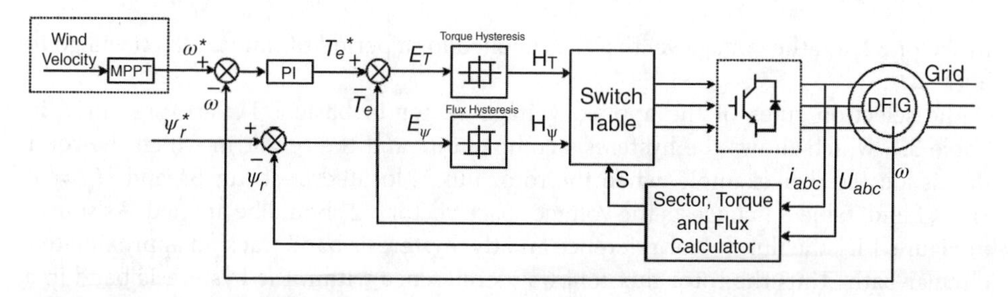

Figure 3.15 DTC system of DFIG.

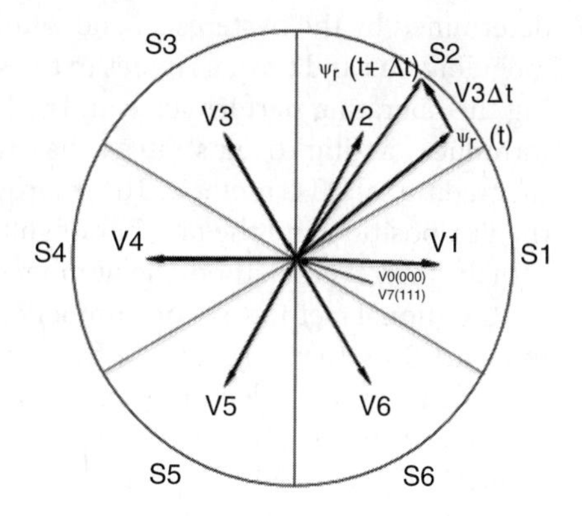

Figure 3.16 Flux sectors and voltage vectors in DTC.

Table 3.1 Switching table in conventional DTC method.

H_ψ	H_T	Sector					
		S1	S2	S3	S4	S5	S6
$H_\psi = 1$	$H_T = 1$	V2	V3	V4	V5	V6	V1
	$H_T = 0$	V6	V1	V2	V3	V4	V5
$H_\psi = 0$	$H_T = 1$	V3	V4	V5	V6	V1	V2
	$H_T = 0$	V5	V6	V1	V2	V3	V4

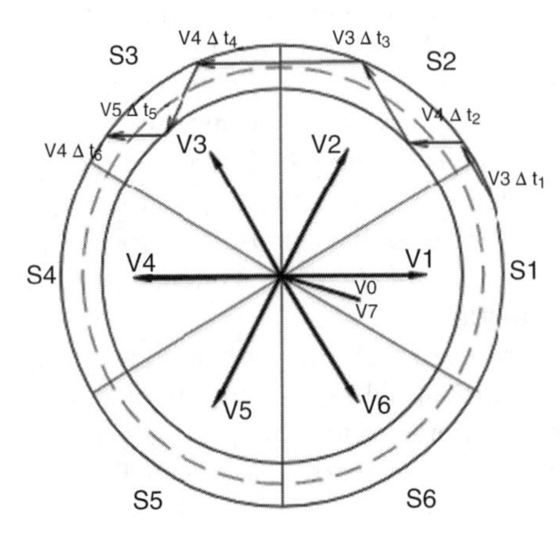

Figure 3.17 Flux trajectory in DTC.

in sector S2 and the voltage vector V3 is applied for a period of time $\Delta t, \Psi_r(t)$ will go to $\Psi_r(t+\Delta t)$.

The selection rules of the applied voltage vector in basic DTC are presented in Table 3.1, which shows the hysteresis comparators and the sector in which the rotor flux is located. For example, when the rotor flux is located in sector S1 and $H_\psi = 1$, $H_T = 1$,and Table 3.1 suggests the voltage space vector V2 should be applied. As shown in Figure 3.17, the flux vector reference and the hysteresis band track an approximately circular path. Thus the rotor flux follows its reference within the hysteresis band in a zigzag trajectory.

It can be seen that the flux trajectory in conventional DTC has relatively large ripples determined by the hysteresis band width. This causes some drawbacks for the system, including flux and torque ripples in the steady state. There are many studies on improving the operating performance and reducing the flux and torque ripples. This section introduces an improved strategy that avoids the drawbacks of the DFIG system using the traditional DTC method. The improved strategy increases the divisional accuracy of the flux position and the number of voltage space vectors.

In the improved method, the number of flux divisions will be modified to twice that in the traditional method, so the number of voltage space vectors is also doubled. As shown in Figure 3.18, there are now 12 flux sectors and 12 voltage space vectors. Accordingly, the voltage vector selection rules in the new DTC are as presented in Table 3.2. In a theoretical analysis, the flux trajectory in new DTC strategy will have smaller ripples, as shown in Figure 3.19.

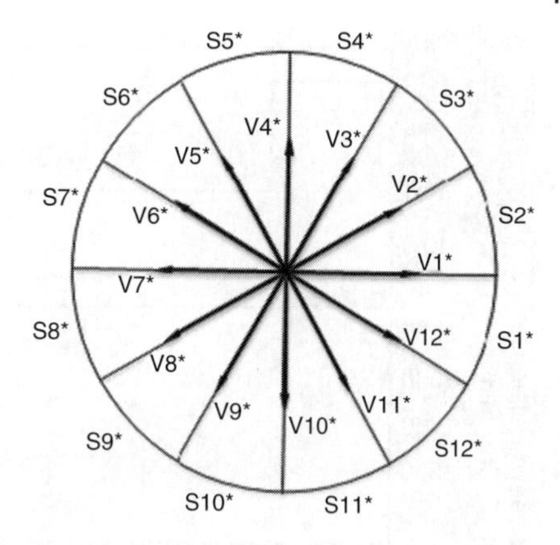

Figure 3.18 Space vectors and flux sectors in new DTC.

Table 3.2 Switching table in new DTC method.

H_ψ	H_T	Sector											
		S1*	S2*	S3*	S4*	S5*	S6*	S7*	S8*	S9*	S10*	S11*	S12*
$H_\psi = 1$	$H_T = 1$	V3*	V4*	V5*	V6*	V7*	V8*	V9*	V10*	V11*	V12*	V1*	V2*
	$H_T = 0$	V10*	V11*	V12*	V1*	V2*	V3*	V4*	V5*	V6*	V7*	V8*	V9*
$H_\psi = 0$	$H_T = 1$	V4*	V5*	V6*	V7*	V8*	V9*	V10*	V11*	V12*	V1*	V2*	V3*
	$H_T = 0$	V9*	V10*	V11*	V12*	V1*	V2*	V3*	V4*	V5*	V6*	V7*	V8*

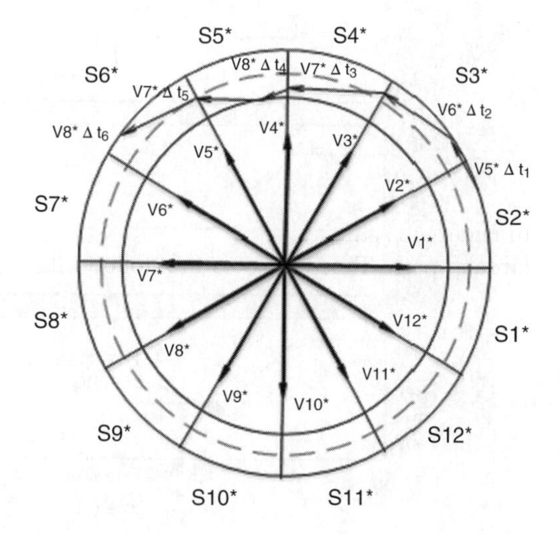

Figure 3.19 Flux trajectory in new DTC.

Simulation results confirm the theoretical analysis. Simulations of the conventional and new DTC methods for a DFIG-based wind generation system were carried out in a Matlab/Simulink environment. A 50 kVA DFIG was used, and the frequency of the AC grid was 50 Hz. The synchronous speed of the rotor was 1500 rpm. The DC-link voltage was 690 V. The simulation results are shown in Figure 3.20. As the wind velocity changes, the rotor speed follows the reference that is determined by the MPPT strategy to capture

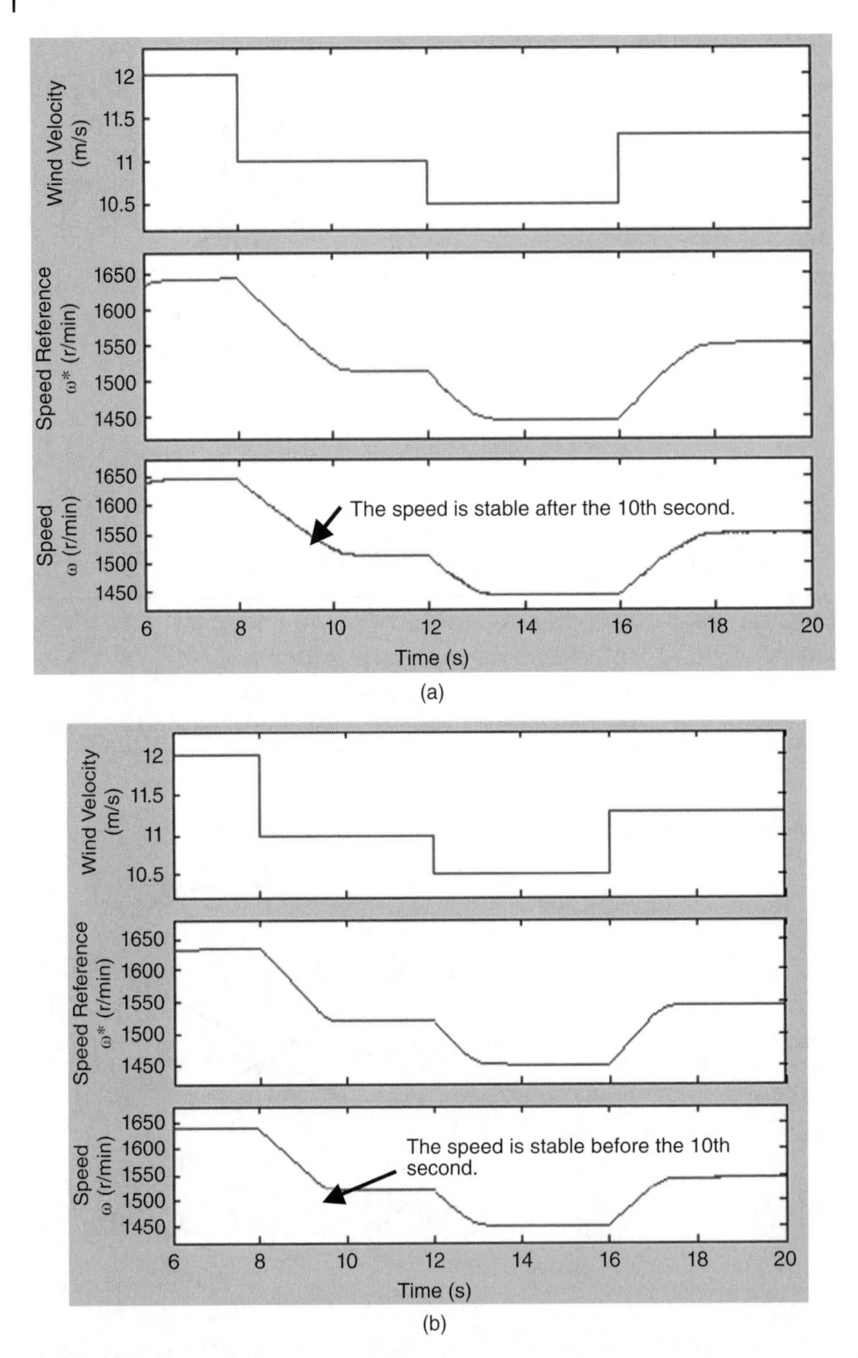

Figure 3.20 Simulation results: (a) conventional DTC; (b) new DTC.

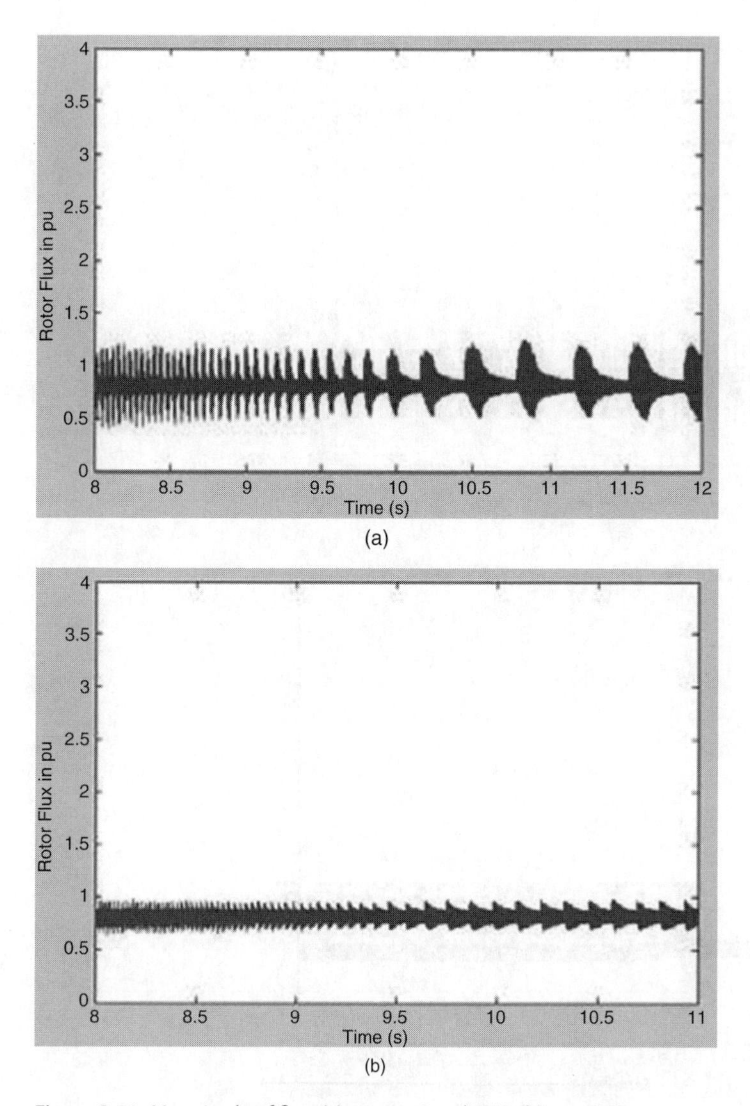

Figure 3.21 Magnitude of flux: (a) conventional DTC; (b) new DTC.

the maximum wind power. The rotor speed in the new DTC system has a faster response than the conventional system: as the wind velocity changes in the eighth second, the new DTC method gives a steady-state rotor speed earlier than the conventional DTC method.

Figure 3.21 shows that the rotor flux ripple using the new method is smaller than when using the conventional method. Both parts of the figure show an average amplitude of the rotor flux of 0.814 pu. The maximum deviation of the rotor flux in Figure 3.21a is 0.435 pu and the maximum deviation of the rotor flux in Figure 3.21b is 0.155 pu. That means the DFIG system using the new DTC method has reduced the deviation of the rotor flux by 64.4%.

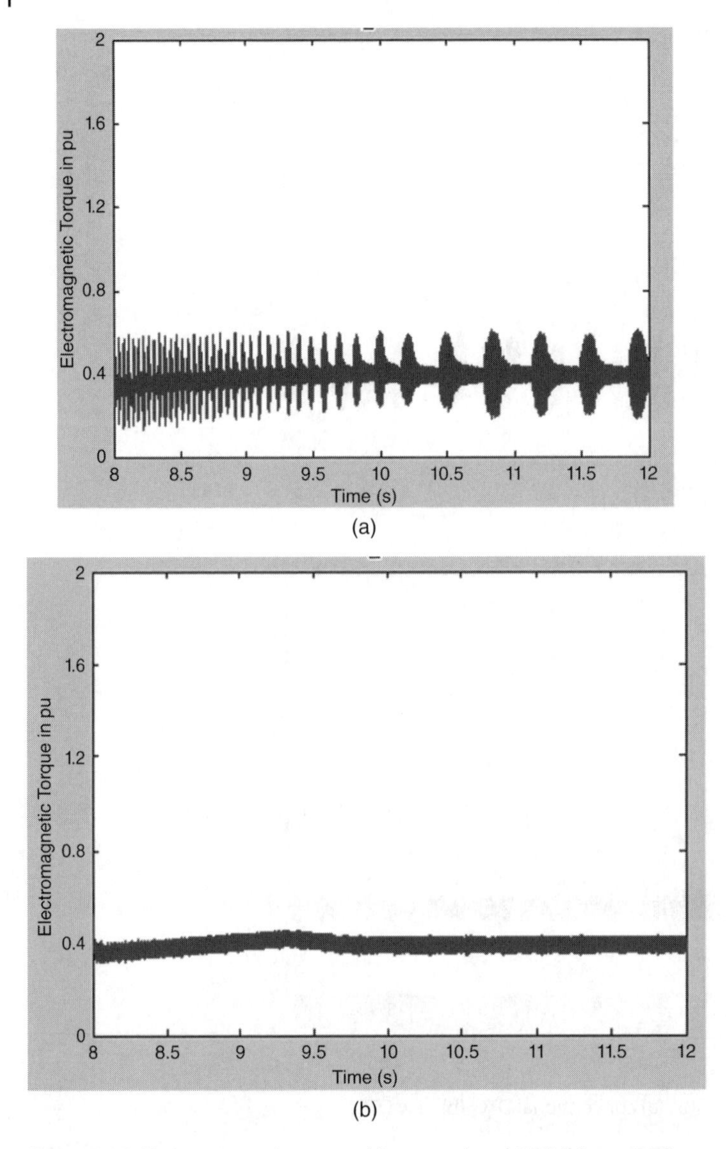

Figure 3.22 Electromagnetic torque: (a) conventional DTC; (b) new DTC.

Figure 3.22 shows that the electromagnetic torque ripple using the new method is smaller than when using the conventional method. The average amplitude of the electromagnetic torque is 0.4, but the maximum deviation of the electromagnetic torque in Figure 3.22a is 0.211 pu and the maximum deviation of the electromagnetic torque in Figure 3.22b is 0.042 pu. That means the new DTC method has reduced the deviation of the electromagnetic torque by 80.1%.

One important performance measure of a DFIG system is the voltage of DC-link between the rotor-side converter and the grid-side converter, as shown is Figure 3.13. It can be seen from Figure 3.23 that the DC-link voltage ripple of the new DTC method is

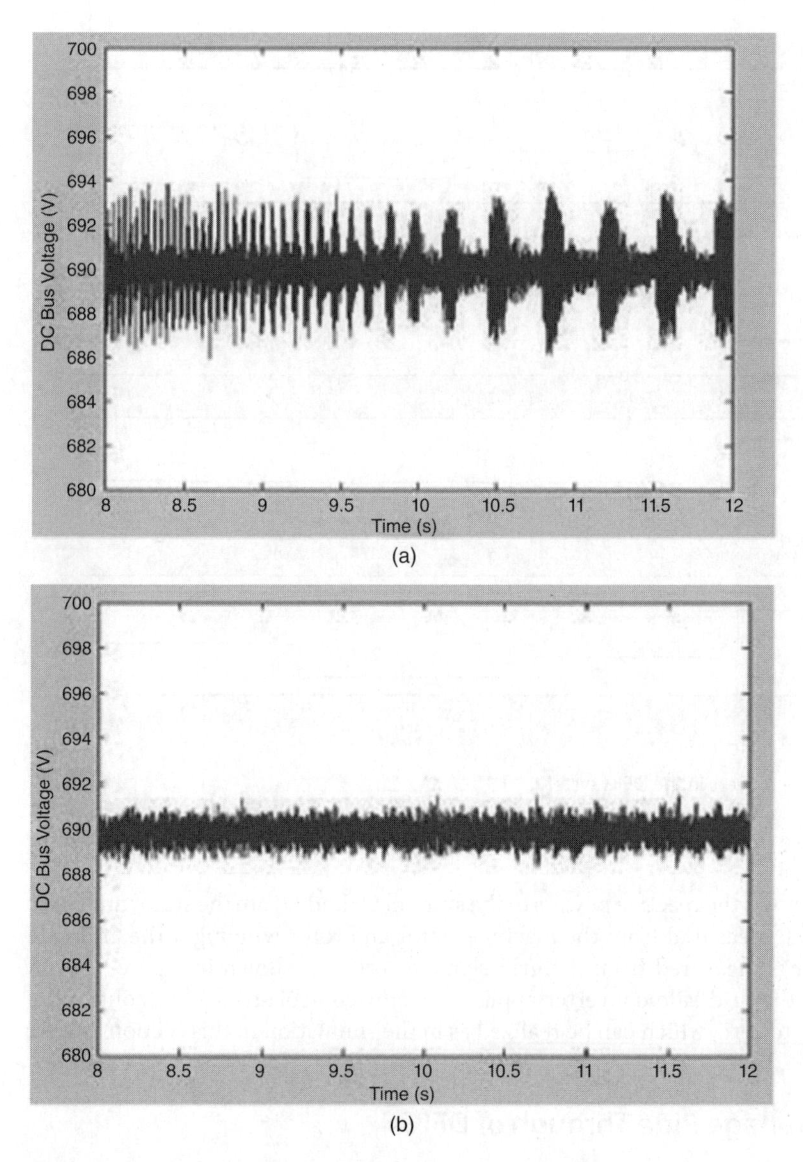

Figure 3.23 DC-link voltage: (a) conventional DTC; (b) new DTC.

smaller than that of the conventional DTC method. The average DC bus voltage is 690 V which is the same point the simulation system sets. The maximum deviation of the DC bus voltage in Figure 3.23a is 3.744 V and the maximum deviation of the DC bus voltage in Figure 3.23b is 1.500 V. That means the new DTC method has reduced the deviation of the DC bus voltage by 59.9%.

Figure 3.24 shows the active powers of the system. P_1 and P_2 are the stator and rotor active powers, measured from the machine stator and rotor windings. The theoretical mechanical active power is calculated as $P_1 - P_2$ and the actual mechanical active power P_m is measured from the rotation shaft that connects the wind turbine and the DFIG.

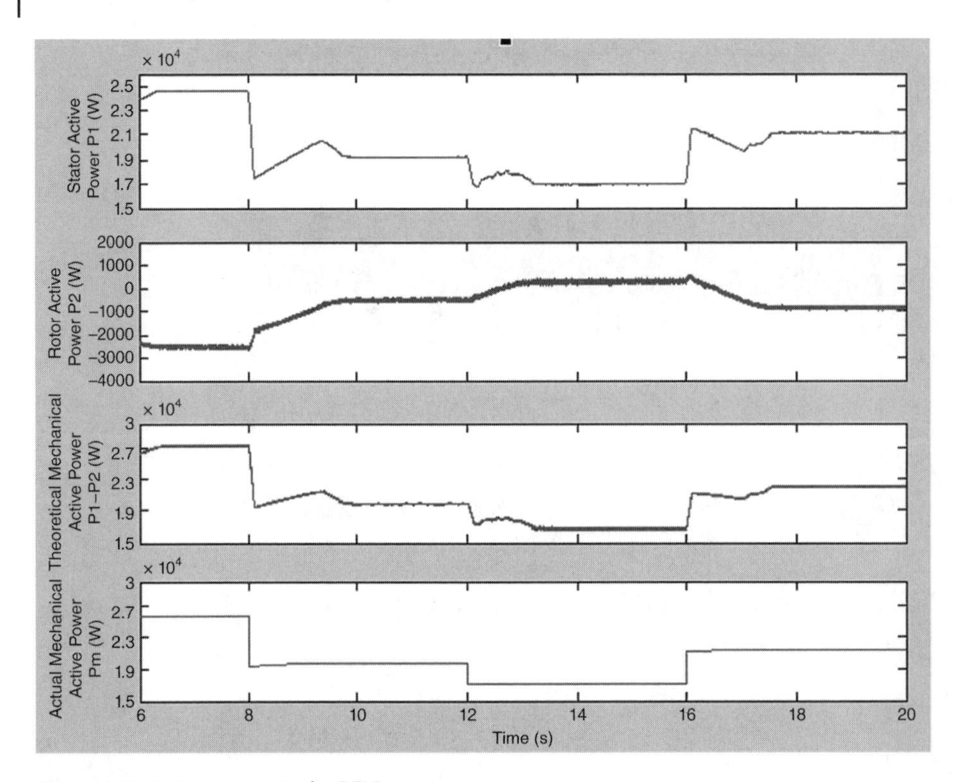

Figure 3.24 Active powers in the DFIG system.

From the Figure 3.24 it can be seen that the actual mechanical active power is close to its theoretical value.

Figure 3.25 shows the reactive powers of the system. Q_1 and Q_2 are the stator and rotor reactive powers, measured from the machine stator and rotor windings. The grid-side reactive power is measured from the grid-side converter, as shown in Figure 3.25. In many studies, the grid-side converter applies a vector control strategy to control the reactive power to zero, which can be realized as in the simulation in this section as well.

3.5 Low-voltage Ride Through of DFIGs

A major drawback of DFIGs is their low-voltage ride through (LVRT) capability during grid faults. Faults in the power system can cause a voltage dip (or "voltage sag") at the connection point of the wind turbine. The dip of the terminal voltage will result in an increase of the current in the rotor circuit of the DFIG. It will also cause over-currents and over-voltages to power-electronic converters, which might damage them [15–17].

The voltage dip is normally short, say 0.5–30 cycles; the grid voltage drops from 1 pu to 0.1–0.9 pu. Conventional power plants can ride through this short period and provide active and reactive regulation support to maintain power system stability after the fault. Unlike conventional power plants, DFIGs would be disconnected in the event of a grid failure, so as to protect the converters. If all wind turbines were disconnected in the event of a grid fault, they would not be able to support the voltage and the frequency

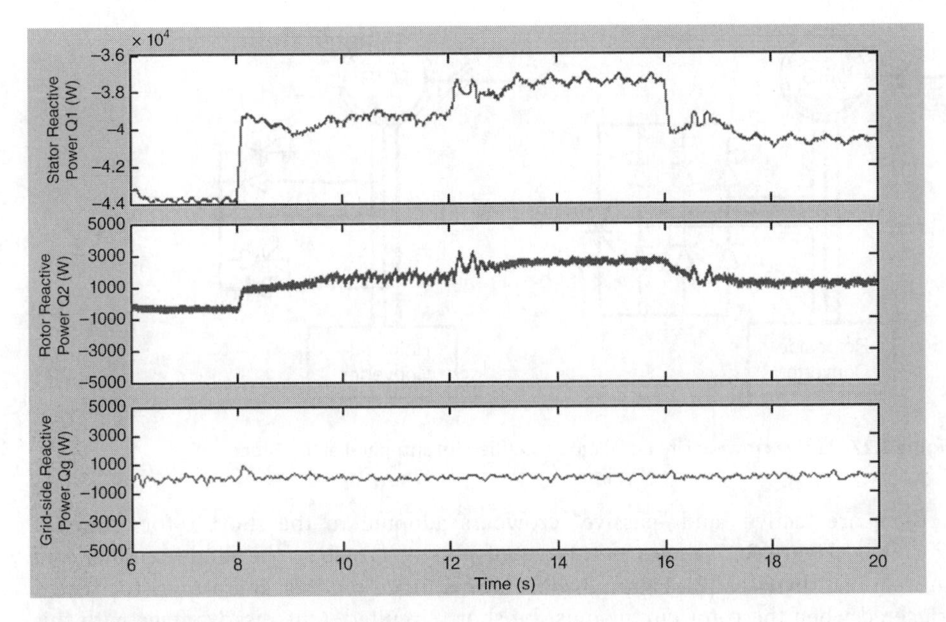

Figure 3.25 Reactive powers in the DFIG system.

of the grid after the fault has been cleared. After a DFIG wind turbine has been disconnected, it takes some time before it can be reconnected to the grid. This can cause major problems for system stability. Nowadays, with a large number of variable-speed wind turbines being connected to power grids, the interactions of the turbines and the grid are becoming increasingly important. A large proportion of the turbines need to stay connected to the grid in the event of a grid fault. They should supply active and reactive power for frequency and voltage support immediately after the fault has been cleared.

A possible solution that is sometimes used is to short circuit the rotor windings of the DFIG with so-called "crowbars", as shown in Figure 3.26; the chopper in the figure is for high voltage protection[15].

Figure 3.26 DFIG with crowbar protection.

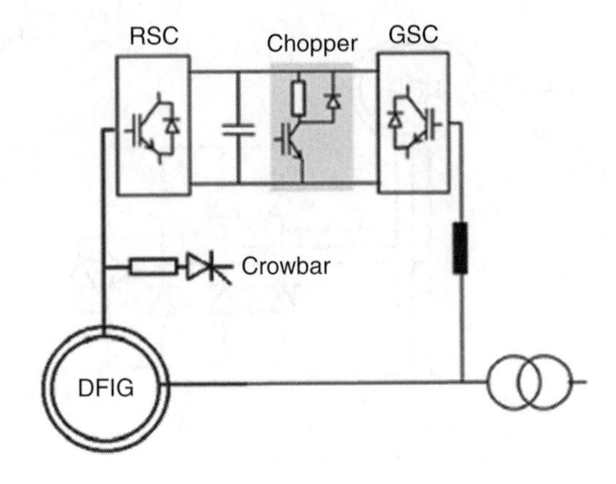

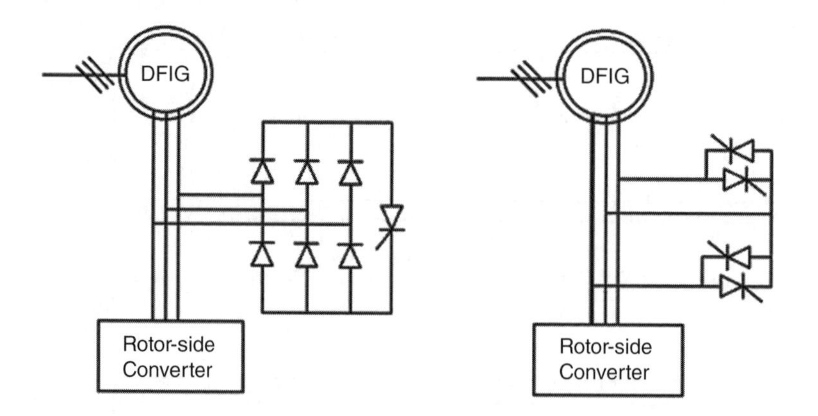

Figure 3.27 Passive crowbar circuit: (a) diode rectifier; (b) anti-parallel thyristor.

There are "active" and "passive" crowbars adopted to the short rotor windings of DFIGs. Figure 3.27 shows two types of passive crowbar: the diode rectifier and anti-parallel thyristor [2]. Figure 3.27a shows a diode rectifier and a thyristor that is triggered when the rotor circuit must be short circuited. One disadvantage with this system is that once the crowbar has been triggered, the turbine must be disconnected from the grid, since the current through the thyristor is a continuous DC current and can only be interrupted if the turbine is disconnected from the grid. Anti-parallel thyristors are used in Figure 3.27b to achieve a quick (within 10 ms) disconnection of the rotor circuit, thereby activating rotor-side converter control as quickly as possible.

Another option is to use an "active" crowbar, which can break the short-circuit current in the crowbar. Figure 3.28 shows the active crowbar circuit. It consists of a diode rectifier and an insulated gate bipolar transistor (IGBT). The crowbar circuit is triggered when the rotor circuit must be short circuited. The "active" crowbar can achieve a very fast (about a few microseconds) disconnection of the rotor circuit by switching off the IGBT immediately.

Besides the crowbar, other proposals, such as STATCOM and SVC, fault current limiters [17], and dynamic voltage restorers, have been proposed for providing LVRT enhancement for DFIG wind turbines. Control methods of the rotor-side converter

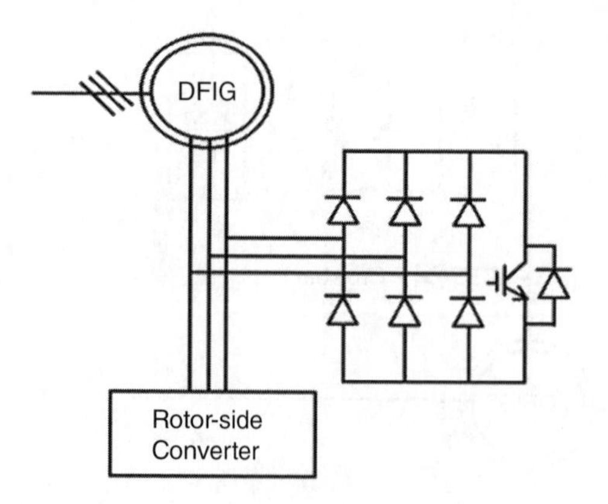

Figure 3.28 Active crowbar circuit.

have also been proposed to meet LVRT requirements, especially for asymmetrical grid fault situations [18].

When the concentration of wind turbines is high, grid operators already have set requirements for the behavior of wind turbines. In 2003, E.On Netz of Germany published the worldwide first LVRT requirement [15]. Instead of disconnecting wind turbines from the grid, the turbines must be able to match certain characteristics. Only when the grid voltage goes below a defined curve (in duration or voltage level) can the turbine be disconnected. In 2007, E.On Netz revised the LVRT requirement so that there were two thresholds: Limit 1 and Limit 2.

Three-phase short circuits or symmetrical voltage drops due to disturbances must not lead to instability or to a disconnection of the generating facility from the network above Limit 1 [16]. Within the area between Limit 1 and Limit 2, all generating facilities should pass through the fault without being disconnected from the network. If a generating facility is not capable of meeting the requirement, it is permissible to shift the limit by agreement with the network operator. Below Limit 2, a short-term disconnection of the generating facility from the network is always permitted.

The LVRT grid code is receiving wide acceptance now and being used as a template for similar requirements in nearly every country with wind generation.

3.6 Conclusions

DFIGs are an attractive and popular method for variable speed wind turbines due to the small size of the converter. In this chapter, the operation principle, PQ control, and DTC of the DFIG were introduced. PQ control is widely used in the practice, giving decoupled active power and reactive power. The DTC does not depend on the motor parameters. The disadvantage of the DFIG is its LVRT capability. However, a crowbar and other methods have been proposed to enable DFIGs to meet grid code requirements. LVRT requirements have been set in the grid codes of many countries where wind power penetration is high.

References

1 Ackermann, T. (2005) *Wind farms in Power Systems.* John Wiley & Sons.

2 Petersson, A. (2005) Analysis, modeling and control of doubly-fed induction generators for wind turbines, PhD thesis, Chalmers University of Technology, Goteborg, Sweden.

3 Wang, Z., Li, G.J., Sun, Y.Z., et al. (2010) Effect of erroneous position measurements in vector-controlled doubly-fed induction generator, *IEEE Transactions On Energy Conversion*, **25**(1), 59–69.

4 Wang, Z., Sun, Y., Li, G., et al. (2010) Magnitude and frequency control of grid-connected doubly fed induction generator based on synchronised model for wind power generation, *IET Renewable Power Generation*, **4** (3), 232–241.

5 Pena, R. Clare, J.C., and Asher, G.M. (1996) Doubly fed induction generator using back-to-back PWM converters and its application to variable-speed wind-energy generation, *IEE Proceedings –Electric Power Applications*, **143** (3), 231–241.

6 Tapia, A. Tapia, G., Ostolaza, J.X., et al. (2003) Modeling and control of a wind turbine driven doubly fed induction generator, *IEEE Transactions on Energy Conversion*, **18** (2), 194–204.
7 Muller, S., Deicke, M., and De Doncker, R.W. (2002) Doubly fed induction generator systems for wind turbines, *IEEE Industry Applications Magazine*, **17** (1), 26–33.
8 Shen, B.K., Mwinyiwiwa, B., Zhang, Y.Z., et al. (2009) Sensorless maximum power point tracking of wind by DFIG using rotor position phase lock loop (PLL), *IEEE Transactions on Power Electronics*, **24** (4), 942–951.
9 Muller, S. Deicke, M., and De Doncker, R.W. (2002) Doubly fed induction generator systems for wind turbines, *IEEE Industry Applications Magazine*, **8** (3), 26–33.
10 Liu, Z. Mohammed, O.A., and Liu, S. (2007) A novel direct torque control of doubly-fed induction generator used for variable speed wind power generation, in *IEEE Power Engineering Society General Meeting*, pp. 1–6.
11 Takahashi, I. and Noguchi, T. (1986) A new quick-response and high efficiency control strategy of an induction motor, *IEEE Transactions on Industry Application*, **IA-22** (5), 820–827.
12 Buja G.S. and Kazmierkowski, M.P. (2004) Direct torque control of PWM inverter-fed AC motors – a survey, *IEEE Transactions on Industrial Electronics*, **51** (4), 744–758.
13 Zarean, N. and Kazemi, H. (2012) A new DTC control method of doubly fed induction generator for wind turbine, in *IEEE 2012 Renewable Energy and Distributed Generation Conference*, pp. 69–74.
14 Zhang Y., Li Z., Wang T., et al. (2011) Evaluation of a class of improved DTC method applied in DFIG for wind energy applications, *Electrical Machines and Systems (ICEMS), 2011 International Conference on. IEEE*, pp. 1–6.
15 Morren, J. and de Haan, S.W.H. (2005) Ride through of wind turbines with doubly-fed induction generator during a voltage dip, *IEEE Transactions on Energy Conversion*, **20** (2), 435–441.
16 Transmission Code 2007. Network and System Rules of the German Transmission System Operators.
17 Guo, W., Xiao, L., Dai, S., Li, Y., et al. (2105) LVRT capability enhancement of DFIG with switch-type fault current limiter, *IEEE Transactions on Industrial Electronics*, **62** (1), 332–342.
18 Geng, H., Liu, C., and Yang. G. (2013) LVRT capability of DFIG-based WECS under asymmetrical grid fault condition, *IEEE Transactions on Industrial Electronics*, **60** (6), 2495–2509.

4

Optimal Control Strategies of Wind Turbines for Load Reduction

Shuju Hu and Bin Song

Institute of Electrical Engineering, Chinese Academy of Sciences

4.1 Introduction

A modern large-scale wind turbine is a complex multi-body system including blades, wind wheel, drivetrain, generator, nacelle, tower and other components. With wind turbines becoming large scale and lightweight [1], the flexibility of components is increasing and the natural frequency of the system is gradually reducing. Meanwhile, due to non-stable factors, such as wind shear, the tower-shadow effect and turbulence, the dynamic loads on wind turbines are very complex. As wind turbines bear long-term dynamic loads, many components will suffer dynamic stress, thereby causing fatigue damage [2]. Fatigue damage is one of the main forms of failure of wind turbine mechanical parts and structural components, so it must be of concern during design and operation processes [3]. To improve the reliability and lifetime of wind turbines, loads must be reduced. Therefore, the research on control strategies for wind turbines for load reduction has attracted both domestic and international attention.

Wind turbine load reduction control comprises passive control and active control. Passive control is realized by selection of suitable elastic support components for damping the wind turbine, so that vibrational restriction and load reduction can be achieved. Although the structure of passive control devices is simple, additional cost is required. Active control is realized by control strategies for the wind turbine through its main control system and actuators based on load signals collected by sensors during operation, so that vibration restriction and load reduction can be achieved. Compared to passive control, active control is economic, reliable and adaptable, so there is a great potential for engineering applications.

Currently, active control focuses on the following four main areas:

- tower fore–aft vibration control
- tower lateral vibration control
- individual pitch control (IPC)
- drivetrain torsional vibration control.

Research into tower fore–aft vibration control and tower lateral-vibration control has been well developed in recent years. Therefore, IPC and drivetrain torsional vibration control are the focus of this chapter. Firstly, the dynamic simulation model of a wind

Modeling and Modern Control of Wind Power, First Edition. Edited by Qiuwei Wu and Yuanzhang Sun.

Companion website: www.wiley.com/go/wu/modeling

turbine is established. Then, an IPC strategy and drivetrain torsional vibration control strategy are proposed. Finally, the effectiveness of the control strategies is verified by simulation.

4.2 The Dynamic Model of a Wind Turbine

In order to study the control strategy of wind turbine load reduction, it is necessary to establish the dynamic model of a wind turbine. A wind turbine is a complex energy conversion system. In order to accurately describe its characteristics, the dynamic model of wind conditions, aerodynamics, the wind wheel, drivetrain, the tower and the electrical control need to be established and coupled to obtain a high-order model of the whole system. In this chapter, the dynamic model of a direct-drive wind turbine is built, with a view to load-reduction controller design,

4.2.1 Wind Conditions Model

Accurate modeling of wind conditions is essential for the analysis of the dynamic characteristics of a wind turbine. At present, a detailed description and mathematical expressions for typical wind conditions are given in the international standards, such as the IEC 61400 and GL 2010 standards. In this chapter, according to the IEC61400 standard,the wind conditions model for direct-drive wind turbine is built using IEAwind and Turbsim software. Figure 4.1 shows the normal turbulence model built in Turbsim, in which the average wind speed is 18 m/s, and the turbulence intensity is 16.978%.

4.2.2 Aerodynamic Model

The aerodynamic torque of the wind turbine is calculated using blade element momentum theory. Firstly, the lift force and drag force acting on the radius r and length δr of the blade element in the span-wise direction are analyzed. The sectional view of blade

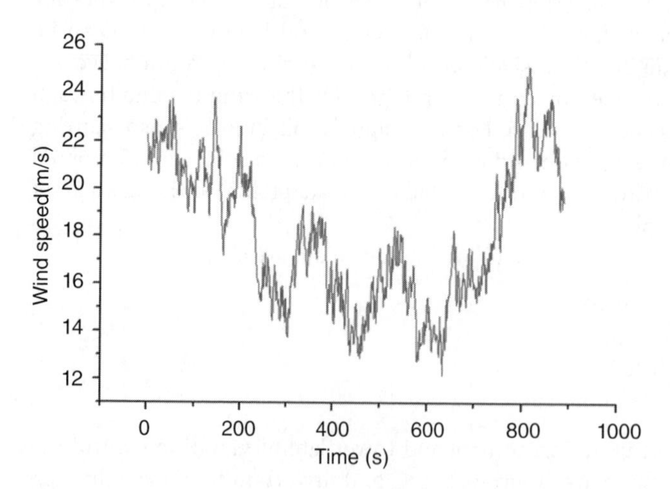

Figure 4.1 The normal turbulence model.

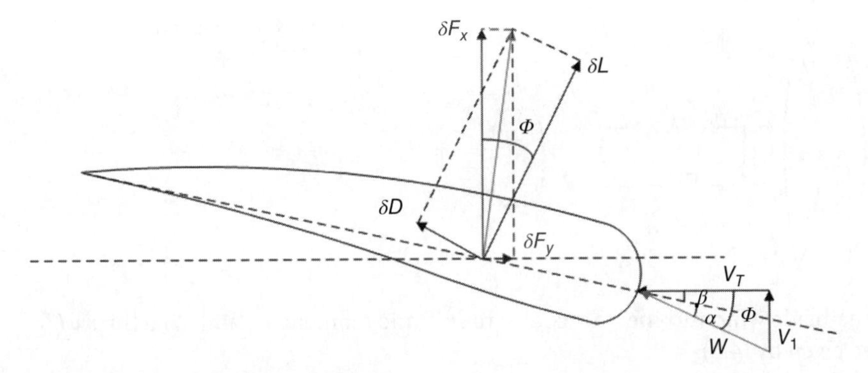

Figure 4.2 Force analysis of blade element.

is shown in Figure 4.2; W represents the synthetic wind speed, α represents the angle of attack, and β represents the pitch angle.

The lift force dL, which is perpendicular to the synthetic wind W, and the drag force dD, which is parallel to the synthetic wind W, are given by (4.1):

$$dL = \frac{1}{2}C_l(\alpha)\rho W^2 l dr$$
$$dD = \frac{1}{2}C_d(\alpha)\rho W^2 l dr \tag{4.1}$$

where ρ represents the air density, l represents the blade chord, and C_l and C_d represent the lift and drag coefficients respectively.

The forces of the blade element on the perpendicular to the blade rotation plane and on the blade rotation plane are given by (4.2):

$$dF_x = dL\cos\phi + dD\sin\phi$$
$$dF_y = dL\sin\phi - dD\cos\phi \tag{4.2}$$

The bending moment that is produced by blade element on the rotor center is given by (4.3):

$$dM_x = rdF_y = r(dL\sin\phi - dD\cos\phi)$$
$$dM_y = rdF_x = r(dL\sin\phi + dD\cos\phi) \tag{4.3}$$

According to (4.1)–(4.3), integrating is done to the entire length of the blade, to get the force F_{xi} and F_{yi} and the bending moment M_{xi} and M_{yi} of a single blade:

$$F_{xi} = \int_0^R \frac{1}{2}\rho W^2(C_l\cos\phi + C_d\sin\phi)l\delta r$$
$$F_{yi} = \int_0^R \frac{1}{2}\rho W^2(C_l\sin\phi - C_d\cos\phi)l\delta r$$
$$M_{xi} = \int_0^R \frac{1}{2}\rho W^2 r(C_l\sin\phi - C_d\cos\phi)l\delta r$$
$$M_{yi} = \int_0^R \frac{1}{2}\rho W^2 r(C_l\cos\phi + C_d\sin\phi)l\delta r \tag{4.4}$$

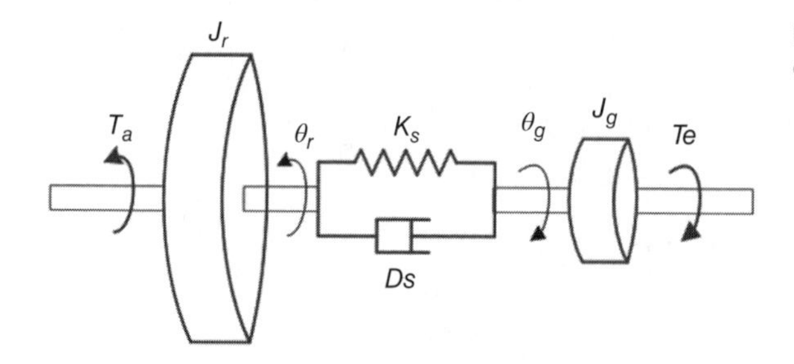

Figure 4.3 The two-mass model of drivetrain .

For three-blade wind turbine, the axial aerodynamic moment T_a and axial thrust F_{ax} of hub are given by (4.5):

$$T_a = \sum_{i=1}^{3} M_{xi}, F_{ax} = \sum_{i=1}^{3} F_{xi} \tag{4.5}$$

4.2.3 Tower Model

The tower dynamic model is simplified as a spring-mass-damper system:

$$M_t \ddot{x}_{fa} + D_t \dot{x}_{fa} + K_t x_{fa} = F_{ax} \tag{4.6}$$

where x_{fa} represents the fore-aft displacement of the tower, M_t represents its equivalent mass, D_t represents the modal damping coefficient of the fore–aft modal, and K_t represents the stiffness coefficient of fore–aft modal.

The natural oscillation angular frequency ω_t and damping ratio ζ_t of the spring-mass-damper system are given by (4.7):

$$\begin{aligned} \omega_t &= \sqrt{\frac{K_t}{M_t}} \\ \zeta_t &= \frac{C_t}{2\omega_t M_t} \end{aligned} \tag{4.7}$$

4.2.4 Drivetrain Model

For direct-drive wind turbines, the drive train can be represented by a two-mass torsional model, as shown in Figure 4.3. The model consists of the wind wheel inertia J_r and the generator inertia J_g, which are connected by a flexible shaft. The flexible shaft can be expressed by the equivalent of a torsional spring K and damping D.

The torsional angular displacement of drive train is defined as $\theta_s = \theta_r - \theta_g$. The torsional angular velocity is defined as $\omega_s = \dot{\theta}_s$. The torque on the shaft is given by (4.8):

$$T_s = K_s \theta_s + D_s \omega_s \tag{4.8}$$

The dynamic equations of the two-mass model are given by (4.9):

$$\begin{cases} J_r \dot{\omega}_r = T_a - T_s - D_r \omega_r \\ J_g \dot{\omega}_g = T_s - T_e - D_g \omega_g \\ \dot{\theta}_r = \omega_r \\ \dot{\theta}_g = \omega_g \end{cases} \tag{4.9}$$

where T_a and T_e represent the aerodynamic torque of the wind turbine and the electromagnetic torque of the generator, respectively, and θ_r and θ_g represent the mechanical angle and electrical angle of the wind turbine, respectively.

4.2.5 Electrical Control Model

Vector control technology based on rotor field orientation is usually applied in permanent magnet synchronous generators. In the dq synchronous rotating coordinate system, the d-axis direction is set to the permanent magnet flux. The dq voltage of the permanent magnet synchronous generator can be deduced as:

$$\begin{cases} u_d = -R_s i_d - L_d \dfrac{di_d}{dt} + \omega_r L_q i_q \\ u_q = -R_s i_q - L_q \dfrac{di_q}{dt} - \omega_r L_d i_d - \omega_r \Psi_f \end{cases} \tag{4.10}$$

where u_d, u_q, i_d, and i_q represent respectively the voltage component and current component of the generator dq-axis, L_d and L_q represent the inductance of the generator dq-axis, R_s represents the stator resistance, and ω_r represents the angular velocity of stator.

The torque can be described as:

$$T_e = p_n[\psi_f i_q + (L_d - L_q) i_d i_q] \tag{4.11}$$

For a salient pole permanent magnet generator, L_d is equal to L_q. Then the torque can be simplified as:

$$T_e = p_n \psi_f i_q \tag{4.12}$$

The electromagnetic torque of the generator of the wind turbine is controlled by the converter. The response characteristic of the electromagnetic torque of the generator can be expressed by a one-order inertia link:

$$T_e = \frac{T_e^*}{(1 + \tau_e s)} \tag{4.13}$$

where T_e^* represents the given value of generator torque and τ_e represents the time constant of the converter.

4.2.6 Wind Turbine Dynamic Model

To build the wind turbine dynamic model based on the wind conditions model, aerodynamic model, drivetrain model, tower model and electrical control model, the combined modeling technique is used, as shown in Figure 4.4. The aerodynamic model, mechanical structure model and electrical control model are built using Aerodyn, FAST, and Matlab/Simulink software, respectively. The dynamic model of the wind turbine is obtained by dynamically linking these packages.

4.3 Wind Turbine Individual Pitch Control

Wind turbine IPC is a new control technology, and is based on collective pitch control technology. In recent years, the capacity and size of wind turbines has been increasing,

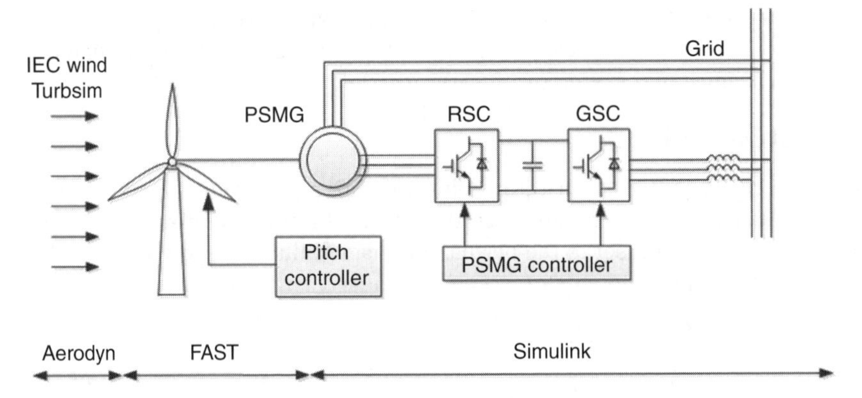

Figure 4.4 Combined model based on Aerodyn, FAST, and Simulink.

leading to large imbalances in the loads on wind wheels, and thereby causing fatigue damage of critical components [4]. IPC will solve this problem because each blade can be controlled separately to reduce the unbalanced loads on the wind wheel. Due to its effectiveness and economy, the approach has become a major development goal for large-scale wind turbines. IPC technology is introduced in detail in this section.

4.3.1 Control Implementation

A typical IPC system based on root load measurements is shown in Figure 4.5. The system consists of a blade root load sensor, an azimuth measuring device, the main control system, and pitch actuators. The load and azimuth signals obtained by the root load measuring device and the azimuth measuring device, respectively, are transmitted to the main control system. An individual pitch control algorithm is built into the programmable logic controller (PLC). Then pitch angle commands are sent to the three separate pitch actuators. Thus independent pitch action is achieved.

4.3.2 Linearization of the Wind Turbine Model

The wind turbine model established in the last section is a non-linear model with strong coupling, which is not suitable for IPC. Therefore, it is necessary to linearize the wind turbine model, as shown in Figure 4.6.

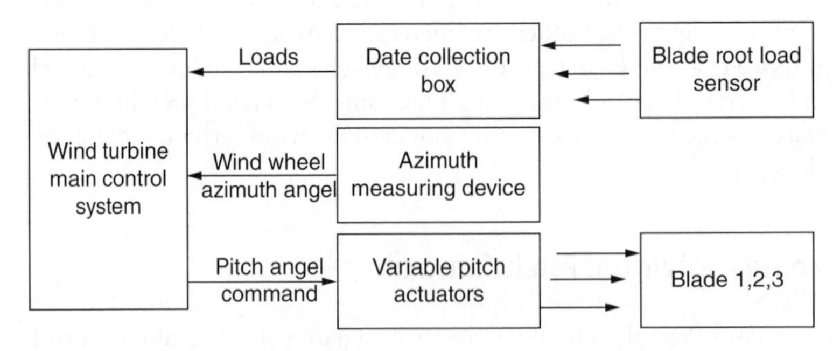

Figure 4.5 Individual pitch control system based on root load measurements.

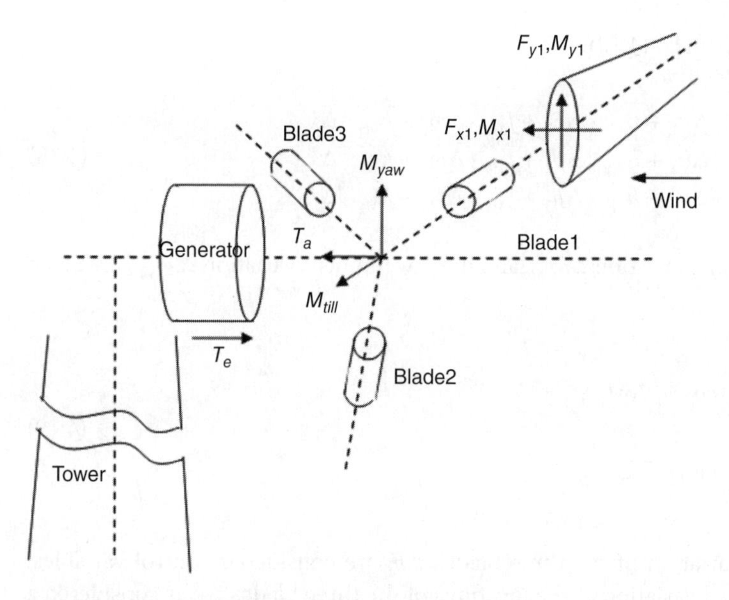

Figure 4.6 The simplified wind turbine model.

First, based on the linear blade element momentum theory, the blade is regarded as a whole. The action point of aerodynamic force is assumed to be located at a position 2R/3 from the blade root. The loads on each blade are related to the effective wind speed V_i, the wind wheel rotational speed ω_r, and the pitch angle β_i. The aerodynamic equation after linearization is given by (4.14):

$$\begin{cases} \Delta F_{xi} = k_{Fx} \cdot \Delta V_i + h_{Fx} \cdot \Delta \beta_i + f_{Fx} \cdot \Delta \omega_r \\ \Delta F_{yi} = k_{Fy} \cdot \Delta V_i + h_{Fy} \cdot \Delta \beta_i + f_{Fy} \cdot \Delta \omega_r \\ \Delta M_{xi} = k_{Mx} \cdot \Delta V_i + h_{Mx} \cdot \Delta \beta_i + f_{Mx} \cdot \Delta \omega_r \\ \Delta M_{yi} = k_{My} \cdot \Delta V_i + h_{My} \cdot \Delta \beta_i + f_{My} \cdot \Delta \omega_r \end{cases} \tag{4.14}$$

In (4.14), the gain in the front of deviation indicates the sensitivity coefficient of aerodynamic loads.

By considering the influence of tower motion in the front and rear directions, the effective wind speed acting on the blades can be expressed as:

$$V_i = v_i - \dot{x}_{fa} \tag{4.15}$$

where v_i represents the actual wind speed acting on the each blade.

According to (4.6), the local equation of the tower motion is given by (4.16):

$$M_t \Delta \ddot{x}_{fa} + C_t \Delta \dot{x}_{fa} + K_t \Delta x_{fa} = \Delta F_{ax} \tag{4.16}$$

According to (4.9), the local equation of wind wheel rotation is given by (4.17):

$$J_r \Delta \dot{\omega}_r = \Delta T_a - \Delta T_s - D_r \Delta \omega_r \tag{4.17}$$

The drivetrain torque T_s is approximately constant if the wind turbine is operating above the rated wind speed. Therefore, $\Delta T_s = 0$, and the bending moment variation of blade root ΔM_{yi} is given by (4.18):

$$\begin{cases} \Delta M_{y1} = k_{My} \cdot \Delta v_1 + h_{My} \cdot \Delta\beta_1 + f_{My} \cdot \Delta\omega_r - k_{My} \cdot \Delta\dot{x}_{fa} \\ \Delta M_{y2} = k_{My} \cdot \Delta v_2 + h_{My} \cdot \Delta\beta_2 + f_{My} \cdot \Delta\omega_r - k_{My} \cdot \Delta\dot{x}_{fa} \\ \Delta M_{y3} = k_{My} \cdot \Delta v_3 + h_{My} \cdot \Delta\beta_3 + f_{My} \cdot \Delta\omega_r - k_{My} \cdot \Delta\dot{x}_{fa} \end{cases} \tag{4.18}$$

The pitching moment variation ΔM_{tilt} and the yaw moment variation ΔM_{yaw} are given by (4.19):

$$\begin{cases} \Delta M_{tilt} = \sum_{i=1}^{3} (\cos\psi_i \cdot \Delta M_{yi}) \\ \Delta M_{yaw} = \sum_{i=1}^{3} (\sin\psi_i \cdot \Delta M_{yi}) \end{cases} \tag{4.19}$$

The pitch angle variation of the three blades $\Delta\beta_i$ are considered control variables. The actual wind speed variation at the locations of the three blades Δv_i is considered a disturbance variation. The speed variation of the wind wheel $\Delta\omega_r$, the pitching moment ΔM_{tilt} and the yaw moment ΔM_{yaw} are all considered as output variables. Therefore, the state equation of the wind turbine is developed from (4.14-4.19) as:

$$\begin{bmatrix} \Delta\dot{\omega}_r \\ \Delta\dot{x}_{fa} \\ \Delta\ddot{x}_{fa} \end{bmatrix} = \begin{bmatrix} \frac{3f_{Mx} - D_r}{J_r} & 0 & -\frac{3k_{Mx}}{J_r} \\ 0 & 0 & 1 \\ 0 & -\frac{K_{tw}}{M_t} & -\frac{D_{tw} + 3k_{Fx}}{M_t} \end{bmatrix} \begin{bmatrix} \Delta\omega_r \\ \Delta x_{fa} \\ \Delta\dot{x}_{fa} \end{bmatrix} \begin{bmatrix} \frac{h_{Mx}}{J_r} & \frac{h_{Mx}}{J_r} & \frac{h_{Mx}}{J_r} \\ 0 & 0 & 0 \\ \frac{h_{Fx}}{M_t} & \frac{h_{Fx}}{M_t} & \frac{h_{Fx}}{M_t} \end{bmatrix} \begin{bmatrix} \Delta\beta_1 \\ \Delta\beta_2 \\ \Delta\beta_3 \end{bmatrix} + \begin{bmatrix} \frac{k_{Mx}}{J_r} & \frac{k_{Mx}}{J_r} & \frac{k_{Mx}}{J_r} \\ 0 & 0 & 0 \\ \frac{k_{Fx}}{M_t} & \frac{k_{Fx}}{M_t} & \frac{k_{Fx}}{M_t} \end{bmatrix} \begin{bmatrix} \Delta v_1 \\ \Delta v_2 \\ \Delta v_3 \end{bmatrix} \tag{4.20}$$

The output equation is given by (4.21):

$$\begin{bmatrix} \Delta\omega_r \\ \Delta M_{tilt} \\ \Delta M_{yaw} \end{bmatrix} = \begin{bmatrix} 1 & 0 & 0 \\ 0 & 0 & 0 \\ 0 & 0 & 0 \end{bmatrix} \cdot \begin{bmatrix} \Delta\omega_r \\ \Delta x_{fa} \\ \Delta\dot{x}_{fa} \end{bmatrix} + \begin{bmatrix} 0 & 0 & 0 \\ h_{My}\cos\psi_1 & h_{My}\cos\psi_2 & h_{My}\cos\psi_3 \\ -h_{My}\sin\psi_1 & -h_{My}\sin\psi_2 & -h_{My}\sin\psi_3 \end{bmatrix} \cdot \begin{bmatrix} \Delta\beta_1 \\ \Delta\beta_2 \\ \Delta\beta_3 \end{bmatrix} + \begin{bmatrix} 0 & 0 & 0 \\ k_{My}\cos\psi_1 & k_{My}\cos\psi_2 & k_{My}\cos\psi_3 \\ -k_{My}\sin\psi_1 & -k_{My}\sin\psi_2 & -k_{My}\sin\psi_3 \end{bmatrix} \cdot \begin{bmatrix} \Delta v_1 \\ \Delta v_2 \\ \Delta v_3 \end{bmatrix} \tag{4.21}$$

As can been seen from (4.21), the pitch moment and yaw moment are both influenced by pitch angle and wind speed, and the degree of influence is constantly changing with the azimuth changes. In order to simplify the controller design, a Coleman transformation is carried out, and the state equation (4.20) is transformed into:

$$\begin{bmatrix}\Delta\dot{\omega}_r\\ \Delta\dot{x}_{fa}\\ \Delta\ddot{x}_{fa}\end{bmatrix} = \begin{bmatrix}\dfrac{3f_{Mx}-D_r}{J_r} & 0 & -\dfrac{3k_{Mx}}{J_r}\\ 0 & 0 & 1\\ 0 & -\dfrac{K_t}{M_t} & -\dfrac{D_t+3k_{Fx}}{M_t}\end{bmatrix}\begin{bmatrix}\Delta\omega_r\\ \Delta x_{fa}\\ \Delta\dot{x}_{fa}\end{bmatrix} + \begin{bmatrix}\dfrac{3h_{Mx}}{J_r} & 0 & 0\\ 0 & 0 & 0\\ \dfrac{3h_{Fx}}{M_t} & 0 & 0\end{bmatrix}\begin{bmatrix}\Delta\beta_0\\ \Delta\beta_c^{(1)}\\ \Delta\beta_s^{(1)}\end{bmatrix}$$
$$+\begin{bmatrix}\dfrac{3k_{Mx}}{J_r} & 0 & 0\\ 0 & 0 & 0\\ \dfrac{3k_{Fx}}{M_t} & 0 & 0\end{bmatrix}\begin{bmatrix}\Delta v_0\\ \Delta v_c^{(1)}\\ \Delta v_s^{(1)}\end{bmatrix} \tag{4.22}$$

The output equation (4.21) is transformed into:

$$\begin{bmatrix}\Delta\omega_r\\ \Delta M_{yc}^{(1)}\\ \Delta M_{ys}^{(1)}\end{bmatrix} = \begin{bmatrix}1&0&0\\0&0&0\\0&0&0\end{bmatrix}\cdot\begin{bmatrix}\Delta\omega_r\\ \Delta x_{fa}\\ \Delta v_{fa}\end{bmatrix} + \begin{bmatrix}0&0&0\\0&h_{My}&0\\0&0&h_{My}\end{bmatrix}\cdot\begin{bmatrix}\Delta\beta_0\\ \Delta\beta_{yc}^{(1)}\\ \Delta\beta_{ys}^{(1)}\end{bmatrix} + \begin{bmatrix}0&0&0\\0&k_{My}&0\\0&0&k_{My}\end{bmatrix}\cdot\begin{bmatrix}\Delta v_0\\ \Delta v_{yc}^{(1)}\\ \Delta v_{ys}^{(1)}\end{bmatrix} \tag{4.23}$$

There is the following relationship between $\Delta M_{yc}^{(1)}$, $\Delta M_{ys}^{(1)}$ and M_{tilt}, M_{yaw}:

$$\begin{cases} M_{tilt} = 1.5M_{yc}^{(1)}\\ M_{yaw} = 1.5M_{ys}^{(1)}\end{cases} \tag{4.24}$$

By means of a coordinate transformation, the time-variant equation of the wind turbine is transformed into a time invariant equation, and there is no coupling among the output variables. Therefore, the control can be carried out according to three independent SISO systems. ω_r is realized through collective pitch control, so the control method of $\Delta M_{yc}^{(1)}$ and $\Delta M_{ys}^{(1)}$ is introduced specifically in this section.

4.3.3 Controller Design

1p Individual Pitch Control

Because there is a large 1p component of the wind wheel imbalance load, the 1p individual pitch control method is first proposed as a way to reduce the 1p component.

By carrying out local linearization, the wind wheel imbalance loads can be expressed as (4.25):

$$\begin{cases} \Delta M_{yc}^{(1)} = h_{My}\cdot\Delta\beta_c^{(1)} + k_{My}\cdot\Delta v_c^{(1)}\\ \Delta M_{ys}^{(1)} = h_{My}\cdot\Delta\beta_s^{(1)} + k_{My}\cdot\Delta v_s^{(1)}\end{cases} \tag{4.25}$$

where $\Delta\beta_c^{(1)}$ and $\Delta\beta_s^{(1)}$ represent control variables, and $\Delta v_c^{(1)}$ and $\Delta v_s^{(1)}$ represent disturbance variables. Therefore, $\Delta M_{yc}^{(1)}$ and $\Delta M_{ys}^{(1)}$ can be reduced by restraining $\Delta v_c^{(1)}$ and $\Delta v_s^{(1)}$. With the same forms and completed decoupling of $\Delta v_c^{(1)}$ and $\Delta v_s^{(1)}$, the controllers can be designed in the same way. In this chapter, the 1p individual pitch control strategy is proposed, as shown in Figure 4.7.

The detailed design process is as follows. First, the loads of the three blades and the azimuth angle of the wind wheel are measured and transformed into $\Delta M_{yc}^{(1)}$ and $\Delta M_{ys}^{(1)}$ through a Coleman transformation. Then the 3p components of $\Delta M_{yc}^{(1)}$ and $\Delta M_{ys}^{(1)}$ are

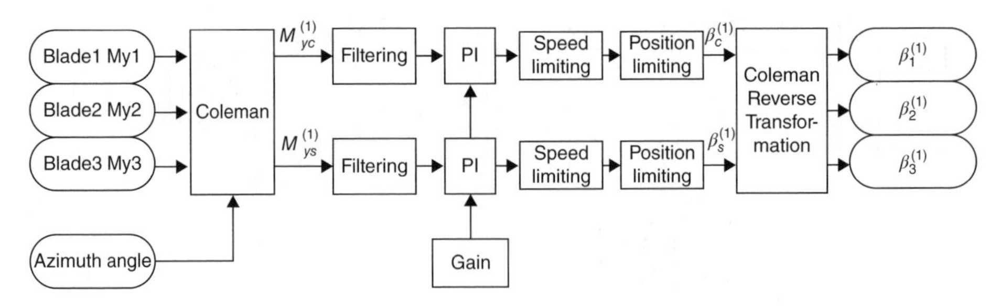

Figure 4.7 The control diagram block of IPC-1p.

removed through a band stop filter. In order to ensure the overall gain of system is unchanged, a gain is added to the PI controller, which takes the average of the three blades' pitch angles feedback as coefficients. Finally, through a Coleman reverse transformation, the outputs $\Delta\beta_c^{(1)}$ and $\Delta\beta_s^{(1)}$ are turned back to $\beta_1{}^{(1)}$, $\beta_2{}^{(1)}$, and $\beta_3{}^{(1)}$ as inputs of the variable pitch actuator. To avoid any instability of $\Delta\beta_c^{(1)}$ and $\Delta\beta_s^{(1)}$, position limiting and rate limiting are also needed.

2p Individual Pitch Control

The 1p individual pitch controller can reduce 1p component of blade loads. However, there is still a high proportion of 3p component, which is mainly caused by 2p cyclic blade loads. Therefore, 2p individual pitch control is designed to reduce 2p cyclic loads. First, the following Coleman transformation is carried out as (4.26):

$$\begin{bmatrix} M_{yc}^{(2)} \\ M_{ys}^{(2)} \end{bmatrix} = \frac{2}{3} \begin{bmatrix} \cos 2\psi_1 & \cos 2\psi_2 & \cos 2\psi_3 \\ \sin 2\psi_1 & \sin 2\psi_2 & \sin 2\psi_3 \end{bmatrix} \begin{bmatrix} M_{y1} \\ M_{y2} \\ M_{y3} \end{bmatrix} \tag{4.26}$$

Through IPC-1p control, the 1p component of the blade loads has been eliminated. Then, the 2p component of the blade loads can be extracted by (4.26), and an IPC-2p controller can be designed to control $\Delta M_{yc}^{(2)}$ and $\Delta M_{ys}^{(2)}$ to the target value. Referring to the IPC-1p controller, the control diagram block of IPC-2p is as shown in Figure 4.8.

Furthermore, it can be expanded to np ($n = 3,4\ldots$) individual control as required. However, as n increases, the pitch angle variation frequency of each closed-loop output also increases, which means a higher action frequency of the variable pitch actuator is required.

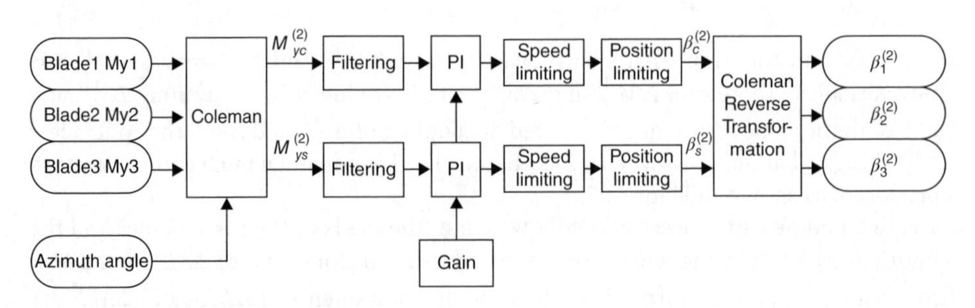

Figure 4.8 The control diagram block of IPC-2p

4.3.4 Simulation Analysis

In order to verify the effectiveness of IPC-1p and IPC-2p control strategies, the wind turbine dynamic simulation is carried out under conditions of wind turbulence. The average wind speed is 18 m/s, and the longitudinal, transverse, and vertical turbulence intensities are 16.97%, 13.29%, and 9.44% respectively. The wind shear coefficient is 0.2. The simulation results are shown in Figure 4.9.

This shows that the blade root loads can be reduced by 25–26% and hub loads by 9–20%, with drivetrain torque increased by 2% through IPC-1p control. The blade root loads can be reduced by 27–29% and hub loads by 14–24%, with drivetrain torque increased by 3%, through IPC-1p+2p control.

Figure 4.10 shows the spectrum of the blade root bending moment. It can be seen the 1p component (0.3 Hz) of the blade root loads can be weakened significantly through IPC-1p control. In addition, the 2p component (0.6 Hz) of blade root loads can be further reduced through IPC-2p control.

Figure 4.11 shows the simulation waveform of the generated power using different pitch control methods. It can be seen that there is no significant effect on output power caused by individual pitch control.

4.4 Drivetrain Torsional Vibration Control

The wind turbine drivetrain is a large-inertia and low-stiffness system, with a low resonant frequency. Low-frequency resonance may easily occur under the influence of external disturbances. Because of the small damping coefficient, it is not easy to restrict resonance, which will seriously affect the dynamic loads of drivetrain components, thus accelerating damage to them. Moreover, resonance also causes power pulsation of the generator and endangers power quality. Therefore, it is necessary to restrain the torsional vibration of large-scale wind turbine drivetrains [5-9]. In this section, a torsional vibration control strategy based on the linear-quadratic-Gaussian method (LQG) is introduced.

4.4.1 LQG Controller Design

The control diagram block of LQG is shown in Figure 4.12. The LQG control system consists of optical state feedback and a Kalman state estimator. Dynamic damping can be applied to the drivetrain using the state feedback controller K, thereby restraining the torsional vibration of drivetrain. The detailed design process of LQG controller will be described in the following paragraphs.

According to Figure 4.3, the torsional angular velocity ω_s, the torsional angle θ_s, and the generator angular velocity ω_g are considered state variables. The electromagnetic torque T_e is considered a control variable. The aerodynamic torque T_a is considered a disturbance variable. Therefore, the state equation of the drivetrain model can be established as follows:

$$\begin{bmatrix} \dot{\omega}_s \\ \dot{\theta}_s \\ \dot{\omega}_g \end{bmatrix} = \begin{bmatrix} -\dfrac{D_s(J_r+J_g)}{J_rJ_g} & -\dfrac{K_s(J_r+J_g)}{J_rJ_g} & 0 \\ 1 & 0 & 0 \\ \dfrac{D_s}{J_g} & \dfrac{K_s}{J_g} & 0 \end{bmatrix} \begin{bmatrix} \omega_s \\ \theta_s \\ \omega_g \end{bmatrix} + \begin{bmatrix} \dfrac{1000}{J_g} \\ 0 \\ -\dfrac{1000}{J_g} \end{bmatrix} T_e + \begin{bmatrix} \dfrac{1000}{J_r} \\ 0 \\ 0 \end{bmatrix} T_a \tag{4.27}$$

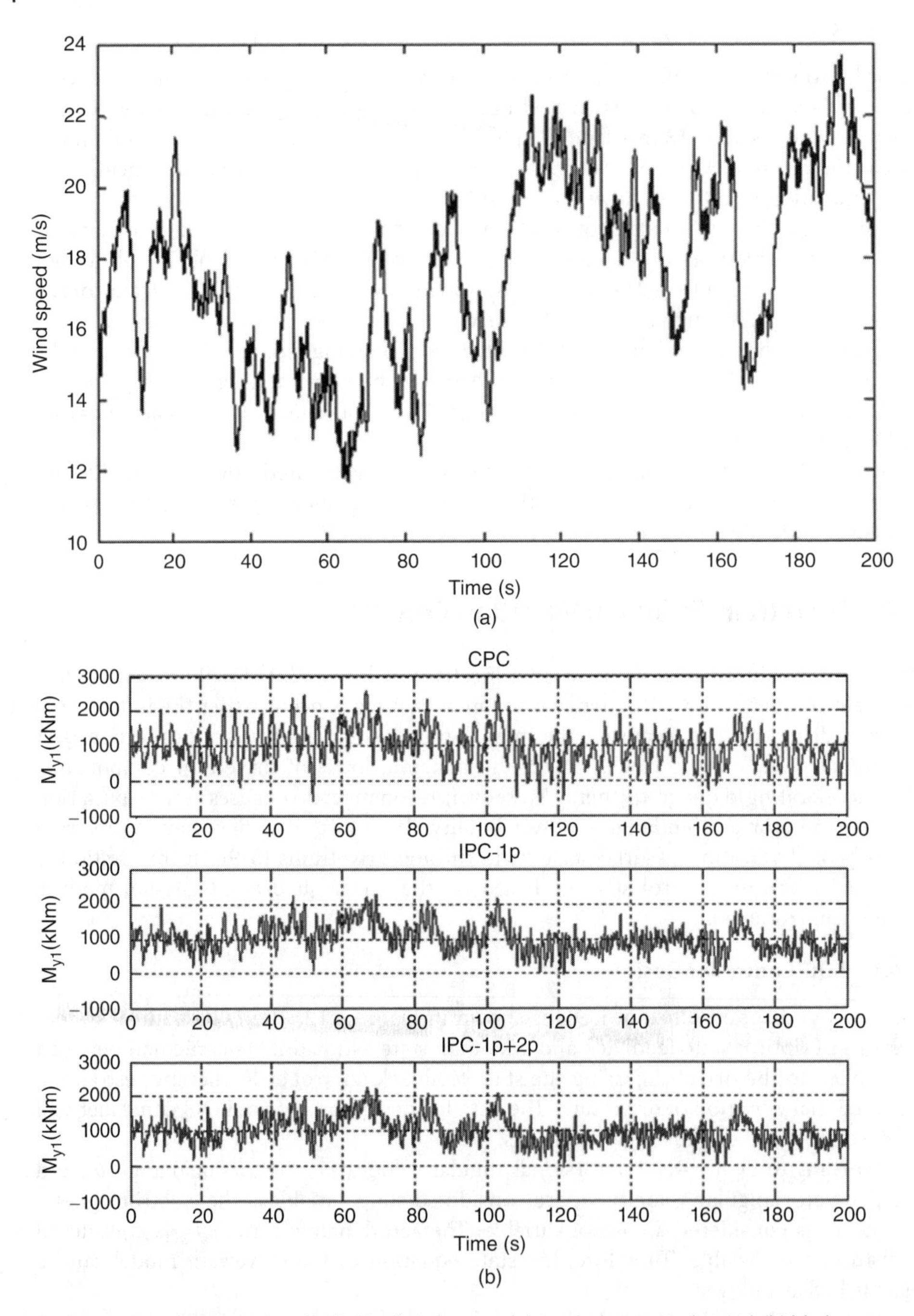

Figure 4.9 Simulation results of individual pitch control: (a) turbulence wind speed; (b) blade root bending moment M_{y1}; (c) wind wheel pitch moment M_{titl}; (d) wind wheel yawing moment M_{yaw}; (e) drivetrain torque T_s.

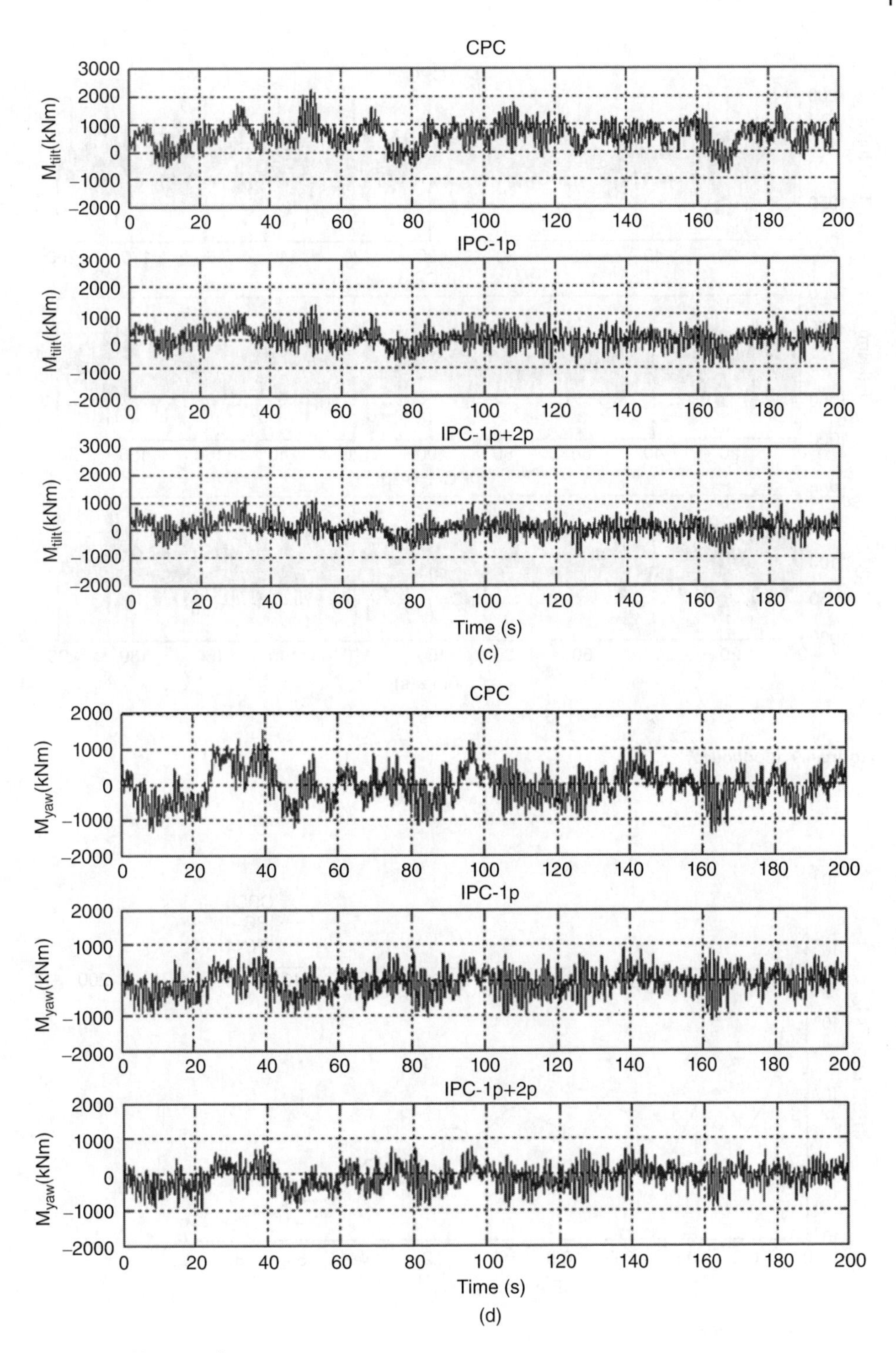

Figure 4.9 (*Continued*)

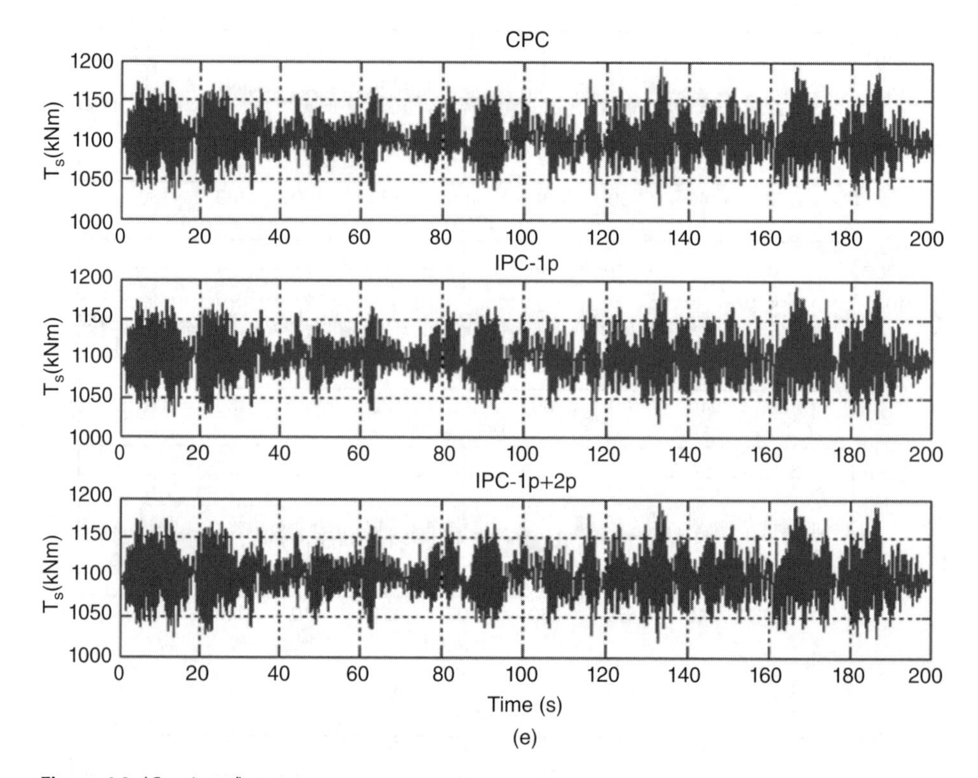

Figure 4.9 (*Continued*)

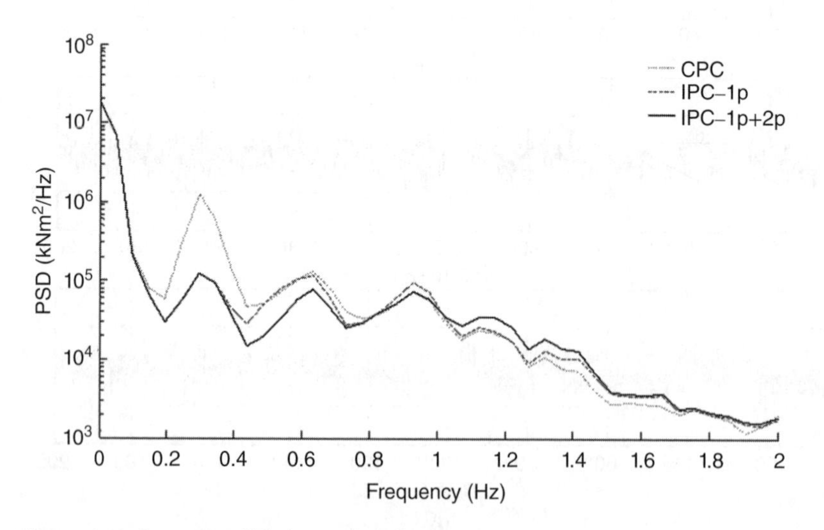

Figure 4.10 Spectrum of blade root bending moment.

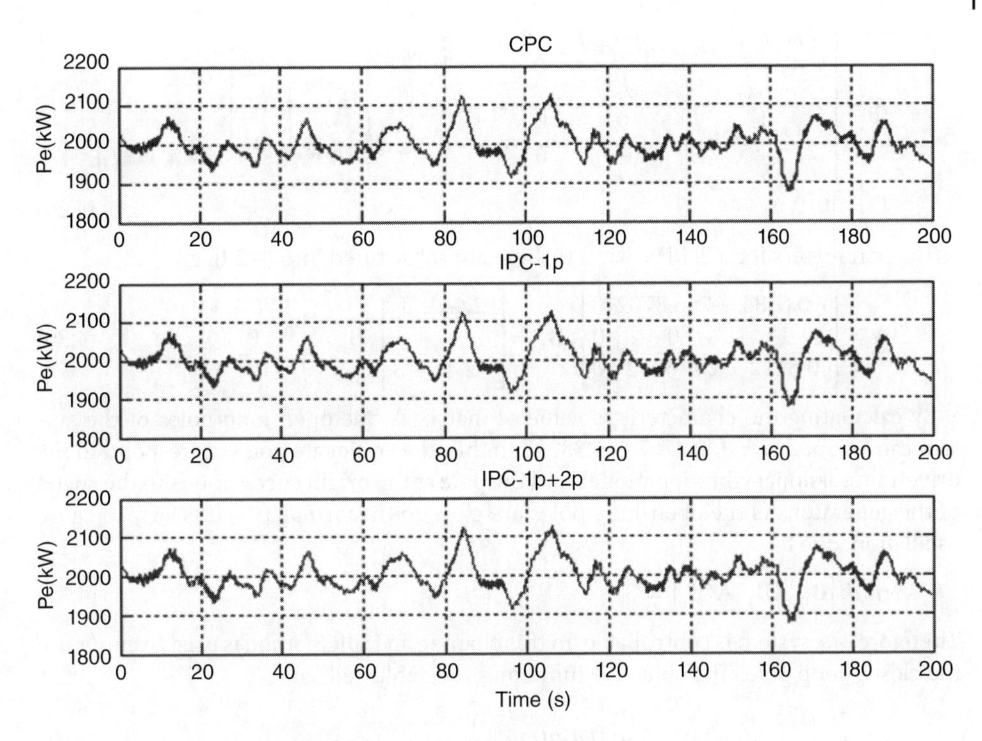

Figure 4.11 Simulation waveform of generation power

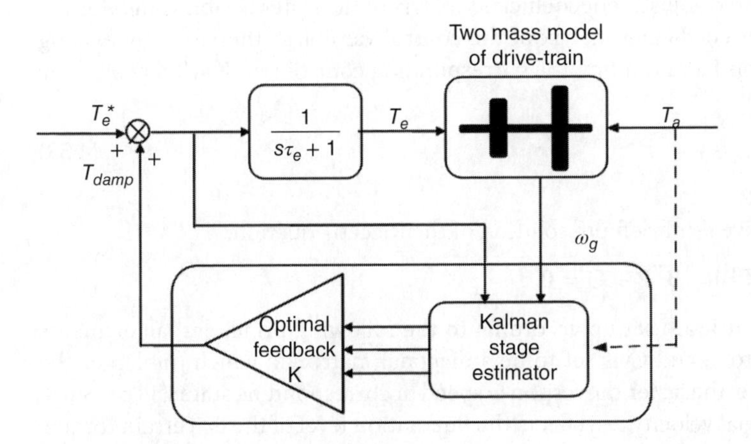

Figure 4.12 The control diagram block for LQG.

The above equation can be described as follows:

$$\begin{cases} \dot{\mathbf{x}} = \mathbf{A}\mathbf{x} + \mathbf{B}_1 u_1 + \mathbf{B}_2 u_2 \\ \mathbf{y} = \mathbf{C}\mathbf{x} \end{cases} \tag{4.28}$$

where $\mathbf{x} = \left[\omega_s \ \theta_s \ \omega_g\right]^T$, $u_1 = T_e^*$, and $u_2 = T_a$.

$$\mathbf{A} = \begin{bmatrix} -\dfrac{D_s(J_r + J_g)}{J_r J_g} & -\dfrac{K_s(J_r + J_g)}{J_r J_g} & 0 \\ 1 & 0 & 0 \\ \frac{D_s}{J_g} & \frac{K_s}{J_g} & 0 \end{bmatrix}, \mathbf{B}_1 = \begin{bmatrix} \frac{1000}{J_g} \\ 0 \\ -\frac{1000}{J_g} \end{bmatrix}, \mathbf{B}_2 = \begin{bmatrix} \frac{1000}{J_r} \\ 0 \\ 0 \end{bmatrix}$$

$$\mathbf{C} = \begin{bmatrix} 0 & 0 & 1 \end{bmatrix} \tag{4.29}$$

The parameters for a 2-MW wind turbine are substituted into (4.29), giving:

$$\mathbf{A} = \begin{bmatrix} -0.6534 & -266.8702 & 0 \\ 1 & 0 & 0 \\ 0.6112 & 249.6202 & 0 \end{bmatrix}, \mathbf{B}_1 = \begin{bmatrix} 2.4e-3 \\ 0 \\ -2.4e-3 \end{bmatrix}, \mathbf{B}_2 = \begin{bmatrix} 1.7e-4 \\ 0 \\ 0 \end{bmatrix} \tag{4.30}$$

By calculating the characteristic value of matrix **A**, the open-loop poles of the system can be obtained: 0,-0.3267±16.3329i. In this, the conjugate poles correspond to the drivetrain torsional vibration model, and the pole at the origin corresponds to the speed of the generator. As the open-loop poles are close to the virtual axis, it is easy to cause oscillation, due to:

$$rank\begin{bmatrix} \mathbf{B}_1 & \mathbf{AB}_1 & \mathbf{A}^2\mathbf{B}_1 \end{bmatrix} = 3 \tag{4.31}$$

Therefore, the system is controllable. In this chapter, an LQR method is used to configure the closed-loop pole. The objective function **J** is established as:

$$\mathbf{J} = \frac{1}{2}\int_{t_0}^{\infty} [\mathbf{x}^T(t)\mathbf{Q}\mathbf{x}(t) + \mathbf{u}^T(t)\mathbf{R}\mathbf{u}(t)]dt \tag{4.32}$$

where **Q** is the weighted adjustment coefficient matrix of the state variables, and **R** is the weighted adjustment coefficient matrix of the control variables. Therefore, by making the objective function **J** a minimum, the corresponding control rate **K** will become the optimal control rate.

$$\mathbf{u}(t) = -\mathbf{K}\mathbf{x}(t) \tag{4.33}$$

where $\mathbf{K} = -\mathbf{R}^{-1}\mathbf{B}^{\mathrm{T}}\mathbf{P}$.

P is the only positive semi-definite solution of the Riccati equation:

$$\mathbf{PA} + \mathbf{A}^{\mathrm{T}}\mathbf{P} - \mathbf{PBR}^{-1}\mathbf{B}^{\mathrm{T}}\mathbf{P} + \mathbf{Q} = 0 \tag{4.34}$$

The matrices **Q** and **R** are set up according to the following principles: all elements of matrix **R** are set to 1, and **Q** is set to be a diagonal matrix, in which the larger the diagonal elements are, the faster the response speed of corresponding state will be. Since the drivetrain torsional velocity ω_s reflects the fluctuation level of the drivetrain torque, the weight of the first line diagonal element Q1 should be larger than that of the second and third line diagonal elements.

When **Q** = diag(2e6, 1, 0.1), the closed loop poles are 0, -1.7284±16.2445i. The coordinate position of the closed-loop poles is shown in Figure 4.13. Thus it can be seen that with LQR closed-loop feedback, the conjugate poles move to the left half plane, and the damping ratio of the torsional mode increases to five times the original value. If Q1 continues to increase, although the damping ratio will be increased, which can speed up the convergence rate of torsional model, it will lead to a larger damping torque, which will cause overload of the converter and generator.

Figure 4.13 The closed-loop poles of system.

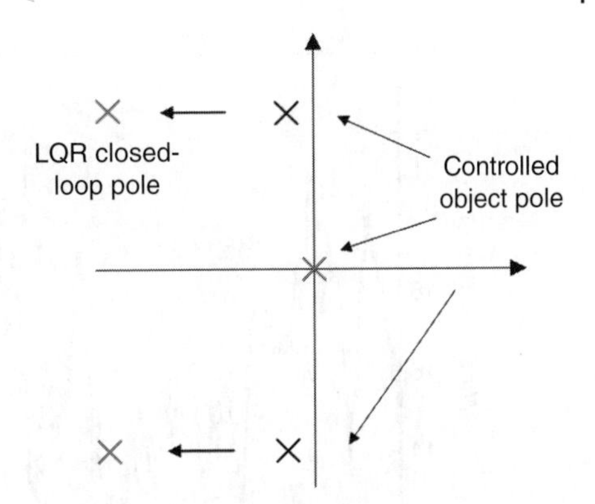

The practical application of control techniques is usually limited by problems in obtaining measurements of the states as needed in the controller. Extra sensors and measurements can add considerable cost and complexity to a wind turbine, while errors in the measurement of these states can cause poor controller behavior. Thus a state estimator is employed, where unknown states are determined from just one measured generator speed. In order to achieve good state estimation despite having uncertainty in the model and measurement noise, a discrete-time Kalman filter is used:

$$\begin{cases} \dot{\hat{\mathbf{x}}}(t) = \mathbf{A}\hat{\mathbf{x}}(t) + \mathbf{B}\mathbf{u}(t) + \mathbf{L}(\mathbf{y}(t) - \hat{\mathbf{y}}(t)) \\ \hat{\mathbf{y}}(t) = \mathbf{C}\hat{\mathbf{x}}(t) \end{cases} \tag{4.35}$$

where, **L** represents the optical state estimation:

$$\mathbf{L} = \mathbf{Y}\mathbf{C}^T\mathbf{V}^{-1} \tag{4.36}$$

Y can be obtained by solving the Riccati equation:

$$\mathbf{Y}\mathbf{A}^{\mathrm{T}} + \mathbf{A}\mathbf{Y} - \mathbf{Y}\mathbf{C}^{\mathrm{T}}\mathbf{V}^{-1}\mathbf{C}\mathbf{Y} + \mathbf{W} = 0 \tag{4.37}$$

where **W** and **V** are respectively the system noise variance and measurement noise variance, both determined by experiment. Here take W = diag(2e-7, 1e-7, 2e-6),V=1e-6. Then **L** can be obtained from (4.36).

4.4.2 Simulation Analysis

In order to verify the control effect of the LQG control strategy on the basis of IPC, a wind turbine dynamic simulation is carried out under the conditions of turbulent wind. The average wind speed is 18 m/s, and the longitudinal, transverse, and vertical turbulence intensities are 16.97%, 13.29%, and 9.44% respectively. The wind shear coefficient is 0.2.

The simulation results of LQG damping control is shown in Figure 4.14. Part (a) shows the turbulent wind speed, (b) shows the value of electromagnetic torque, which is a constant value in the condition of undamped control, but includes a tiny disturbance in the condition of LQG control to realize the adding of resistance to drivetrain. Part (c) shows the simulation waveform of the blade root bending moment, (d) shows the drivetrain torque and (e) the drivetrain torque spectrum.

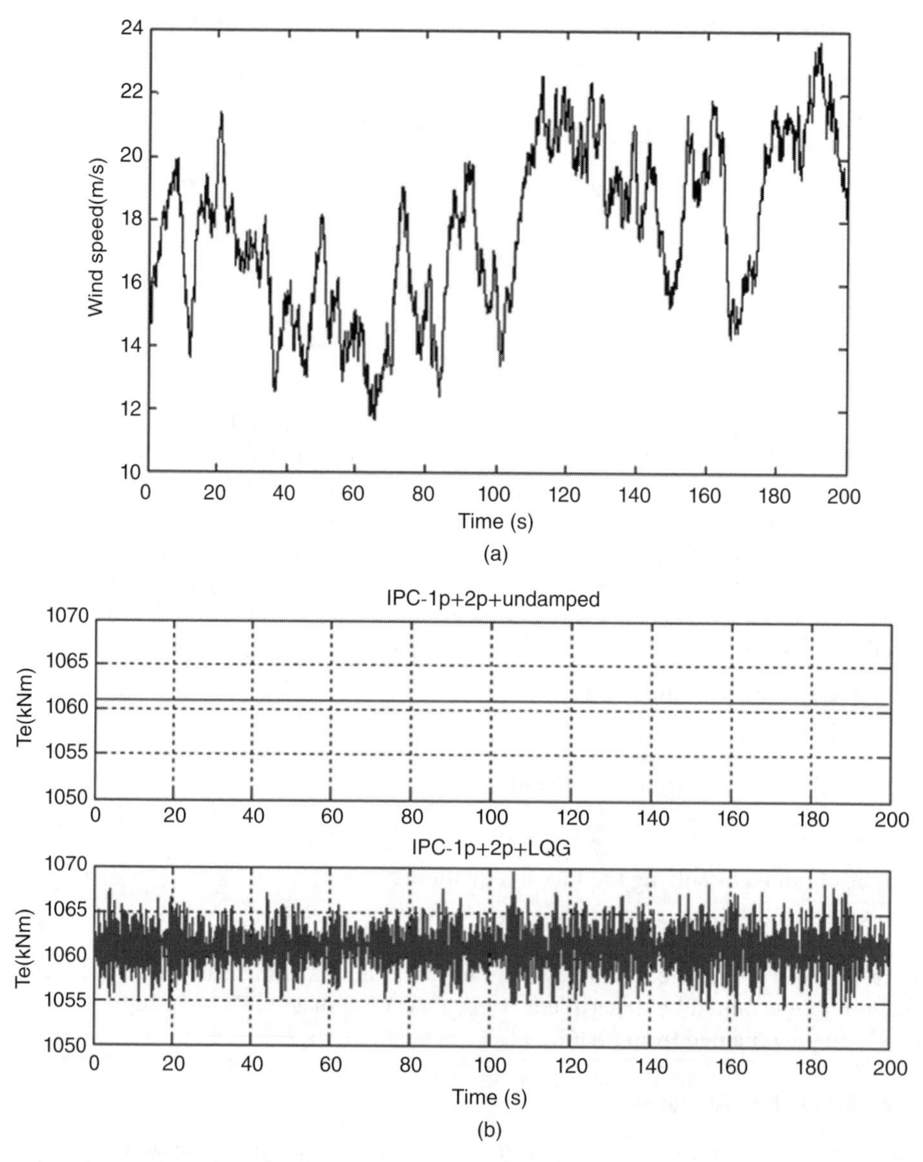

Figure 4.14 Simulation results of LQG control: (a) turbulent wind speed; (b) electromagnetic torque; (c) blade 1 bending moment; (d) drivetrain torque; (e) drivetrain torque spectrum.

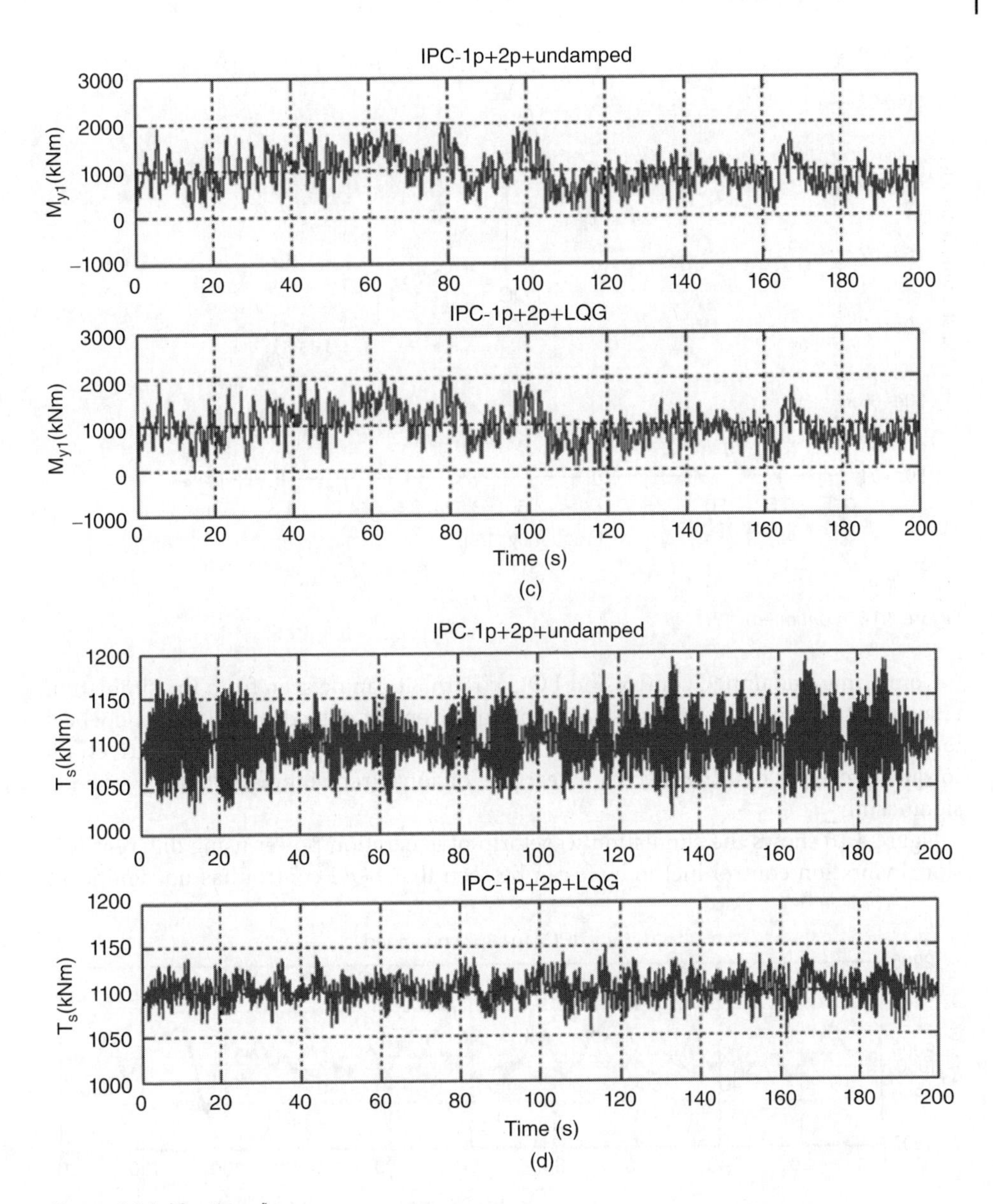

Figure 4.14 *(Continued)*

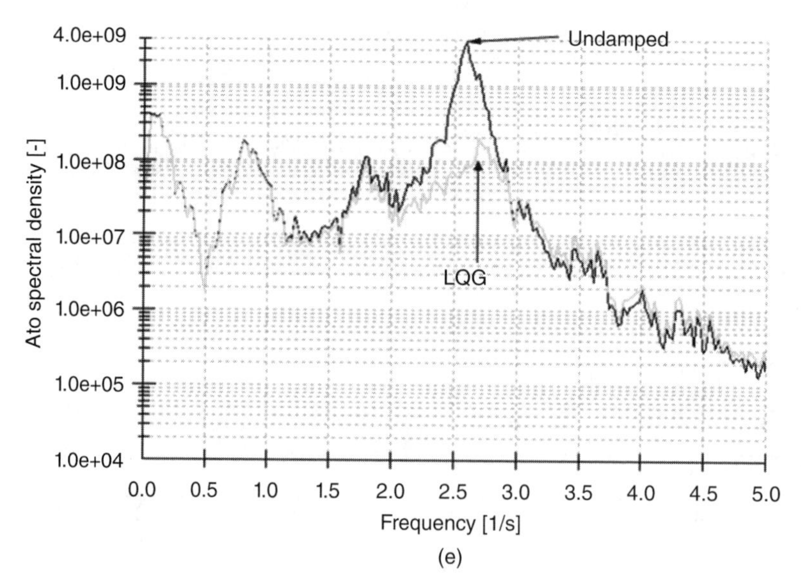

Figure 4.14 (*Continued*)

Comparing undamped control and LQG control, it can be seen from the simulation results that although there is no significant improvement of blade root loads and wheel loads, the drivetrain torque is reduced by 50%. From the spectrum of the drivetrain torque we can see that the spectrum near the torsional frequency (2.6 Hz) is weakened significantly

Figure 4.15 shows the simulation waveform of generation power using different torsional vibration control methods. It can be seen that LQG control has no significant

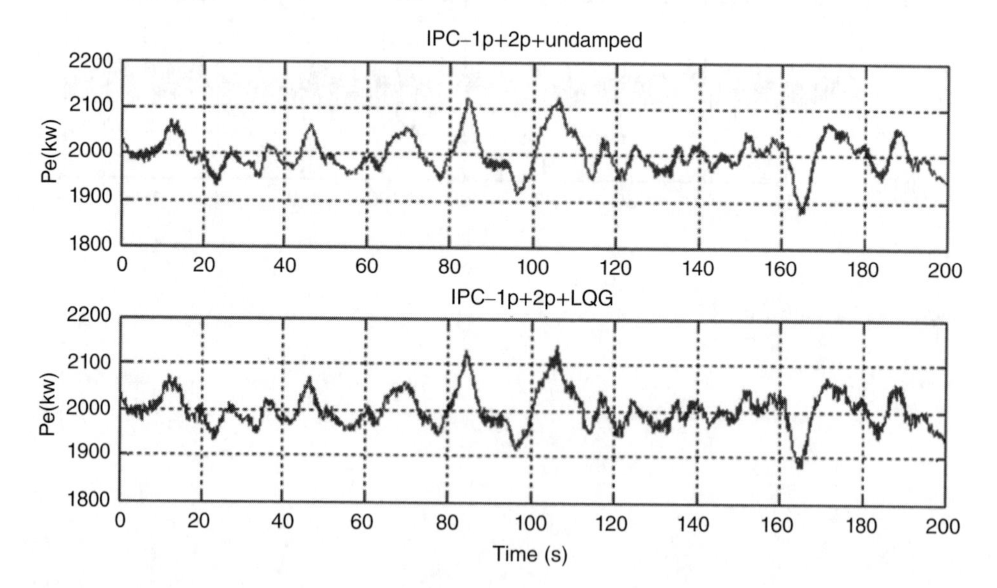

Figure 4.15 The simulation waveform of generated power.

effect on output power. This indicates that the LQG control strategy can effectively reduce wind turbine loads and thus improve the reliability and lifetime of wind turbines without affecting power generation.

4.5 Conclusion

In this chapter, the load optimization control strategies of large-scale wind turbines was examined. First, the dynamic model for permanent magnet direct-driven wind turbines was developed with wind condition, aerodynamics, tower, drivetrain and electric control system considered. Then, based on blade root load measurements and linear quadratic-Gaussian control theory, the individual pitch control and drivetrain torsion vibration control strategies were proposed. Finally, the effectiveness of these strategies was validated by simulation. The simulation results show that the proposed strategies can effectively reduce wind turbine loads and thus improve the reliability and lifetime of wind turbines while not affecting power generation. Moreover, they can be easily applied to commercial wind turbines.

References

1 CWEA (2014) *Statistics of China wind power installed capacity in 2014.* Chinese Wind Energy Association.

2 Ying, Y. and Xu, G.D. (2011) Development of pitch control for load reduction on wind turbines. *Journal of Mechnical Engineering*, **47** (16), 106–111.

3 Darrow, J., Johnson, K., and Wright, A. (2011) Design of a tower and drive train damping controller for the three-bladed controls advanced research turbine operating in design-driving load cases. *Wind Energy*, **14**(4), 571–601.

4 Jin, X., Li, L., He, Y.L., and Yang, X.G. (2015) Independent pitch control strategy analysis of MW wind turbine. *Journal of ChongqingUniversity*, **38** (1), 61–67.

5 Mandic, G., Nasiri, A., Muljadi, E., and Oyague, F. (2012) Active torque control for gearbox load reduction in a variable-speed wind turbine. *IEEE Transactions on Industrial Applications*, **48** (6), 2424–2432.

6 Camblong, H., Vechiu, I., Etxeberria, A., and Martínez, M.I. (2014) Wind turbine mechnical stresses reduction and contribution of frequency regulation. *Control Engineering Practice*, **30**, 140–149.

7 Xu, H., Hu, S.S., Song, B.,and Xu, H.H. (2014) Optimal control of DFIG wind turbines for load reduction under symmetrical grid fault. *Automation of Electric Power System*, **38** (11), 20–26.

8 Xu, H. (2014) Research on optimal control strategies of large scale wind turbine for load reductiond. University of Chinese Academy of Sciences.

9 Xu, H., Xu, H.H., Chen, L., and Wenske, J. (2013) Active damping control of DFIG wind turbines during fault ride through. *2013 3rd International Conference on Electric Power and Energy Conversion Systems*, Istanbul, Turkey, October 2-4.

5

Modeling of Full-scale Converter Wind Turbine Generator

Yongning Chi[1], Chao Liu[1], Xinshou Tian[1], Lei Shi[2], and Haiyan Tang[1]

[1] *China Electric Power Research Institute, Beijing, China*
[2] *NARI Technology Co. Ltd, Jiangsu, China*

5.1 Introduction

The full-scale converter (FSC) wind turbine generator (WTG) is connected to the grid through full rating AC/DC/AC converters. The FSC-WTG consists of the wind turbine, generator and FSC [1], and the FSC system can be further divided into the generator-side converter, DC link, and the grid-side converter. The output of the FSC-WTG is rectified by the generator-side converter and then supported by the capacitor, and the energy is fed to the grid through the grid-side converter. The FSC-WTG can achieve a flexible connection with grid through the control over-converter, and in the event of a grid fault, the converter can also greatly improve the reactive current contribution of the WTG to the grid, enhancing the ability of the WTG to deal with grid disturbances [2].

The gearbox is optional for the FSC WTG. There are many types of generator used in FSC-WTGs, such as the squirrel cage induction generator (SCIG), the permanent magnet synchronous generator (PMSG), and the electrical excitation synchronous generator. The basic structure of a FSC-WTG is shown in Figure 5.1. With its excellent operating characteristics, the FSC-WTG has become mainstream, alongside the doubly-fed induction generator (DFIG) WTG. In particular, FSC-WTGs based on SCIG and PMSG technologies have been widely studied and used [1]. They are therefore the focus of this chapter.

FSC-WTGs based on PMSGs have fault ride through (FRT) capability and give good grid connection [3–5]. Depending on whether there is a gearbox, they can be categorised as direct-drive, semi-direct-drive, or high-speed. Direct-drive PMSG-based wind turbines use multipolar PMSG, omitting the gearbox, so the generator is directly driven by the wind turbine. The application of PMSG improves the system efficiency, and the omission of the gearbox reduces the system fault rate. High-power multipolar PMSG has shortcomings, such as the large diameter of the machine, and difficulties in design, manufacture, installation, and maintenance. Semi-direct-drive PMSG-based wind turbines make use of a gearbox of low gearing ratio, and the rotational speed of the PMSG is 120–450 rpm. The increase of speed may reduce the volume and cost of the permanent magnet generator and facilitate design, manufacture, installation, and maintenance. However, the existence of the gearbox may still lead to a deterioration of

Modeling and Modern Control of Wind Power, First Edition. Edited by Qiuwei Wu and Yuanzhang Sun.

Companion website: www.wiley.com/go/wu/modeling

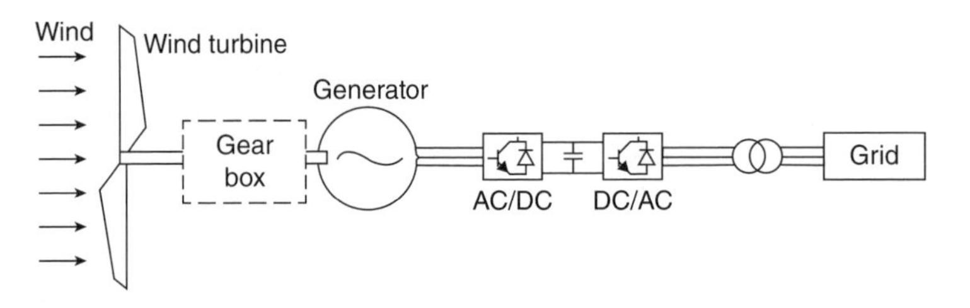

Figure 5.1 Structure diagram of FSC-WTG.

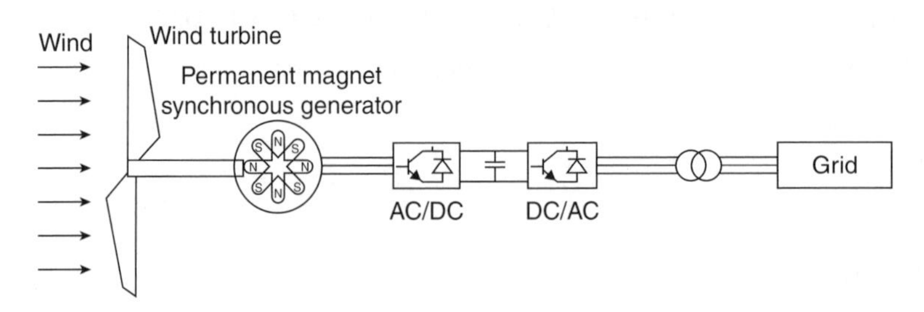

Figure 5.2 Structure diagram of direct-drive PMSG-based wind turbine.

reliability. The high-speed PMSG-based wind turbine is basically similar to induction generator based wind turbines, with the high-speed PMSG together with a gearbox of higher gearing ratio facilitating the production, installation, operation, and maintenance, and improving the efficiency. However, the gearbox is a problem. Direct-drive PMSG-based wind turbines are the most widely used. A direct-drive PMSG-based wind turbine is shown in Figure 5.2.

Due to the limited speed range of traditional wind turbine generators, a gearbox is adopted for the traditional WTG system to increase the speed of the wind turbine. However, the gearbox generates noise, increases the fault rate, and reduces the utilization of wind power. The direct-drive WTG system is set up so that the wind turbine directly drives the low-speed synchronous AC machine to generate electrical energy, and is directly connected to the grid through the FSC, greatly improving the system efficiency [6]. With direct-drive technology, a gearbox is not necessary between the wind turbine and the AC generator, therefore reducing the maintenance interval, reducing noise due to the gearbox, and increasing the efficiency at the low wind speeds. The number of poles of the rotor of a low-speed AC generator is much higher than that of an ordinary AC synchronous generator, and always above 30 pairs. Therefore, the excircle of the rotor and the inner diameter of the stator are greatly increased, but the axial length is relatively short, in a disk shape. The permanent magnet machine significantly simplifies the system control structure and reduces the volume and mass of the generator. The PMSG is excited by the permanent magnet, and there is no exciting winding on the rotor. As such, the exciting winding does not have copper loss, and the efficiency is higher than an electrically excited generator with the same capacity. There is no slip ring on the rotor, which ensures safe and reliable operation. In the wind power system, the AC current from the generator is converted by the rectifier and the inverter, and is then connected to grid. The DC link has a large shunt capacitor, which can maintain a constant DC bus voltage.

The gearbox, which has a high fault rate and cost, is eliminated from the direct-drive PMSG-based wind turbine, reducing the operating and maintenance costs. However, the direct connection between the wind turbine and the generator rotor results in the generator converting all the rotor torque into electric energy. To make up the speed, the generator radius must be increased, and the volume increase may cause huge problems for transport and installation. In addition, the direct connection between the rotor and the generator may increase the impulse load of the blade and the fault rate of the generator. Moreover, the cost of the direct-drive PMSG-based wind turbines may be influenced by the price of permanent magnet materials.

Because of these shortcomings of direct-drive PMSG-based wind turbines, induction generators driven by full-scale converter induction generators (FSC-IGs) are being widely studied and used. Currently, the slip ring and brush on the doubly-fed machine seriously restricts the power increase, and the high cost, large volume and complex manufacturing process of permanent magnet machines limit the increase of power per unit. Therefore, the squirrel cage asynchronous machine has evident advantages in terms of cost and reliability. In addition, the technology of the high-voltage squirrel cage machine is mature, further improving the power per unit for WTGs.

The FSC-IG WTG makes use of SCIGs, which have a simple structure, and are cheap and can be used in harsh environments. They also have technically mature gearboxes to augment the speed of the wind turbine, reducing the volume of the generator and reducing costs of manufacture, transport, and maintenance, and giving better performance. The topological structure of a wind turbine based on an FSC-IG is shown in Figure 5.3.

With such mature technology, induction generators can easily generate up to MW level power, and are suitable for the high-power wind power applications. The generator stator is connected to the grid through the FSC. Maximum power point tracking (MPPT) and flexible connection can be achieved with the control over-converter, ensuring excellent performance. The generator is isolated from the grid, giving strong FRT capability and good grid connections. However, the adoption of the FSC increases the hardware costs.

Meanwhile, due to the low efficiency of the induction generator, a gearbox with a high gearing ratio is always required, thus increasing the system fault rate. Nevertheless, the induction generator is advanced in terms of technology, robustness, and maintenance. Therefore, it is perfectly suited to offshore wind farms. Many large-scale offshore wind farms projects are interested in FSC-IG based wind turbines. Currently, they have been adopted by Siemens and they account for a considerable proportion of offshore wind farms.

This chapter describes the operating characteristics of the FSC-WTG, and the FSC-WTG is modeled, including the shaft model, generator model, and the FSC model (including the generator-side converter model, DC-link model and grid-side converter model). The stability of grid-connected wind farms based on FSC-WTGs is analyzed against a test wind farm based on direct-drive PMSG-based wind turbines.

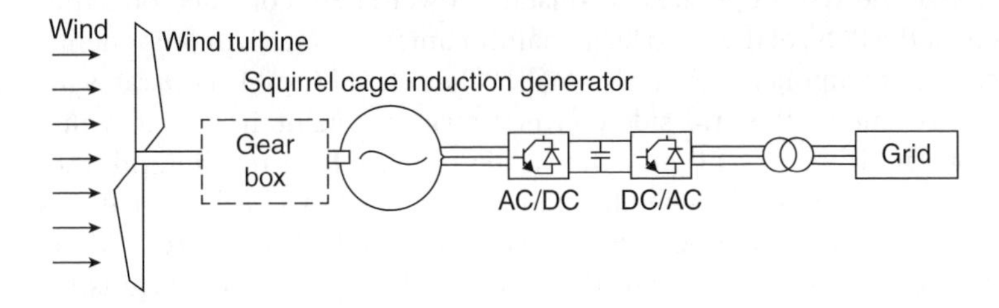

Figure 5.3 Structure diagram of FSC-IG based wind turbine.

5.2 Operating Characteristics of FSC-WTGs

The operating characteristics of wind farms depend on the WTGs, and directly affect grid operations. The operating characteristics of WTGs must therefore be analyzed before there are any grid-connected operations. In addition, the protection devices of grid-connected WTGs must be selected, based mainly on the transient characteristics of the WTGs during fault periods. This must also be carefully considered during the planning and design of grid-connected wind farms. This section describes the operating characteristics and transient characteristics of FSC-WTGs.

During normal operation, the control targets of the generator-side converters (generator active power, AC voltage) and the grid-side converters (DC bus voltage, grid-side reactive power) of FSC-WTGs can be independently achieved. The change of WTG active power may influence the DC bus voltage, and the change of reactive power of the grid-side converter will mainly affect the AC voltage of the grid-side converter [7–9]. In the event of a fault, if the WTG does not trip due to protective action and continues its grid-connected operation, the WTG terminal voltage and active power will quickly decrease, and the AC current of the grid-side converter will sharply increase at the instant of a short-circuit fault. During the fault, the DC bus voltage will experience a gradual increase, and after clearing the fault, the WTG terminal voltage will quickly increase and return to the level before the fault, the active power of the WTG will gradually recover, and the DC bus voltage will continue to decrease. A fault on the grid-side has only a limited effect on the generator-related variables because the generator is decoupled from the grid by the FSC [10–13].

The following analyses all assume that the WTGs do not trip due to the protective action. The control strategies of the grid-side converters of the FSC-WTGs involve DC bus voltage/reactive power and DC bus voltage/AC-side voltage of the converter. These two control strategies are called the constant reactive power control mode and the constant voltage control mode. In the event of a grid fault, the transient characteristics of the point of connection (POC) of the wind farm are described with both control modes of the WTG below. The FSC-WTG operates under the constant power factor control mode; that is, the reactive power of the grid-side converter is 0: power factor is 1 and the DC voltage is 1.0 pu. The current at the POC of the wind farm quickly increases at the instant a fault occurs and reaches a peak soon afterwards; then the short-circuit current decays, stabilizes and maintains a constant value until the fault is cleared. This is due to the buffer to the grid fault provided by the FSC, which reduces the influence of the fault on the generator. During the fault, the reactive component of the grid-side converter AC current changes, and so does the corresponding reactive power of the WTG. During the short circuit, the short-circuit current at the POC of the wind farm mainly depends on the active power to the grid from the wind farm before fault and the voltage at the POC of the wind farm. When the WTG operates in constant power factor control mode, the short-circuit current at the POC of the wind farm mainly contains the active component, with the reactive current component relatively small. When the FSC-WTG operates in constant voltage control mode, the grid-side converter tries to maintain the AC voltage at 1.0 pu and the DC voltage is 1.0 pu. After a three-phase fault on the grid, the short-circuit current at the POC of the wind farm increases, reaches a peak soon afterwards, and maintains a constant short-circuit current until the fault is cleared. After the fault is cleared, the short-circuit current at the POC of the wind farm begins to decay, and the current returns to the stable level before the fault after a certain time. The WTG terminal voltage rapidly drops at the instant of the grid fault, but in constant voltage control mode, the terminal voltage increases constantly; the WTG terminal voltage increases quickly at the instant of the fault clearance, and reaches a peak soon after the fault. The terminal voltage will recover to the set point with the action of the control system. With the action of the constant AC voltage controller, the reactive current component of the grid-side converter gradually increases, and the reactive power to the grid provided by the wind farm increases correspondingly. At the instant of the fault clearance, the reactive current component reaches a peak, and so does the reactive power of the WTG; in the event of a fault, the active current component begins to increase to maintain the balance of active power. After fault clearance, the active and reactive current components of the grid-side converter recover to their stable levels before the fault.

5.3 FSC-WTG Model

The FSC-WTG model includes a wind turbine model, shaft model, generator model and FSC model. The wind turbine model of the FSC-WTG is the same as that for the DFIG WTG, so is not described here.

5.3.1 Shaft Model

The FSC-WTG based on a SCIG has a gearbox, but the direct-drive PMSG-based wind turbine does not. Therefore, the shaft models with and without a gearbox are described in this section.

Shaft Model with Gearbox

The mechanical drivetrain of a wind turbine delivers most of the wind power captured to the generator. Under normal operations, the variable-speed WTG achieves a decoupling of the mechanical and electrical parts through its decoupling control. With a proper control strategy, the harmonics generated by torsional vibration of the shaft system are almost imperceptible. However, in the event of a serious fault in the grid – say a short-circuit fault – the flexibility of the wind turbine drivetrain may lead to a dynamic change of speed.

Depending on the nature and purpose of the study, different modeling methods may be used for the WTG drivetrain. Generally, WTG drivetrain models can be classified as concentrated single-mass models, two-mass models, and three-mass models. In a concentrated single-mass model, the wind turbine blade, hub, gearbox, and generator rotor are equivalent to a mass. When considering the flexibility of the wind turbine, the wind turbine and generator rotors are taken as separate masses, thus establishing the two-mass model. In addition, if the flexibility of the blade is considered, the blade is also taken as a separate mass, to create a three-mass model.

In transient stability analysis of power systems, the drivetrain is always described by a two-mass dynamic model [14–16]. As shown in Figure 5.4, the two-mass model of the drivetrain is composed of the low-speed shaft, gearbox, and high-speed shaft. The blade and low-speed shaft are treated as a mass, and the gearbox and high-speed shaft are treated as a completely rigid shaft.

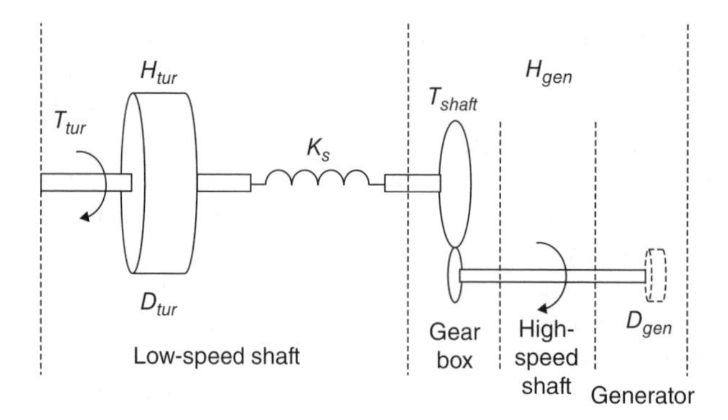

Figure 5.4 Two-mass model of transmission chain.

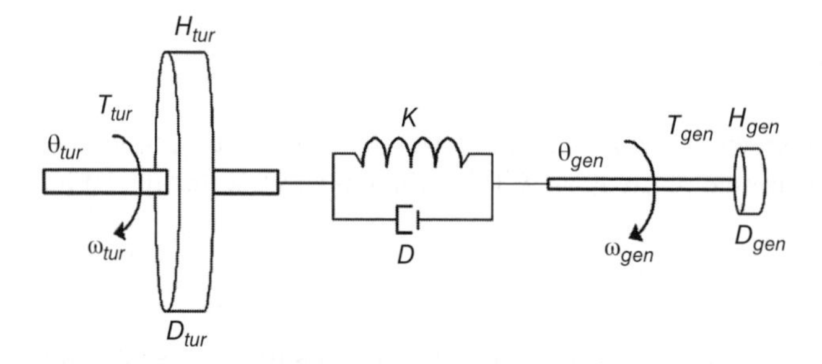

Figure 5.5 Drive system two-mass spring and damper model.

Shaft Model without Gearbox

A direct-drive PMSG-based wind turbine is equipped with a multipole PMSG without a gearbox, and the generator is directly driven by the wind turbine. As the PMSG has several poles, the diameter of machine is larger, and the inertia of generator is much larger than that of a generator with two or four poles. In addition, the increasing number of poles may reduce the rigidity of the generator shaft, which means that any change of the mechanical torque will lead to an obvious change of phase angle of the multipole generator system. A suitable model should be established to study the dynamic characteristics.

Most studies have shown that the two-mass model can exactly reflect the operating characteristics of the WTG shaft, so it has been widely used. The two-mass shaft model of the WTG is shown in Figure 5.5, where the bigger mass corresponds to the inertia H_{tur} of the wind turbine rotor, and the smaller mass corresponds to the inertia H_{gen}. of the generator.

The shaft connecting the wind turbine and PMSG is represented by a rigidity coefficient K and a damping coefficient D. In the two-mass motion equation, viscous friction of the two-mass model is expressed by the damping coefficients D_{tur} of the wind turbine and the damping coefficients D_{gen} of the generator.

Mathematical Model of Shaft

The two-mass model can represent the operating characteristics of shaft models with or without gearboxes, and the mass model can be expressed by (5.1).

$$\begin{cases} 2H_{tur}\dfrac{d\omega_{tur}}{dt} = T_{tur} - K\theta_s - D_{tur}\omega_{tur} \\ 2H_{gen}\dfrac{d\omega_{gen}}{dt} = K\theta_s - T_{gen} - D_{gen}\omega_{gen} \\ \dfrac{d\theta_s}{dt} = \omega_0(\omega_{tur} - \omega_{gen}) \end{cases} \tag{5.1}$$

where H_{tur} and H_{gen} is the inertial time constants of the wind turbine and generator respectively, K is the rigidity coefficient of the shaft (kgm^2/s^2), D_{tur} and D_{gen} are the damping coefficients (Nm/rad) of the wind turbine rotor and the generator rotor respectively, θ_S is the angular displacement (rad) between the two masses, T_{tur} and T_{gen}, refer to the mechanical torque of wind turbine and the electromagnetic torque of the generator respectively, ω_{tur} and ω_{gen} are the speed of the wind turbine and generator rotor respectively, and ω_0 is the synchronous speed (rad/s).

The relative angular displacement θ_S between two masses can be expressed by (5.2):

$$\theta_s = \theta_{tur} - \theta_{gen} \tag{5.2}$$

where θ_{tur} and θ_{gen} respectively refer to the angular displacements (rad) of the wind turbine rotor and generator rotor.

5.3.2 Generator Model

Currently, FSC-WTGs based on SCIGs and PMSGs are widely used. The SCIG and PMSG models are described here.

SCIG

In the FSC-IG, the SCIG is vital, and a proper mathematical model is necessary for the study of the dynamic and static characteristics and the control technology. To simplify the analysis, it is usually assumed that the surface of the stator and rotor of the three-phase asynchronous machine is smooth and that the three-phase winding is symmetrical. Magnetic saturation, eddy currents, and iron core loss are ignored.

The SCIG has a high-order characteristic, and is non-linear and strongly coupled. The mathematical model of the machine may be effectively simplified by an *abc*/*dq* coordinate transformation. With this transformation, the mathematical model of the SCIG under the *dq* synchronously rotating coordinate system can be obtained, with the equations for the voltage and flux linkage as follows:

$$\begin{cases} u_{sd} = R_s i_{sd} + \dfrac{d\psi_{sd}}{dt} - \omega_0\psi_{sq} \\ u_{sq} = R_s i_{sq} + \dfrac{d\psi_{sq}}{dt} + \omega_0\psi_{sd} \\ u_{rd} = R_r i_{rd} + \dfrac{d\psi_{rd}}{dt} - \omega_s\psi_{rq} = 0 \\ u_{rq} = R_r i_{rq} + \dfrac{d\psi_{rq}}{dt} + \omega_s\psi_{rd} = 0 \end{cases} \tag{5.3}$$

$$\begin{cases} \psi_{sd} = L_s i_{sd} + L_m i_{rd} \\ \psi_{sq} = L_s i_{sq} + L_m i_{rq} \\ \psi_{rd} = L_r i_{rd} + L_m i_{sd} \\ \psi_{rq} = L_r i_{rq} + L_m i_{sq} \end{cases} \tag{5.4}$$

where ω_0 is the speed of the synchronously rotating coordinate system, $\omega_s = \omega_0 - \omega_r$, ω_r is the rotor speed, L_s and L_r are the inductances of the stator and rotor respectively, L_m is magnetic inductance, u_{sd} and u_{sq} are the dq components of the stator output voltage, and i_{sd} and i_{sq} are the dq components of the stator output current.

In different coordinate systems – abc reference frame, $\alpha\beta$ reference frame, dq reference frame – electromagnetic torque T_e corresponds to different expressions, and the most common expression is as follows:

$$T_e = 1.5n_p \frac{L_m}{L_r}(\psi_{rd} i_{sq} - \psi_{rq} i_{sd}) \tag{5.5}$$

where T_e (Nm) is the electromagnetic torque and n_p is the number of pole-pairs of the machine.

Equations 5.3–5.5 form the mathematical model of an induction generator in the dq reference frame, and the equivalent circuit at the dq reference frame is shown in Figure 5.6.

PMSG

The PMSG is excited by permanent magnets. To establish the mathematical model in the dq synchronously rotating reference frame, it is first assumed that [17]:

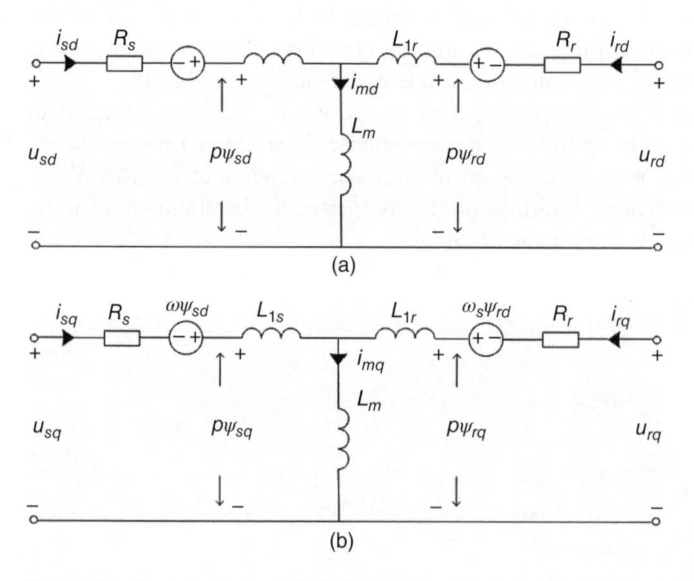

Figure 5.6 dq equivalent models of SCIG: (a) equivalent circuit of d-$axis$: (b) equivalent circuit of q-axis.

- the saturation of the machine iron core is ignored
- the eddy current and hysteretic losses in the machine are excluded
- flux in the stator winding generated by permanent magnets is in a sinusoidal distribution.

Based on these assumptions, the expressions for the components of the stator voltage at the d and q axes are:

$$\begin{cases} u_{sd} = r_s \cdot i_{sd} + \dfrac{d\psi_{sd}}{dt} - \omega_e \cdot \psi_{sq} \\ u_{sq} = r_s \cdot i_{sq} + \dfrac{d\psi_{sq}}{dt} + \omega_e \cdot \psi_{sd} \end{cases} \tag{5.6}$$

where i_{sd}, i_{sq}, u_{sd}, and u_{sq} are the current and voltage of the d and q shafts respectively, r_s is the stator resistance, ω_e is the electrical angular frequency of the generator, $\omega_e = n\omega_{gen}$, ω_{gen} is speed of the generator rotor, n is the number of pole-pairs of the generator rotor, and ψ_{sd} and ψ_{sq} are the flux linkages of the d and q axes respectively.

The expressions for the components of the stator flux linkage at the d and q axes are:

$$\begin{cases} \psi_{sd} = L_d \cdot i_{sd} + \psi_{PM} \\ \psi_{sq} = L_q \cdot i_{sq} \end{cases} \tag{5.7}$$

where ψ_{PM} is the flux linkage generated by the permanent magnet. Substituting (5.7) into (5.6), the following equation can be derived:

$$\begin{cases} \dfrac{di_{sd}}{dt} = \dfrac{1}{L_d}\left(u_{sd} - r_s \cdot i_{sd} + \omega_e \cdot L_q \cdot i_{sq}\right) \\ \dfrac{di_{sq}}{dt} = \dfrac{1}{L_{sq}}\left(u_{sq} - r_s \cdot i_{sq} - \omega_e \cdot (L_d \cdot i_{sd} + \psi_{PM})\right) \end{cases} \tag{5.8}$$

The electromotive force (EMF) of the *q-axis* is defined as $e_q = \omega_e \psi_{PM}$, and the EMF of the *d-axis* $e_d = 0$. As the rotor of the PMSG has a symmetrical structure, it is assumed that the d and q axes have the same inductive reactance; that is, $L_d = L_q = L$. Then (5.8) can be rewritten as:

$$\begin{cases} \dfrac{di_{sd}}{dt} = -\dfrac{1}{L} \cdot r_s \cdot i_{sd} + \omega_e \cdot i_{sq} + \dfrac{1}{L} \cdot u_{sd} \\ \dfrac{di_{sq}}{dt} = -\dfrac{1}{L} \cdot r_s \cdot i_{sq} - \omega_e \cdot i_{sd} - \dfrac{1}{L} \cdot \psi_{PM} + \dfrac{1}{L} \cdot u_{sq} \end{cases} \tag{5.9}$$

Based on the above equation, the equivalent circuit of the PMSG on the dq synchronously rotating reference frame is as shown in Figure 5.7.

The active power of the PMSG is:

$$P_S = \frac{3}{2}(u_{sd} \cdot i_{sd} + u_{sq} \cdot i_{sq}) \tag{5.10}$$

The reactive power of the PMSG is:

$$Q_S = \frac{3}{2}(u_{sq} \cdot i_{sd} - u_{sd} \cdot i_{sq}) \tag{5.11}$$

The electromagnetic torque of the PMSG is:

$$T_e = \frac{3}{2}n \cdot \mathrm{Im}\left[\overline{\psi}_s^* \overline{i}_s\right] = \frac{3}{2}n\left[\psi_{sd}i_{sq} - \psi_{sq}i_{sd}\right]$$
$$= \frac{3}{2}n\left[(L_d - L_q)i_{sd}i_{sq} + \psi_{PM}i_{sq}\right] \tag{5.12}$$

Based on the above assumptions, the torque equation can be simplified as:

$$T_e = \frac{3}{2}n \cdot \psi_{PM} \cdot i_{sq} \tag{5.13}$$

5.3.3 Full-scale Converter Model

There are many kinds of converters. Converters can be classified as AC-DC-AC converters and AC-AC converters in terms of the main circuit structure of the converter. They can be classified as voltage converters and current converters in terms of the nature of variable-frequency power source. AC-DC-AC dual PWM converters have drawn a great deal of attention because of their favorable transmission characteristics, high power levels, and the small harmonics of the grid-side current. An FSC-WTG comprises a three-phase PWM rectifier and a three-phase PWM inverter. The rectifier and the inverter are of the same structure, and make up a back-to-back FSC. An insulated gate bipolar transistor (IGBT) is used as the power electronic component. A simplified structure diagram and equivalent circuit of the FSC are shown in Figure 5.8 [18], showing the generator-side converter, reactor, grid-side converter, and DC link.

For the DC link, the total energy flowing into the DC-link capacitor must be equal to the sum of the output energy of the capacitor and its energy loss. Then:

$$I_v - I_d = C\frac{dU_{dc}}{dt} \tag{5.14}$$

where the current injecting to the grid-side converter, $I_d = \dfrac{u_{cq}i_{cq} + u_{cd}i_{cd} - P_{c,loss}}{U_{dc}}$ and the current out of the generator-side converter $I_v = \dfrac{u_{sq}i_{sq} + u_{sd}i_{sd} - P_{s,loss}}{U_{dc}}$.

Substituting I_d and I_v into (5.14):

$$\frac{dU_{dc}}{dt} = -\frac{1}{CU_{dc}}\left((u_{cq}i_{cq} + u_{cd}i_{cd}) + (u_{sq}i_{sq} + u_{sd}i_{ss}) - (P_{s,loss} + P_{c,loss})\right) \tag{5.15}$$

In (5.15), subscript ‘c’ means the grid-side converter, u_{cd}, u_{cq}, i_{cd}, and i_{cq} refer to the components of AC voltage and the current of the grid-side converter in the *d* and *q* axes,

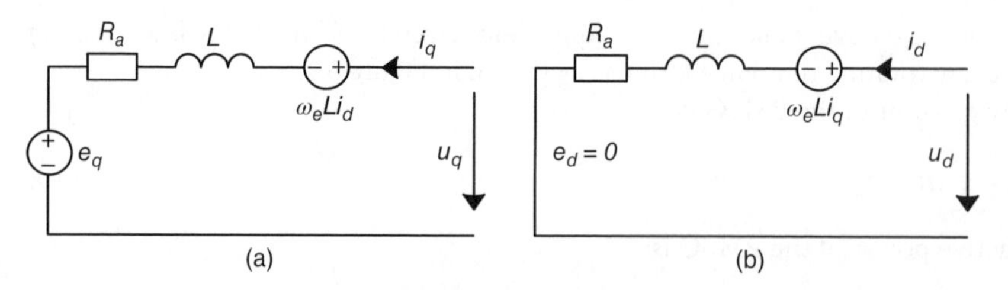

Figure 5.7 PMSG equivalent circuit: (a) Equivalent circuit of *q* shaft; (b) Equivalent circuit of *d* shaft.

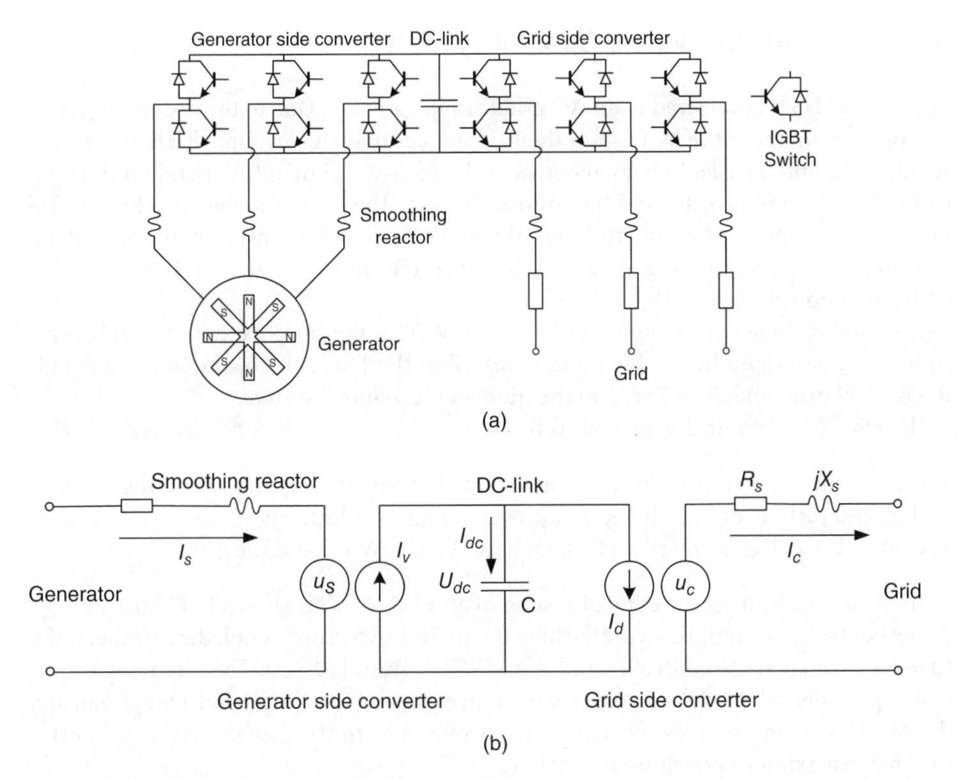

Figure 5.8 Simplified structure and equivalent circuit diagram of full scale converter: (a) simplified structure diagram; (b) equivalent circuit.

respectively, and $P_{s,loss}$ and $P_{c,loss}$ are the active losses of the generator-side converter and the grid-side converter, respectively.

The active power P_C and reactive power Q_C of the grid-side converter are expressed as:

$$\begin{cases} P_C = \dfrac{3}{2}(u_{cq}i_{cq} + u_{cd}i_{cd}) \\ Q_C = \dfrac{3}{2}(u_{cq}i_{cd} - u_{cd}i_{cq}) \end{cases} \tag{5.16}$$

In steady-state operation, it is assumed that the converter loss $P_L = P_{s,loss} + P_{c,loss}$ is zero, then $P_S = P_C$, and the DC voltage U_{dc} is constant at the initial value of the DC voltage. When a disturbance occurs in the system, there will be a current $I_{dc} = I_v - I_d$ through the DC capacitor, and the DC voltage begins to fluctuate. In this case, the DC voltage U_{dc} can be calculated by:

$$U_{dc}(t) = \sqrt{U_{dc}{}^2(0) + \frac{2}{C}\int_0^t (P_S(\tau) - P_C(\tau)) \cdot d\tau} \tag{5.17}$$

where C is the DC capacitance.

Switching loss is a major loss of converters, which can be replaced by a resistor between the poles of the DC link. As the loss is relatively small, it can be ignored, and so can the loss of the series resistor of the converter.

5.4 Full Scale Converter Control System

The FSC-WTG is connected to an AC grid through an FSC. Due to the decoupling control of the converter, the WTG and the grid are completely decoupled. Therefore, the steady-state and transient characteristics of the FSC-WTG mainly depend on the control system of the converter and the control strategy. The control system of a FSC-WTG principally comprises the wind turbine control system and the converter control system, the latter consisting of the generator-side converter control system and the grid-side converter control system [19].

The wind turbine control system of an FSC-WTG is the same as that of a DFIG one, so it is not described here. This section describes the FSC control system, and details the control principle of an FSC and the model of its control system.

The FSC-WTG control system aims to:

- control the active power from the generator to track the optimal operating point of the wind turbine or limit its output power at high wind speeds
- control the reactive power exchanged between the WTG and the grid.

The block diagram of the control system of an FSC-WTG is shown in Figure 5.9.

The control system of the wind turbine comprises two cross-coupled controllers: the speed and pitch angle controller and the MPPT system [20–22]. The pitch angle controller provides the pitch angle control reference value β^{ref} to the pitch drive system, and the MPPT system provides the active power reference to the generator-side converter to achieve maximum power point tracking.

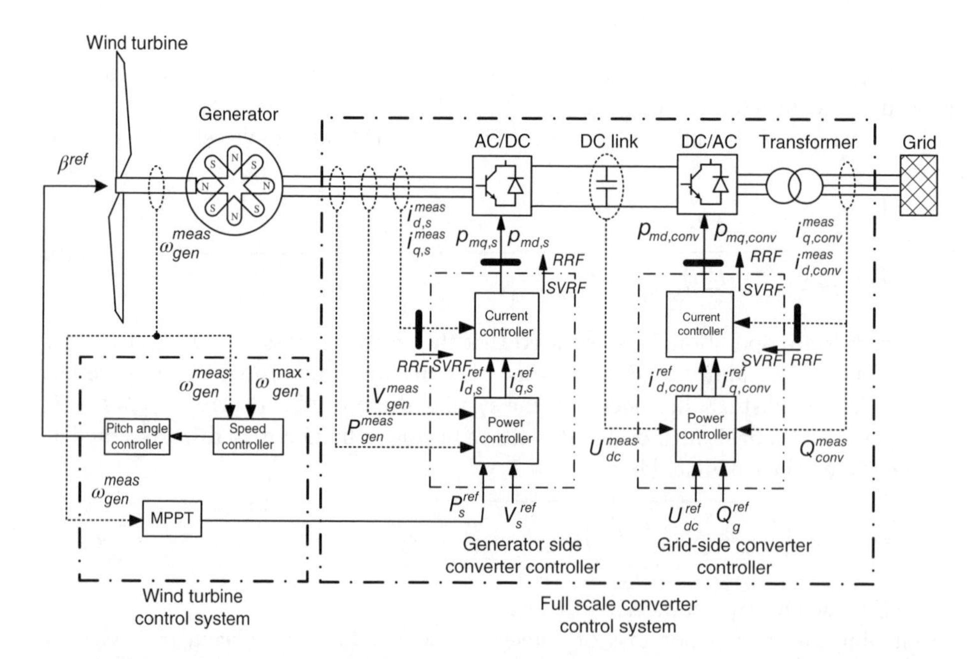

Figure 5.9 Control system of an FSC-WTG. SVRT: stator voltage reference frame; (RRF) rotor reference frame (RRF)

The FSC control system is for the decoupling control of active and reactive power of the WTG, and includes two decoupling control loops: the generator-side converter control system and the grid-side converter control system. The grid-side converter works as an inverter, and its function is to convert DC into AC of a fixed frequency. The generator-side converter is used for rectifying and may comprise electronic components without turn-off capability, such as diodes, or components with turn-off capability, such as IGBTs and gate turn-off thyristors. When the generator-side converter has turn-off capability the generator will be optimally controlled.

5.4.1 Control System of Generator-side Converter

The control system model of a generator-side converter is certainly related to the type of the connected generator. The control systems of generator-side converters will be described for the widely used SCIG and PMSG systems.

Control System of SCIG Side Converter

To achieve control of active and reactive decoupling of the generator, and obtain a better dynamic performance, the generator-side converter is generally controlled according to the rotor flux orientation technology. The rotor flux orientation means that the generator rotor flux direction coincides with the d-axis of the synchronously rotating reference frame. The q-axis component of the rotor flux is zero; that is, $\psi_{rd} = |\psi_r|$ and $\psi_{rq} = 0$. Flux is eliminated through the combination of (5.3) and (5.4) to obtain the voltage equation of the SCIG:

$$\begin{cases} u_{sd} = R_s i_{sd} + \sigma L_s \dfrac{di_{sd}}{dt} + \dfrac{L_m}{L_r}\dfrac{d\psi_{rd}}{dt} - \omega_0 \sigma L_s i_{sq} \\ u_{sq} = R_s i_{sq} + \sigma L_s \dfrac{di_{sq}}{dt} + \omega_0 \left(\sigma L_s i_{sd} + \dfrac{L_m}{L_r}\psi_{rd} \right) \end{cases} \tag{5.18}$$

The given value ψ_{rd} of the rotor flux is usually set as the rated value, and the actual value is generally calculated by the following equation:

$$\psi_{rd} = \frac{L_m}{1 + \tau_r p} i_{sd} \tag{5.19}$$

$\psi_{rq} = 0$ is put into the torque equation (5.5):

$$T_e = n_p \frac{L_m}{L_r} \psi_{rd} i_{sq} \tag{5.20}$$

The rotor flux angle θ_r can be calculated as follows:

$$\theta_r = \int (\omega_r + \omega_s) dt \tag{5.21}$$

$$\omega_s = \frac{L_m i_{sq}}{\tau_r \psi_{rd}} \tag{5.22}$$

where $\sigma = 1 - L_m^2/(L_r L_s)$ is the magnetic leakage coefficient, τ_r is a time constant, and $\sigma\omega_0 L_s i_{sq}$ and $\sigma\omega_0 L_s i_{sd}$ are decoupled items of the stator voltage.

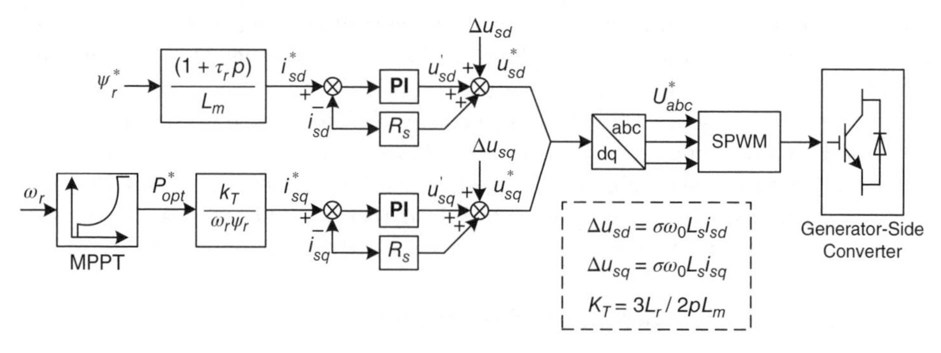

Figure 5.10 Block diagram of generator-side inverter control.

Based on (5.19) and (5.20), and under the *dq* rotating coordinate system, according to the rotor flux vector orientation, the rotor flux is only related to the *d*-axis component i_{sd} of the stator current, and in case of a constant rotor flux, the electromagnetic torque is only related to the *q*-axis component i_{sq} of the stator current. Therefore, the flux and torque of the generator can be controlled by altering the *d*- and *q*-axis components i_{sd} and i_{sq} of the generator stator current to achieve decoupled control of active and reactive power. The block diagram for control of a SCIG side converter is shown in Figure 5.10.

As shown in Figure 5.10, the generator-side converter has a dual closed-loop controller structure. For the outer loop, the reference value of the components of the stator current at the *dq* axes is obtained with the rotor flux and electromagnetic torque, and is adjusted through the proportional integral (PI) of the current inner loop plus the decoupled compensation items. Through the Parker inverse transformation, reference values of stator voltage u^*_{sd} and u^*_{sq} will lead to the reference value U^*_{abc} of the three-phase stator voltage and thus the control signals of the generator-side converter through the PWM link, ultimately achieving MPPT control.

Control System of PMSG Side Converter

For the generator-side converter, three different control strategies may be used, as shown in Figure 5.11:

Maximum torque control Under the RRF, the electromagnetic torque equation of the PMSG is as follows:

$$\begin{aligned} T_e &= \frac{3}{2}n \cdot Im[\overline{\psi}_s^* \bar{i}_s] = \frac{3}{2}n \cdot [\psi_{sd} i_{sq} - \psi_{sq} i_{sd}] \\ &= \frac{3}{2}n \cdot [(L_d - L_q) i_{sd} i_{sq} - \psi_{PM} i_{sd}] \\ &= \frac{3}{2}n \cdot [\psi_{PM} i_{sq}] \end{aligned} \tag{5.23}$$

As shown in (5.23), the electromagnetic torque of the PMSG only depends on the *q*-axis component of the stator current. If the stator current is only controlled by the *q*-axis component, the PMSG will provide the maximum electromagnetic torque. In

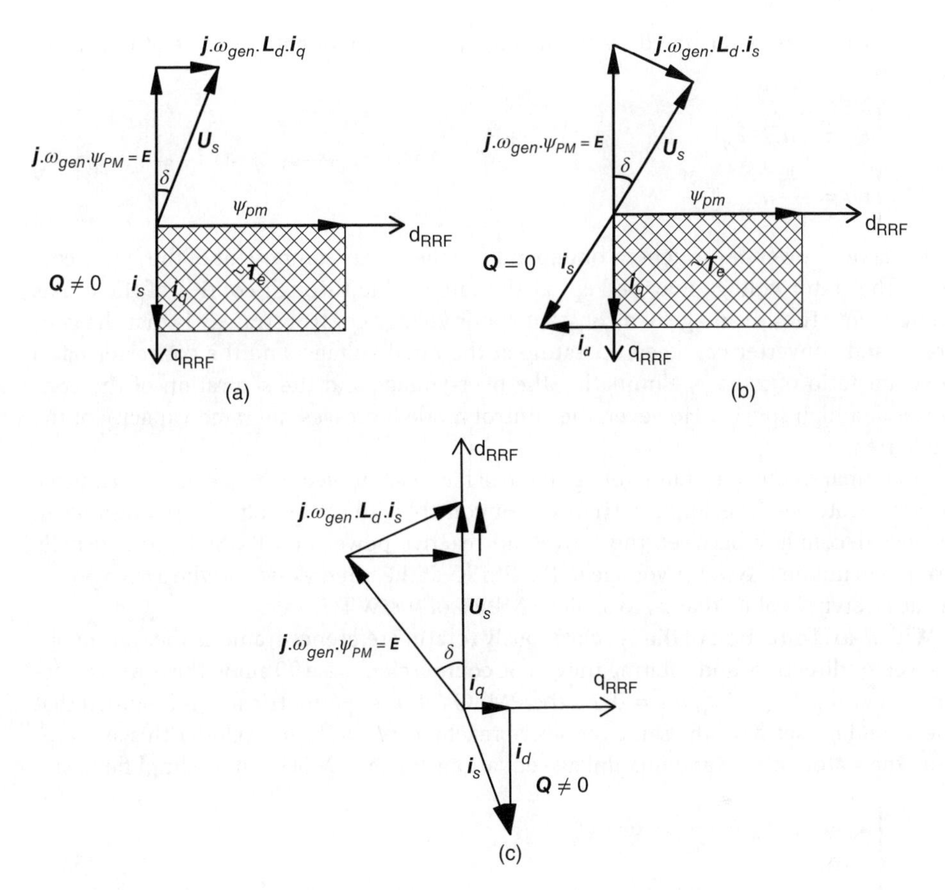

Figure 5.11 Vector chart for three generator-side converter control strategies: (a) maximum torque (b) unit power factor: (c) constant stator voltage.

this control mode, the reactive power of the generator is not 0, and the rated capacity of the converter is increased.

Unit power factor control If the d-axis component of the stator current is controlled to adjust the reactive power required for the generator, the generator will be able to run at a unit power factor ($Q = 0$). This control strategy can reduce the rated capacity of the converter. However, the stator voltage will vary with the generator speed, which may lead to overvoltage of the converter and overspeed of the generator rotor. Since the reactive power of the generator is not necessarily related to the reactive power exchanged between the WTG and the grid, the control strategy for the reactive power of the generator has little significance.

Constant stator voltage control To avoid overvoltage of the converter caused by overspeed of the generator, a control strategy of directly controlling the generator stator voltage can be adopted. The control objective is achieved at the stator voltage reference frame (SVRF); that is, the direction of the d-axis is taken as the vector direction of the stator voltage. In this case, there is no coupling between the active power and reactive power, which are determined by the d- and q-axis components of the stator current respectively:

$$\begin{cases} P_S = \frac{3}{2} u_{sd} \cdot i_{sd} \\ Q_S = -\frac{3}{2} u_{sd} \cdot i_{sq} \end{cases} \tag{5.24}$$

The active power of the generator depends on the d-axis component of the stator current. The stator voltage may be kept at the rated value through control of the q-axis component. The advantage of a constant stator voltage control strategy is that the generator and converter can keep operating at the rated voltage, and the converter has a constant ratio of u_{dc}/u_{ac}, eliminating the over-voltage and the saturation of the converter at a high speeds. However, the control mode increases the rated capacity of the converter.

In summary, constant stator voltage control has been proved to be the best of the three control strategies. The stator current vector control with stator voltage orientation can achieve decoupling between the active and reactive power of a PMSG, and is usually used to maintain the stator voltage of the PMSG at the rated value and the active power at the reference value; that is, to achieve MPPT of the WTG.

With d-axis direction of the synchronously rotating reference frame as the stator voltage vector direction, and rotating the q-axis counterclockwise 90 along the d-axis direction, we will get $u_{sd} = u_s$, $u_{sq} = 0$. As the PMSG rotor is symmetrical, it is assumed that the d- and q-axes have the same inductive reactance: $L_d = L_q = L$. Under this assumption, the stator voltage and flux linkage equations for the PMSG can be simplified as:

$$\begin{cases} u_{sd} = r_s \cdot i_{sd} + \frac{d\psi_d}{dt} - \omega_e \cdot \psi_q = u_s \\ u_{sq} = r_s \cdot i_{sq} + \frac{d\psi_q}{dt} + \omega_e \cdot \psi_d = 0 \end{cases} \tag{5.25}$$

$$\begin{cases} \psi_{sd} = L \cdot i_{sd} + \psi_{PM} = 0 \\ \psi_{sq} = L \cdot i_{sq} = -\frac{u_s}{\omega_s} \end{cases} \tag{5.26}$$

Equations (5.25) and (5.26) show:

$$\begin{cases} u_{sd} = r_s \cdot i_{sd} + L \cdot \frac{di_{sd}}{dt} - \omega_e \cdot L \cdot i_{sq} \\ u_{sq} = r_s \cdot i_{sq} + L \frac{di_{sq}}{dt} + \omega_e \cdot (L \cdot i_{sd} + \psi_{PM}) \end{cases} \tag{5.27}$$

The active and reactive power of the PMSG can be simplified as:

$$\begin{cases} P_S = \frac{3}{2} u_s \cdot i_{sd} \\ Q_S = -\frac{3}{2} u_s \cdot i_{sq} \end{cases} \tag{5.28}$$

Therefore, the active and reactive currents are completely decoupled in the SVRF, and the corresponding control voltage vectors are coupled.

Two new inputs can be defined as follows:

$$\begin{cases} u'_{sd} = u_{sd} + \omega_e \cdot L \cdot i_{sq} \\ u'_{sq} = u_{sq} - \omega_e \cdot L \cdot i_{sd} \end{cases} \tag{5.29}$$

Substituting (5.29) into (5.17), we can get:

$$\begin{cases} u'_{sd} = r_s \cdot i_{sd} + L \cdot \dfrac{di_{sd}}{dt} \\ u'_{sq} = r_s \cdot i_{sq} + L\dfrac{di_{sq}}{dt} + \omega_e \cdot \psi_{PM} \end{cases} \tag{5.30}$$

From this, decoupled control of current can be achieved through feedforward compensation; that is, feedforward inputs $\omega_e L i_{sq}$ and $-\omega_e L i_{sd}$ are added to the d- and q-axes current controllers.

For the PWM converter, the relationship between the AC and DC voltages can be expressed as follows:

$$\begin{cases} u_{sd} = \dfrac{\sqrt{3}}{2\sqrt{2}} p_{md} U_{dc} \\ u_{sq} = \dfrac{\sqrt{3}}{2\sqrt{2}} p_{mq} U_{dc} \end{cases} \tag{5.31}$$

where P_{md} and P_{mq} are the d- and q-axis components of the converter modulation factor P_m. As such, the decoupling of active and reactive power of the generator can be achieved through control of the modulation factor of the converter.

The block diagram of the generator-side converter control system is shown in Figure 5.12. It aims to change the active power sent to the FSC from the PMSG according to the optimum power curve and maintain the stator voltage at its reference value. The generator-side converter control system is divided into the inner-loop controller and the outer-loop controller. The inner-loop controller is used to control the d- and q-axis components of the generator-side converter current; the outer-loop controller controls the active power input to the converter from the generator and the stator voltage of the PMSG. The reference values of the current components are determined by the outer-loop active power and voltage controller; the reference value of the active power depends on the MPPT characteristics, with the goal of making the generator at any time operate at the optimal operating speed – power curve, ensuring optimal $C_p(\lambda)$. The reference value of the stator voltage is generally set to its rated value.

5.4.2 Grid-side Converter Control System

The grid-side converter is controlled using voltage-oriented vector control to achieve decoupled control of the active and reactive power between the grid-side converter and the grid. It is primarily used for regulating the DC voltage of the DC link, and the reactive power between the WTG and the grid. Its control objectives are:

- keep the DC voltage constant
- keep the reactive power exchanged between the grid-side converter and the grid at the reference value.

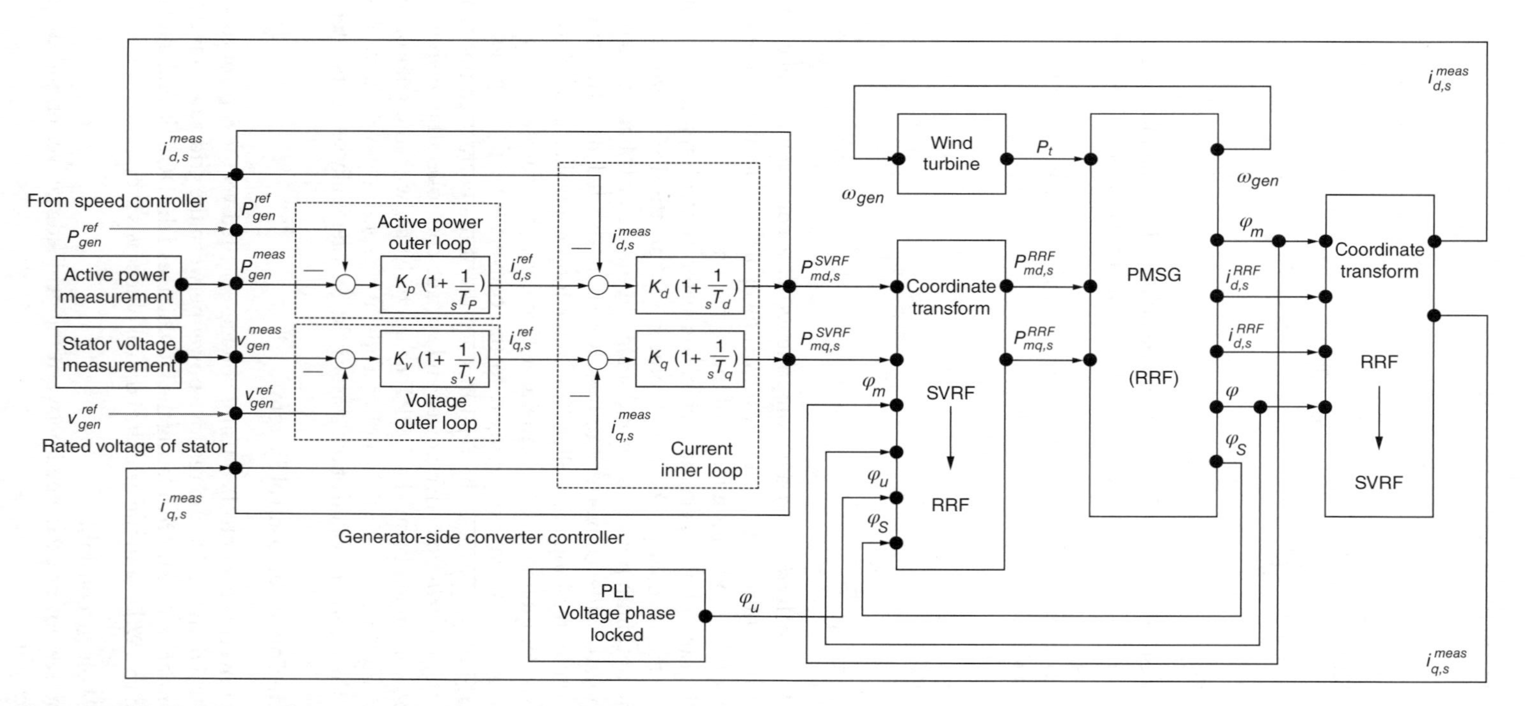

Figure 5.12 Block diagram of generator-side converter control system.

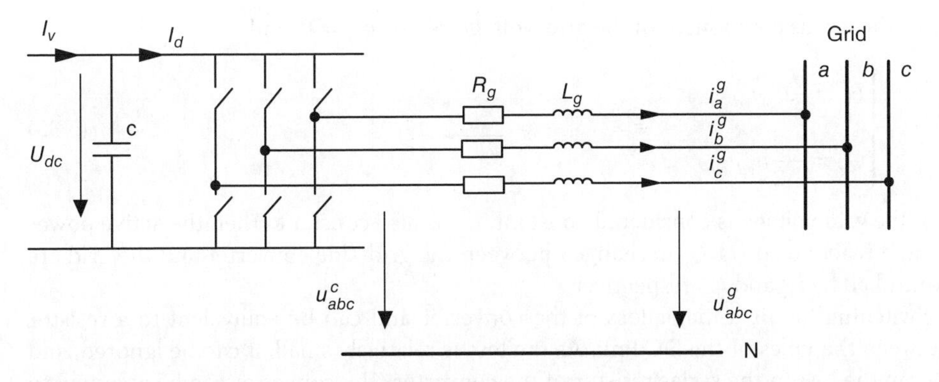

Figure 5.13 Structure diagram of PWM voltage source converter at grid-side.

The structure diagram of the PWM voltage source converter at the grid-side is shown in Figure 5.13.

Figure 5.13 shows:

$$\begin{bmatrix} u_{ga} \\ u_{gb} \\ u_{gc} \end{bmatrix} = R_g \begin{bmatrix} i_{ga} \\ i_{gb} \\ i_{gc} \end{bmatrix} + L_g \frac{d}{dt} \begin{bmatrix} i_{ga} \\ i_{gb} \\ i_{gc} \end{bmatrix} + \begin{bmatrix} u_{ca} \\ u_{cb} \\ u_{cc} \end{bmatrix} \tag{5.32}$$

where u_{gabc} is the three-phase grid voltage (kV), u_{cabc} is the three-phase grid-side converter voltage (kV), i_{ga}, i_{gb}, and i_{gc} are the three-phase grid-side converter currents (kA), R_g and L_g are series resistance (Ω) and inductance (H) of the grid-side converter, I_d and I_v are the DC currents (kA) of the grid-side and the generator, and C is the DC-link capacitance.

Through a PARK transformation of (5.32), we can get the voltage equation in the dq synchronously rotating frame:

$$\begin{cases} u_{gd} = R_g i_{gd} + L_g \dfrac{d}{dt} i_{gd} - \omega_0 L_g i_q^g + u_{cd} \\ u_{gq} = R_g i_{gq} + L_g \dfrac{d}{dt} i_{gq} + \omega_0 L_g i_{gd} + u_{cq} \end{cases} \tag{5.33}$$

where i_{gd}, i_{gq}, u_{gd}, and u_{gq} are the d- and q-*axis* components of the grid current and voltage, and u_{cd} and u_{cq} are the d- and q-*axis* components of the converter voltage at the grid-side.

The active power P_g and reactive power Q_g exchanged between the grid-side converter and the grid can be expressed as:

$$\begin{cases} P_g = \dfrac{3}{2}(u_{gd} i_{gd} + u_{gq} i_{gq}) \\ Q_g = \dfrac{3}{2}(u_{gq} i_{gd} - u_{gd} i_{gq}) \end{cases} \tag{5.34}$$

In this section, grid-voltage-oriented vector control is adopted, with the d-axis direction as the grid-voltage direction and the q-axis rotating 90° along the d shaft. In this case, the q-axis component of the grid-voltage vector $u_{gq} = 0$, and:

$$\begin{cases} P_g = \frac{3}{2} u_{gd} i_{gd} \\ Q_g = -\frac{3}{2} u_{gd} i_{gq} \end{cases} \tag{5.35}$$

If the grid voltage is considered constant, u_{gd} is also constant. Then the active power P_g and reactive power Q_g exchanged between the grid-side converter and the grid are controlled by i_{gd} and i_{gq}, respectively.

Switching loss is a major loss of the converter and can be equivalent to a resistor between the poles of the DC link. As the loss is relatively small, it can be ignored, and so can the loss in the series resistor of the converter. The converter can be assumed to be an ideal converter, and the following equation can be derived:

$$U_{dc} I_{dc} + \sqrt{3} \mathrm{Re}(\dot{U}_{ac} I_{ac}^*) = 0 \tag{5.36}$$

In other words:

$$U_{dc} I_d = \sqrt{3} u_{gd} i_{gd} = P_g \tag{5.37}$$

The relation between the AC and DC voltages of the PWM converter are rewritten as follows:

$$\begin{cases} u_{gd} = \frac{\sqrt{3}}{2\sqrt{2}} P_{md} U_{dc} \\ u_{gq} = \frac{\sqrt{3}}{2\sqrt{2}} P_{mq} U_{dc} \end{cases} \tag{5.38}$$

where P_{md} and P_{mq} are the PWM modulation factors of the grid-side converter:

$$I_d = \frac{3}{2\sqrt{2}} P_{md} i_{cd} \tag{5.39}$$

For the capacitor C of the DC loop, the following formula holds:

$$C \frac{dU_{dc}}{dt} = I_d - I_v \tag{5.40}$$

We can get the relationship of the DC voltage U_{dc} and the grid current active component i_{gd} as:

$$C \frac{dU_{dc}}{dt} = \frac{3}{2\sqrt{2}} P_{md} i_{gd} - I_v \tag{5.41}$$

Through a Laplace transform of the above equation with I_v as a disturbance variable, we can get the transfer function with the DC voltage U_{dc} as output and the grid current active component i_{gd} as input:

$$\frac{U_{dc}(s)}{i_{gd}(s)} = \frac{3P_{md}}{2\sqrt{2}Cs} \tag{5.42}$$

Equation 5.42 shows that the DC-link voltage U_{dc} is controlled by i_{gd}. As the converter is ultimately controlled through the voltage, it is necessary to establish the relationship between the inverter voltage u_{cabc} and the current $i_{gd}/\ i_{gq}$. In (5.33), we assume that:

$$\begin{cases} u'_{gd} = R_g i_{gd} + L_g \dfrac{di_{gd}}{dt} \\ u'_{gq} = R_q i_{gq} + L_g \dfrac{di_{gq}}{dt} \end{cases} \tag{5.43}$$

Equation (5.33) can be transformed into:

$$\begin{cases} u_{cd} = u_{gd} - u'_{gd} + \omega_e L_g i_{gq} \\ u_{cq} = -u'_{gq} - \omega_e L_g i_{gd} \end{cases} \tag{5.44}$$

Through the Laplace transform of (5.44), we can get the transfer function with u'_g as the input and the grid current i_g as the output:

$$\frac{i_{gd}(s)}{u'_{gd}(s)} = \frac{i_{gq}(s)}{u'_{gq}(s)} = \frac{1}{L_g s + r_g} \tag{5.45}$$

To establish a relationship between the grid current i_g and the converter voltage u_{cabc}, the feedforward links $\omega_0 L_g i_{gq}$ and $-\omega_0 L_g i_{gd}$ must be introduced to the control system.

The grid-side converter aims to keep the DC link voltage at the preset constant value, and to ensure that the reactive power is the preset reference value according to the requirement of the WTG. In order to take full advantage of the control ability of the converter and generate as much active power as possible, the reactive power exchange between the grid-side converter and the grid is set as 0. However, under certain operating conditions, if the grid needs the WTG to provide reactive power support, the reference value of the reactive power may be set to meet the requirement of the grid. Figure 5.14 shows the grid-side converter system and controller, comprising the following components:

(1) DC voltage measurement module, which measures the DC voltage of the DC link;
(2) converter current measurement module, which measures the AC current of the grid-side converter;
(3) voltage phase-locked loop (PLL), which measures the voltage phase angle of the grid-side converter, and achieves decoupled control of the grid-side converter with grid-voltage-oriented vector control;
(4) power measurement link: measure the reactive power from the grid-side converter;
(5) coordinate transformation module: as the grid-side voltage and current are expressed in the three-phase stationary reference coordinate system, and the grid-side converter is controlled according to the grid-voltage-oriented vector at the grid-voltage reference coordinates, the input/output signals of the grid-side converter controller must be transformed;
(6) grid-side converter module: the IGBT converter model;
(7) grid-side converter controller.

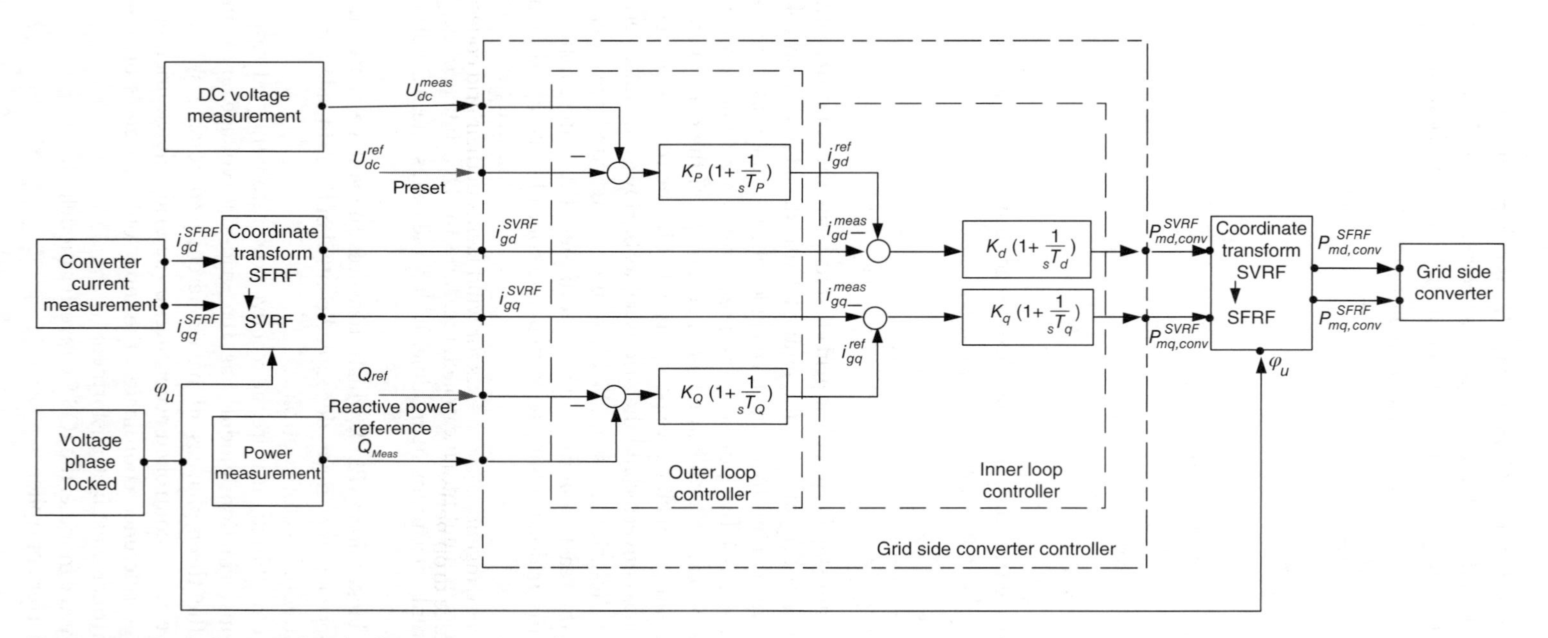

Figure 5.14 Block diagram of grid-side converter control system.

The grid-side converter controller comprises two cascaded PI controllers (an outer-loop controller and an inner-loop controller). The slow outer loop is used to control the DC voltage of the DC link and reactive power of the converter (or the grid-side AC voltage of the converter), and the fast inner loop ensures that the current reaches the reference value determined by the outer loop. The output signals of the grid-side converter controller define the amplitude and phase angle of the output voltage at the AC side of the converter. Under grid-voltage-oriented vector control, the converter current is divided into two mutually perpendicular current components: the d-axis current component is the active current and the q-axis component is the reactive current. The active current at the d-axis is used to control the DC voltage of the DC link, and the reactive current at the q-axis is used to control the reactive power from the converter (or AC voltage at grid-side converter). The converter modulation factor output from the current inner-loop controller of the grid-side converter is in the voltage-oriented reference frame of the grid-side converter, and it must be transformed into the three-phase stationary reference coordinate system through coordinate transformation for the grid-side converter.

5.5 Grid-connected FSC-WTG Stability Control

From the perspective of active power, a grid-connected wind farm is a sending-end system. However, from the perspective of reactive power, the wind farm is still a positive reactive load. Due to the strong correlation between voltage stability and reactive power, voltage collapse caused by wind farms, has the same mechanism as conventional power system voltage instabilities. In addition to frequent small disturbances, large disturbances such as short circuits and disconnections also occur, so it is necessary to study the transient voltage stability of power systems with wind power integrated in the event of a large disturbance fault. In the early development of wind power, as they only accounted for a very small proportion share of grid generation, wind farm were not generally required to participate in power system stability control. In the event of a fault in the grid, WTGs usually tripped to ensure the safety of the wind farm and the grid. With the increasing penetration of wind power in the grid, several countries have specified more stringent requirements for wind farms in their grid codes: during a fault and grid-voltage drop periods, wind farms must be capable of continuously operating for a certain time without tripping from the grid, and must have dynamic reactive power support capability. For example, since 2003, the German grid company E.On has required that, in addition to the grid-connected operation of a wind farm during the recovery period of the grid voltage after the fault, WTGs must dynamically provide reactive power to support the grid voltage to prevent tripping of WTGs due to the low voltage. The ability of a wind farm to maintain uninterrupted operation during a fault period is called low-voltage ride through (LVRT) of the wind farm. Similarly, for the possible overvoltage after a grid fault, high-voltage ride through (HVRT) requirements are also specified for WTGs and wind farms. LVRT and HVRT are collectively known as 'fault ride through' (FRT). It has been confirmed that the FSC-WTG has the excellent performance in this regard [17–20]. This section describes the stability characteristics of steady-state voltage and transient voltage of grid-connected wind farms based on FSC-WTGs.

For variable-speed WTGs based on FSCs, the FSC control system can realize decoupled control of active and reactive power of the generator, and improve the power factor and voltage stability of the wind farm. Therefore, the steady-state and transient stability performance are much better than for wind farms based on constant-speed WTGs. At the same time, the stability of the transient voltage of wind farms based on FSC-WTGs is related to the control strategy of the FSC control system. As the FSC has a certain isolation effect on the grid fault, in the event of a large disturbance fault at the grid-side, such as a three-phase short-circuit, the FSC-WTG is less affected by the grid fault. However, in constant power factor control mode, the WTG cannot provide dynamic reactive power to support the voltage, and the cut-off of the fault line will result in a weaker grid structure, reduced terminal voltage, and incomplete WTG active power transmission. When the DC bus voltage of the FSC rises to the overvoltage limit, protection of all WTGs in the wind farm will act and all WTGs will trip from the grid. The operation of the wind farm and the safety of the grid will be affected. Therefore, additional control is required to improve the transient voltage stability of FSC-WTG.

5.5.1 Transient Voltage Control of Grid-side Converter

Under the DC voltage/reactive power control mode of the grid-side converter, during a voltage drop and the post-fault voltage recovery in the event of a grid fault, the FSC-WTG does not participate in the transient voltage stability control. The FSC is designed to transmit active power to the grid at a fixed frequency; the variables such as the reactive power exchanges between the grid-side converter and the grid must be limited to ensure the AC current of the grid-side converter remains in the permissible range. In fact, if the system requires the WTG to provide reactive power support during a grid fault, an additional voltage controller must be added to the original control system of the grid-side converter to participate in transient voltage stability control. The block diagram for transient voltage control of the grid-side converter is shown in Figure 5.15.

By comparing the voltage reference value U_{grid}^{ref} and the measured voltage value during the fault, the PI controller determines the reactive power reference value Q_{grid}^{ref} that the generator-side converter must send according to the error signal. The actually transmitted reactive power is then adjusted by the inner-loop current controller to help the FSC-WTG to recover the terminal voltage to the reference value after the fault.

5.5.2 Additional DC Voltage Coupling Controller

In the FSC control system described in Section 5.3, the generator-side converter and the grid-side converter have separate control systems. For an actual FSC system, there is a coupling relationship between the converters at both sides. To make the FSC system

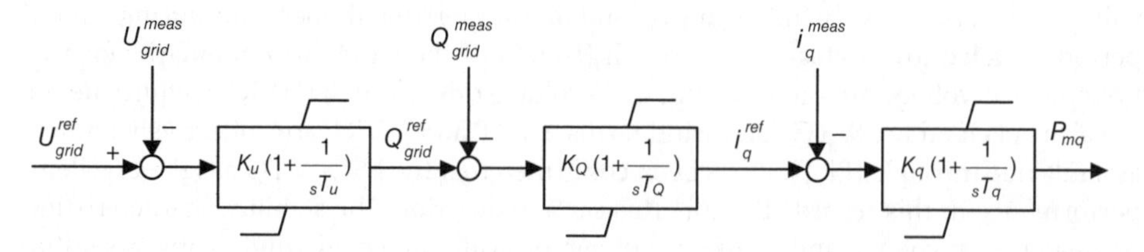

Figure 5.15 Grid-side converter transient voltage controller.

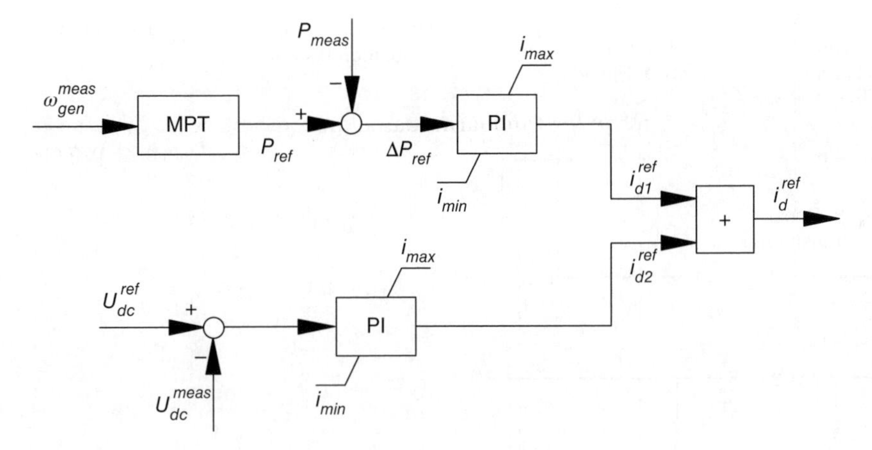

Figure 5.16 Block diagram of DC voltage coupling control system.

operate more stably, an additional DC voltage coupling controller is introduced between the generator-side converter and the original control system of the grid-side converter, and its control block diagram is shown in Figure 5.16. The error signal of the DC voltage generates a reference value of the d-axis component of the stator current by the PI controller. It is superimposed with the reference value of the d-axis component of the stator current generated by the active power control of the original control system, and the sum is the actual reference value of the d-axis current.

During steady-state operation, the DC bus voltage is the rated value, the additional DC voltage coupling controller does not work, and the generator-side converter control system functions to implement MPPT of the WTG. In the event of a grid fault and greater fluctuation of the DC voltage, by comparing the measured and rated values of the DC bus voltage, the reference value of d-axis component of the stator current of the generator-side converter is obtained according to the error signal by the PI controller, and is superimposed with the reference value of d-axis component of the stator current generated by the active power control in the original control system. The generator active power is limited by the inner-loop current control system to achieve coordinated control over the DC bus voltage by the generator-side converter and the grid-side converter until the DC bus voltage regains its rated value and the generator-side converter control system returns to its MPPT function. The use of the DC voltage coupling controller facilitates the transient voltage stability of the WTG.

5.5.3 Simulations

The following simulation analysis is of a regional power system. The local grid and the wind farm are shown in Figure 5.17. Bus 2 of this local grid is the POC of the wind farm; a thermal power plant is connected at Bus 4, with an installed capacity of 2 × 200 MW; Bus 8 is connected to the main grid through dual-circuit 500 kV line. In the test system, the wind farm is an aggregated model. The synchronous generator model of each power plant comprises an excitation system, a prime mover, and a governor system model. The synchronous generator model is the sixth-order model using E''_d, E''_q, E'_d, and E'_q; the load model consists of 50% constant impedance and 50% induction motor model.

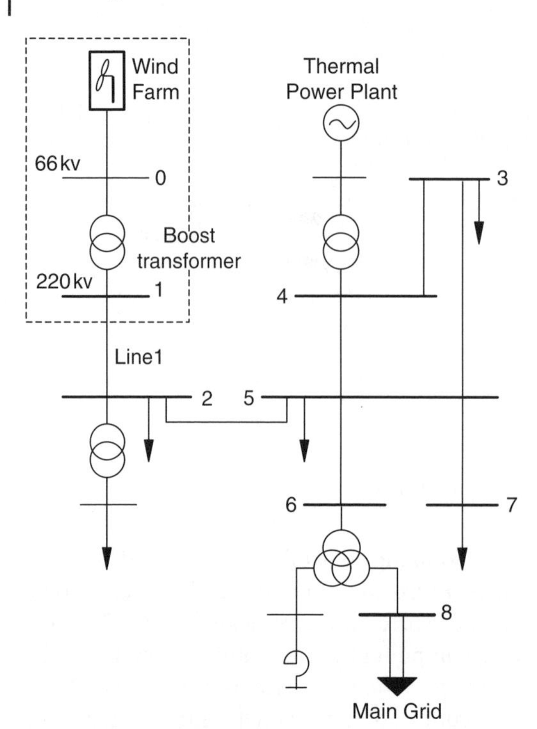

Figure 5.17 Simplified wiring diagram of simulation system.

The wind farm is based on an FSC-WTG, with a total capacity of 100 MW. It is connected to the grid from Bus 1 through a LGJ-2 × 400 single overhead line, which is 70 km in length. A three-phase short-circuit fault occurs in the middle of line 4–5 at 0.1 s of the simulation time and continues for 0.12 s. Then the protection acts to clear the fault. It is assumed that the WTG is able to maintain grid-connected operation in the simulation process, and the FSC-WTG in the wind farm is a direct-drive PMSG-based wind turbine.

Without additional control system

Figure 5.18 shows the changes of various variables of the wind farm after the short-circuit fault with a grounding impedance of 10 Ω.

If the WTG does not make use of any additional control, it can be seen that at the instant of the fault, the terminal voltage of the WTG and active power of the wind farm decline rapidly, the reactive power of the wind farm quickly increases and the DC bus voltage gradually increases. As the mechanical torque is greater than the electromagnetic torque of the generator at the instant of the fault, the speed of the PMSG increases and the AC current of the grid-side converter increases rapidly. At the instant of the fault clearance, the instantaneous DC bus voltage reaches a peak; with the action of the converter control system, various relevant variables are able to recover and are stable after clearing the fault. It can also be seen from the simulation results that as the generator-side converter and the grid-side converter have separate control systems, and the DC link provides certain isolation from the grid fault, which has a limited effect on the generator-side converter and the PMSG. Since the grid fault may cause the DC bus

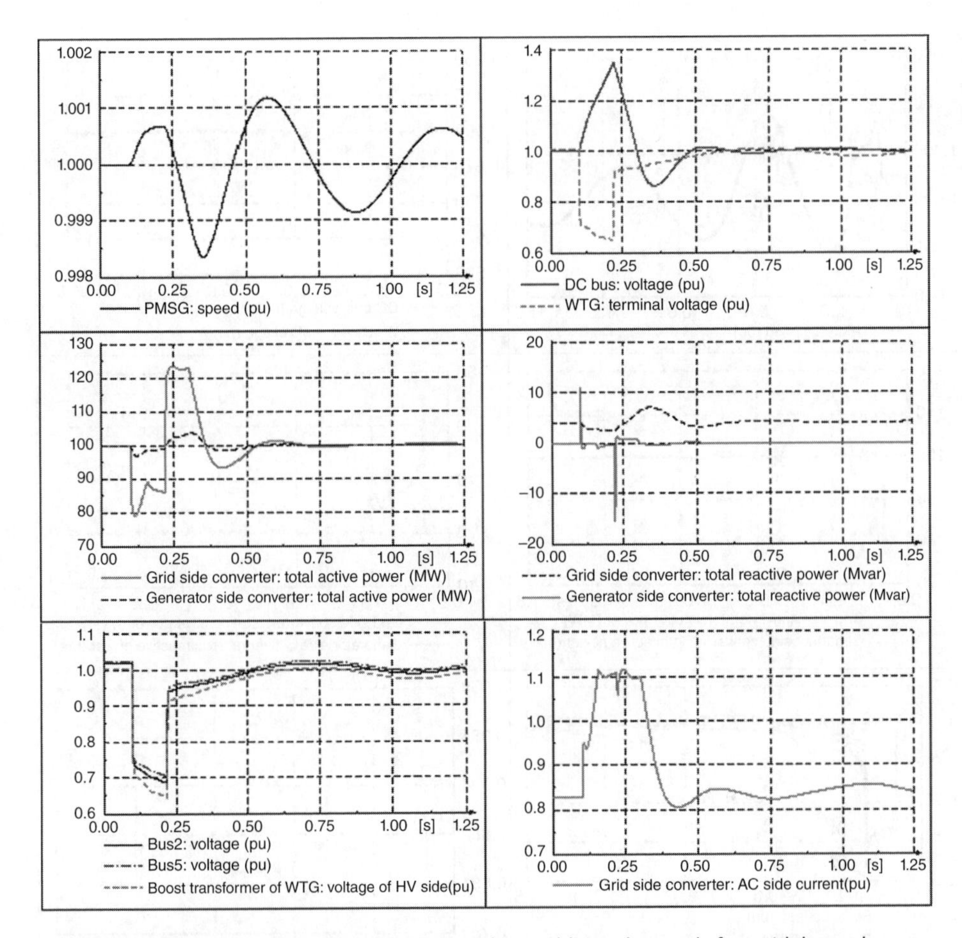

Figure 5.18 Characteristic curve of wind farm without additional control after grid three-phase short-circuit fault (grounding impedance 10 Ω).

voltage to exceed its overvoltage limit, and the wind farm may trip due to the fault, the LVRT function may not be fulfilled, which will directly affect the safety of the grid.

Additional Transient Voltage Control of Grid-side Converter

With the same grid short-circuit fault, the same impedance, but with a grid-side converter using transient voltage control, the changes of variables of the wind farm are as shown in Figure 5.19.

It can be seen from the simulation results that at the instant of the fault, the terminal voltage of the WTG and the active power of the wind farm decline rapidly, and the reactive power of the wind farm quickly increases. As the mechanical torque is greater than the electromagnetic torque of the generator at the instant of the fault, the speed of the PMSG increases. As the grid-side converter operates under in transient voltage control mode, the grid-side converter is able to provide reactive power to support the terminal voltage during the fault and to support the recovery of the grid voltage. The maximum DC bus voltage is certainly improved compared to the WTG without the

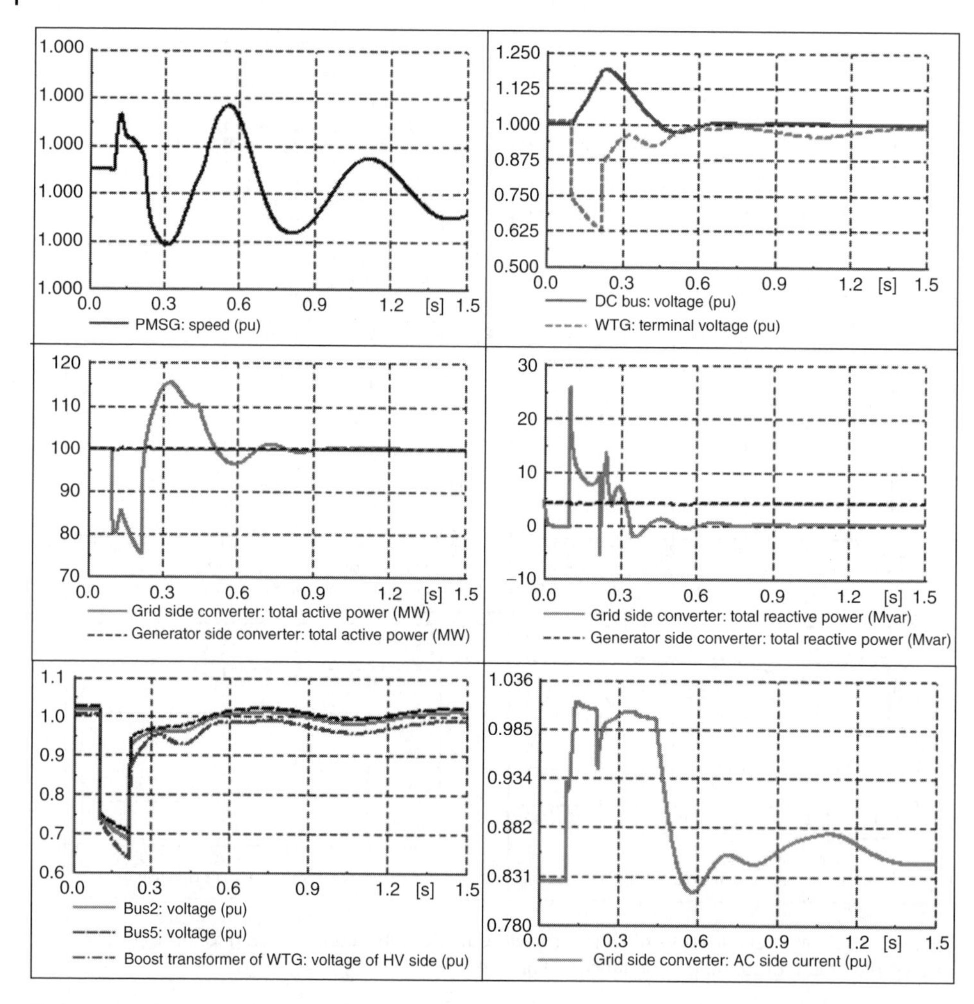

Figure 5.19 Characteristic curve of wind farm with additional transient voltage control at grid three-phase short-circuit fault (grounding impedance 10 Ω).

additional controller; with the additional control system, various relevant variables are able to rapidly decay, and recover to stability after clearing the fault.

Additional DC Voltage Coupling Control System

In the event of the same short-circuit fault of the grid, the same impedance, but with a WTG with an FSC using DC voltage coupling control, the changes of the wind farm variables are as shown in Figure 5.20.

It can be seen from the simulation results that, with the additional DC voltage controller, the control system of the WTG is able to better control the changes of various variables caused by the grid short-circuit fault. The entire wind farm can recover to stability after clearing the fault. At the instant of the fault, the various variables of the WTG and the wind farm vary in the same way, whether or not the DC voltage coupling

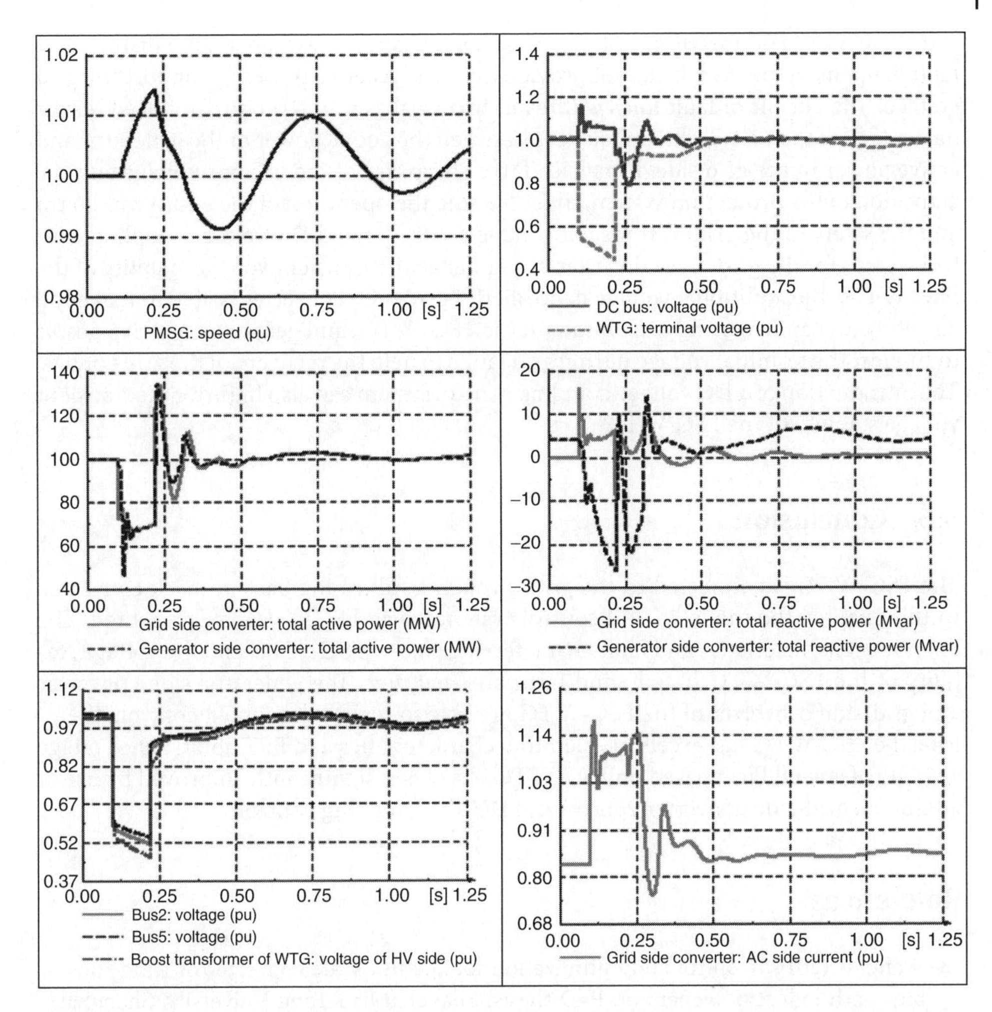

Figure 5.20 Characteristic curve of wind farm with additional DC voltage coupling control after grid three-phase short-circuit fault (Grounding impedance 10 Ω)

control system is used. During the fault, the speed of the PMSG shows an attenuation trend. Through the DC voltage coupling coordination control system, the AC active power of the generator-side converter varies in a similar way to the AC active power of the grid-side converter, and it is possible to control the DC bus voltage within a reasonable range. As the control system of generator-side converter participates in the regulation of the DC bus voltage, the use of a coupling control system increases the influence of the grid fault on the PMSG and the generator-side converter.

In summary, if the grid-side converter of the FSC-WTG adds transient voltage control, the WTG can provide reactive power during a grid fault to support the recovery of the grid voltage. If the grid-side converter of the FSC-WTG makes use of DC voltage coupling control, the DC bus voltage is well controlled, which improves the LVRT of the FSC-WTG.

If the FSC-WTG operates with a general control strategy, when a large disturbance fault happens in the grid, it cannot provide dynamic reactive power to support the grid voltage. The cut-off of fault lines will result in a weaker grid structure, reduced terminal voltage of the WTG, and imbalances between the active power of the generator and active power of the grid side. It may lead to a sharp rise of the DC bus voltage and the activation of the protection system, thus affecting the operation of the entire wind farm and the safety of the grid. A transient voltage controller and DC voltage coupling control system for the grid-side converter can enhance the transient voltage stability of the FSC-WTG. The additional voltage controller of grid-side converter is able to make full use of the dynamic control capabilities of the FSC-WTG, and generates reactive power to support the terminal voltage during grid faults to help the recovery of the grid voltage. The introduction of a DC voltage coupling control system can also improve the transient voltage stability of the FSC-WTG.

5.6 Conclusion

The FSC-WTG is connected to the grid through a full-rating back-to-back converter. In this chapter, the model of and control system for an FSC-WTG are presented. The operating characteristics and fault characteristics of the FSC-WTG and the voltage stability of the FSC-WTG based wind farm are analyzed. The generator-side converter and grid-side converter of the FSC-WTG are able to realize decoupled control. Therefore, the FSC-WTG has excellent operating characteristics and FRT ability. The voltage stability of a wind farm based on FSC-WTGs has been significantly improved by use of additional grid-side converter control and DC-link coupling control.

References

1 Wang, P. (2014) Control and optimization for the full-scale converter of wind turbine with induction generator, PhD thesis, Shanghai Jiao Tong University, Shanghai, China.
2 Chen, Y. (2011) Research on the optimize control technologies of the running quality of wind turbine with full-power converter, PhD thesis, Yanshan University, Qinhuangdao, China.
3 Chinchilla, M., Arnaltes, S., and Burgos, J.C. (2006) Control of permanent-magnet generators applied to variable-speed wind-energy systems connected to the grid, *IEEE Transactions on Energy Conversion*, **21** (1), 130–135.
4 Yin, M., Li, G., Zhang, J., Zhao, W., and Xue, Y. (2007) Modeling and control strategies of directly driven wind turbine with permanent magnet synchronous generator, *Power System Technology*, **31** (15), 61–65.
5 Yan, G., Wei, Z., Mu, G., Cui, Y., Chen, W., and Dang, G. (2009) Dynamic modeling and control of directly-driven permanent magnet synchronous generator wind turbine, *Proceedings of the CSU-EPSA*, **21** (6), 34–39.
6 Chen Y. (2008) Research on full-scale grid-connected power conversion technology for direct-driven wind generation system, PhD thesis, Beijing Jiaotong University, Beijing, China.

7 Hansen, A.D. and Michalke, G. (2008) Modeling and control of variable-speed multi-pole permanent magnet synchronous generator wind turbine, *Wind Energy*, **11** (5), 537–554.

8 Yao, X., Bao, J., Li, W., Liu, Y., and Li, G. (2008)Control strategy and system development of direct drive, VSCF wind turbine, *Journal of Shenyang University of Technology*, **30** (4), 399–403.

9 Yao, J., Liao, Y., Qu, X., and Liu, R. (2008) Optimal wind-energy tracking control of direct-driven permanent magnet synchronous generators for wind turbines, *Power System Technology*, **32** (10), 11–15.

10 Liao, Y., Zhuang, K., and Yao, J. (2012) A control strategy for grid-connecting full power converter of wind power generation system under unbalanced grid voltage, *Power System Technology*, **36** (1), 72–78.

11 Conroy, J.F. and Watson, R. (2007) Low-voltage ride-through of a full converter wind turbine with permanent magnet generator, *IET Renewable Power Generation*, **1** (3), 182–189.

12 Li, J., Hu, S., Kong, D., Xu, H. (2008) Studies on the low voltage ride through capability of fully converted wind turbine with PM SG, *Automation of Electric Power Systems*, **32** (19), 92–95.

13 Geng, Q., Xia, C., Yan, Y., and Chen, W. (2010)Direct power control in constant switching frequency for PWM rectifier under unbalanced grid voltage conditions, *Proceedings of the CSEE*, **30** (36), 79–85.

14 Fu, X., Guo, J., Zhao, D., and Xu, H. (2009) Characteristics and simulation model of direct-drive wind power system, *Electric Power Automation Equipment*, **29** (2), 1–5.

15 Yang, Y., Yi, R., Ren, Z., Liu, X., and Shen, H. (2009) Grid-connected inverter in direct-drive wind power generation system, *Power System Technology*, **33** (17), 157–161.

16 Geng, H., Xu, D., Wu, B., Yang, G. (2009) Control and stability analysis for the permanent magnetic synchronous generator based direct driven variable speed wind energy conversion system, *Proceedings of the CSEE*, **29** (33), 68–75.

17 Tang, R., *Modern Permanent Magnet Machines Theory and Design.* China Machine Press, 1997.

18 Gao, F., Zhou, X., Zhu, N., Su, F., and An, N. (2011) Electromechanical transient modeling and simulation of direct-drive wind turbine system with permanent magnet synchronous generator, *Power System Technology*, **35** (11), 29–34.

19 Zhuang, K. (2012) Research on the key technologies of grid-connected converter for PMSG-based direct-driven wind turbine generation system, PhD thesis, Chongqing University, Chongqing, China.

20 Moor, G.D. and Beukes, H.J. (2004) Maximum power point trackers for wind turbines, in *Proceedings of the 35th Annual IEEE Power Electronics Specialists Conference*, pp. 2044–2049.

21 Chinchilla, M. Arnaltes, S. and Burgos, J.C. (2006) Control of permanent magnet generators applied to variable-speed wind-energy systems connected to the grid, *IEEE Transactions on Energy Conversion*, **21** (1), 130–135.

22 Lin, H. (2012) Research on dynamic equivalent modeling of direct-drive wind farms, PhD thesis, Xinjiang University, Urumqi, China.

6

Clustering-based Wind Turbine Generator Model Linearization

Haoran Zhao and Qiuwei Wu

Technical University of Denmark

In this chapter, a dynamic discrete time piecewise affine (PWA) model of a wind turbine is presented. This can be used for the advanced optimal control of a wind farm, in approaches such as model predictive control (MPC). In contrast to the partial linearization of the wind turbine model at selected operating points, the non-linearities of the wind turbine model are represented by a piecewise static function based on the wind turbine system inputs and state variables. The nonlinearity identification is based on a clustering-based algorithm, which combines clustering, linear identification, and pattern recognition techniques.

6.1 Introduction

Nowadays, the requirements for wind farm controllability specified by system operators have become more stringent, including active power control [1]. Different types of active power control have been specified, including absolute power limitation, delta limitation, and balance control. The modern wind farm is expected to operate like a conventional power plant, and be capable of tracking specific power references.

Due to the rapid development of power electronics, the controllability of modern wind turbines (WTs) has been much improved. According to the distribution algorithm of the wind farm control system, power references are assigned to each turbine. The role of individual wind turbines is as an actuator. Based on the power reference, a pitch angle reference and a generator torque reference are computed by the local wind turbine control and this information is given to the actuating subsystems.

The power-controlled wind turbine model developed by the US National Renewable Energy Laboratory (NREL) is used to represent a variable speed pitch-controlled wind turbine [2]. The model structure is shown in Figure 6.1. This model is highly coupled and non-linear. Conventionally, the wind turbine model is linearized at a specific operating point. However, with changes of power reference and wind speed, the operating point shifts from time to time. A controller designed for a specific operating point cannot guarantee control performance over the whole operating range. For advanced

Modeling and Modern Control of Wind Power, First Edition. Edited by Qiuwei Wu and Yuanzhang Sun.

Companion website: www.wiley.com/go/wu/modeling

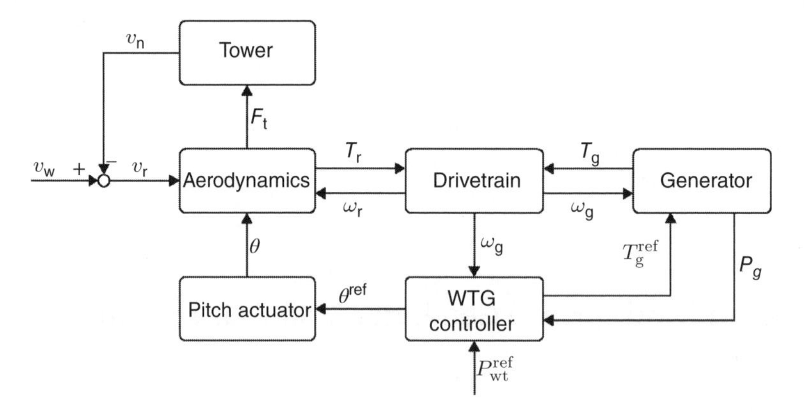

Figure 6.1 Power-controlled wind turbine model.

control strategies, the model should be a discrete-time PWA model, covering the entire operating regime.

Since the state and input variables of the wind turbine system are numerous, the operating points are multidimensional. Conventionally, the non-linearities are linearized by a first-order Taylor series approximation at various operating points [3–5]. In most references, the whole operating regime is partitioned based only on the wind speed. The impacts of other state and input variables are not considered. The main challenge of the identification of PWA models involves the estimation of both the parameters of the affine submodels and the coefficients of the polytopes defining the partition of the regressor set. In this chapter, a novel algorithm, developed by Ferrari-Trecate et al. [6] is used to identify the PWA wind turbine model. This algorithm combines clustering, pattern recognition, and linear identification techniques, the advantage of which is that it transforms the problem of classifying the data into an optimal clustering problem.

This chapter focuses on the identification of a PWA wind turbine model for wind farm control applications. The non-linear control loops, including generator torque control and pitch angle control, are embedded in the model and are identified. Identification was performed in the whole derated operational regime of the wind turbine.

6.2 Operational Regions of Power-controlled Wind Turbines

The wind turbine model developed by NREL consists of several subsystems, including representations of the aerodynamics, drivetrain, tower, generator, pitch actuator and the wind turbine controller. According to the power reference P_{wt}^{ref} generated from the wind farm controller, a pitch reference θ^{ref} and a generator torque reference T_g^{ref} are computed and sent to the pitch controller and the generator torque controller, respectively.

In different wind conditions, the wind turbine control objective varies [7]. When the wind speed v_w is larger than the nominal value, in a full-load regime, the controller aims to limit the captured wind power to the rated power ($P_{wt}^{ref} = P_{rated}$) by regulating the pitch angle to prevent the generator ω_g from overspeeding. In a partial-load regime, the controller aims to extract the maximum available wind power ($P_{wt}^{ref} = P_{wt}^{avi} \leq P_{rated}$) by fixing the pitch angle ($\theta = 0$) and adjust the generator torque to track the optimal rotor speed. The optimal regimes characteristic (ORC) of each wind turbine is plotted in the

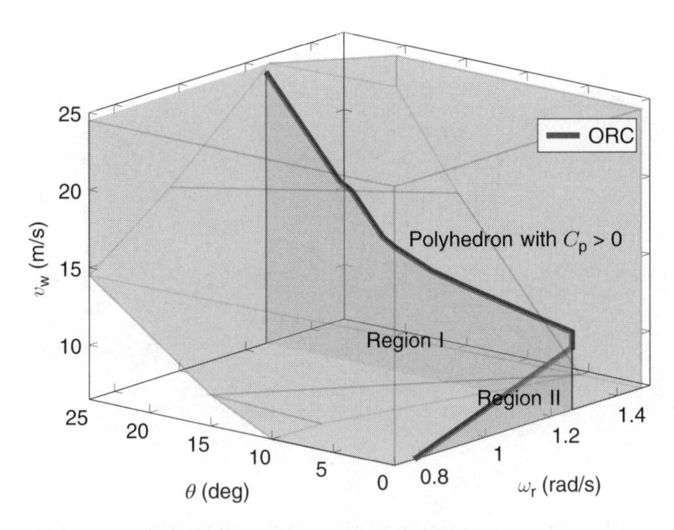

Figure 6.2 Operational regions of a wind turbine.

$\omega_r - \theta - v_w$ space, as illustrated in Figure 6.2. The polyhedron marked in the figure is the area in which the power coefficient $C_p > 0$.

When the wind turbine operates in derated mode – that is, $P_{wt}^{ref} \leq P_{wt}^{avi} \leq P_{rated}$ – the operational region in the steady-state should be below the ORC. According to the activation status of the pitch angle, two regions are defined, as shown in Figure 6.2:

- Region I, where the pitch angle control is activated ($\theta > 0$)
- Region II, where the pitch angle control is deactivated ($\theta = 0$).

6.3 Simplified Wind Turbine Model

The sampling time t_s of the wind farm controller is normally measured in seconds, for example $t_s = 1$ s in the example by Spudić et al. [8]. The fast dynamics related to the electrical system are disregarded in the wind turbine model. Moreover, to reduce the complexity, the oscillations of shaft torsion and tower nodding are also ignored. Accordingly, the simplified NREL wind turbine model is as described by the following equations.

6.3.1 Aerodynamics

The aerodynamic torque T_a and thrust force F_t are the main source of non-linearities, and can be expressed as in (6.1) and (6.2), respectively,

$$T_a = \frac{0.5\pi\rho R^2 v_r^3 C_p(v_r, \omega_r, \theta)}{\omega_r}, \tag{6.1}$$

$$F_t = 0.5\rho R^2 v_r^2 C_t(v_r, \omega_r, \theta), \tag{6.2}$$

where ρ is the air density, R is the blade length, v_r is the effective wind speed on the rotor, and C_p and C_t are the power coefficient and thrust coefficient, respectively. The functions $C_p(\lambda, \theta)$ and $C_t(\lambda, \theta)$ are non-linear ($\lambda = \frac{\omega_r R}{v_r}$), as shown in Figure 6.3.

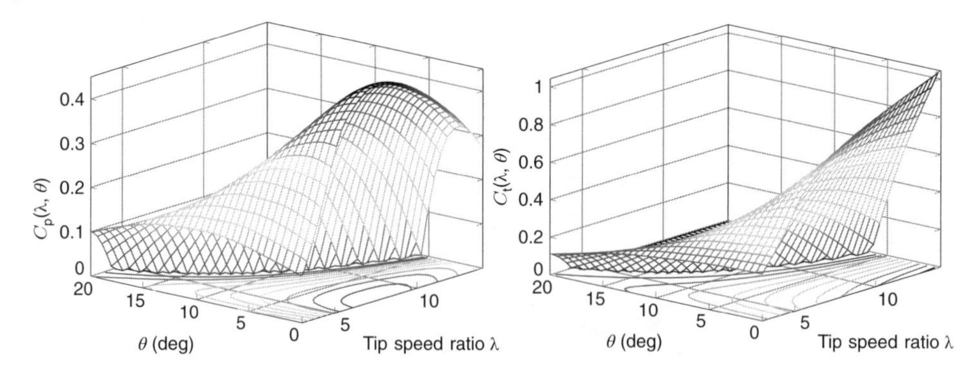

Figure 6.3 $C_p(\lambda, \theta)$ and $C_t(\lambda, \theta)$

6.3.2 Drivetrain

The drivetrain is considered to be rigidly coupled and the single-mass model is used. The rotor inertia J_r and generator inertia J_g are lumped into one equivalent mass J_t. The low-speed shaft motion equations are,

$$J_t = J_r + \eta_g^2 J_g, \tag{6.3}$$

$$\dot{\omega}_r = \frac{1}{J_t}(T_a - \eta_g T_g), \tag{6.4}$$

$$\omega_g = \eta_g \omega_r, \tag{6.5}$$

where η_g is the gear ratio, ω_r and ω_g indicate the speed of the rotor and generator, and J_r and J_g are the inertias of the rotor and generator, respectively.

The shaft torque T_s, twisting the low-speed shaft, is introduced. T_s can be calculated as,

$$T_s = \frac{\eta^2 J_g}{J_t} T_a + \frac{\eta J_r}{J_t} T_g. \tag{6.6}$$

6.3.3 Generator

The vector control scheme is applied in the local torque control loop and ensures a fast and accurate tracking of the electrical torque T_g to the torque reference T_g^{ref}, calculated by,

$$T_g^{\text{ref}} = \frac{P_g^{\text{cmd}}}{\omega_g}, \tag{6.7}$$

where P_g^{cmd} is the power command for the generator. The timescale of the dynamic is normally of the order of milliseconds and therefore can be disregarded,

$$T_g \approx T_g^{\text{ref}}. \tag{6.8}$$

The output power P_g can be derived by,

$$P_g = \mu T_g \omega_g, \tag{6.9}$$

where μ is the generator efficiency. Assuming that μ is well compensated by setting $P_{\mathrm{g}}^{\mathrm{cmd}} = \frac{P_{\mathrm{wt}}^{\mathrm{ref}}}{\mu}$,

$$P_{\mathrm{g}} = P_{\mathrm{wt}}^{\mathrm{ref}}. \tag{6.10}$$

6.3.4 Tower

The relative wind speed on the rotor is affected by the swaying movement of the nacelle. By defining the tower deflection as x_{t}, the effective wind speed on the rotor can be derived as $v_{\mathrm{r}} = v_{\mathrm{w}} - \dot{x}_{\mathrm{t}}$. In this study, the dynamics are disregarded,

$$v_{\mathrm{r}} \approx v_{\mathrm{w}}. \tag{6.11}$$

6.3.5 Pitch Actuator

As proposed by Spudić et al. [3], the dynamics and the non-linearities of the pitch actuator are ignored. The pitch angle θ is regulated by the gain-scheduled PI controller,

$$\theta = -\frac{K_{\mathrm{p}}}{K_{\mathrm{c}}}\dot{\omega}_{\mathrm{f}} - \frac{K_{\mathrm{i}}}{K_{\mathrm{c}}}(\omega_{\mathrm{f}} - \omega_{\mathrm{rated}}), \tag{6.12}$$

$$\dot{\omega}_{\mathrm{f}} = \frac{1}{\tau_{\mathrm{g}}} \cdot \omega_{\mathrm{g}} - \frac{1}{\tau_{\mathrm{g}}} \cdot \omega_{\mathrm{f}}, \tag{6.13}$$

$$K_{\mathrm{c}} = f_{\mathrm{corr}}(P_{\mathrm{wt}}^{\mathrm{ref}}(k-1), \theta(k-1)), \tag{6.14}$$

where K_{p} and K_{i} are the proportional and integral gains of the PI controller, K_{c} indicates the correction factor, which is dependent on $P_{\mathrm{wt}}^{\mathrm{ref}}$ and θ of the last time step ($P_{\mathrm{wt}}^{\mathrm{ref}}(k-1)$, $\theta(k-1)$), ω_{rated} is the rated generator speed and τ_{g} is the time constant of the measurement filter for generator speed ω_{g}.

Based on (6.1)–(6.14), the non-linear dynamics of the wind turbine can be described by:

$$\begin{aligned} \dot{x} &= Ax + Bg, \\ y &= Cx + Dg, \end{aligned} \tag{6.15}$$

with $x = [\omega_{\mathrm{r}}, \omega_{\mathrm{f}}, \theta]'$, $g = [T_{\mathrm{a}}, T_{\mathrm{g}}, F_{\mathrm{t}}, \omega_{\mathrm{rated}}]'$, and $y = [T_{\mathrm{s}}, F_{\mathrm{t}}]'$,

$$A = \begin{bmatrix} 0 & 0 & 0 \\ \frac{\eta_{\mathrm{g}}}{\tau_{\mathrm{g}}} & -\frac{1}{\tau_{\mathrm{g}}} & 0 \\ -\frac{K_{\mathrm{p}}\eta_{\mathrm{g}}}{K_{\mathrm{c}}\tau_{\mathrm{g}}} & \frac{K_{\mathrm{p}} - K_{\mathrm{i}}\tau_{\mathrm{g}}}{K_{\mathrm{c}}\tau_{\mathrm{g}}} & 0 \end{bmatrix}, \quad B = \begin{bmatrix} \frac{1}{J_{\mathrm{t}}} & -\frac{\eta_{\mathrm{g}}}{J_{\mathrm{t}}} & 0 & 0 \\ 0 & 0 & 0 & 0 \\ 0 & 0 & 0 & \frac{K_{\mathrm{i}}}{K_{\mathrm{c}}} \end{bmatrix},$$

$$C = \begin{bmatrix} 0 & 0 & 0 \\ 0 & 0 & 0 \end{bmatrix}, \quad D = \begin{bmatrix} \frac{\eta_{\mathrm{g}}^2 J_{\mathrm{g}}}{J_{\mathrm{t}}} & \frac{\eta_{\mathrm{g}} J_{\mathrm{r}}}{J_{\mathrm{t}}} & 0 & 0 \\ 0 & 0 & 1 & 0 \end{bmatrix}.$$

With sampling time t_s, the corresponding discrete-time form of the system (6.15) can be derived according to the method introduced by Maciejowski et al. [9],

$$x(k+1) = A_d{}'x(k) + B_d{}'g(k),$$
$$y(k) = C_d{}'x(k) + D_d{}'g(k), \tag{6.16}$$

where A'_d, B'_d, C'_d, and D'_d are the discrete forms of A, B, C, and D in (6.15), respectively.

6.4 Clustering-based Identification Method

In this section, the procedure for the clustering-based method is briefly introduced. A non-linear static function $f(x)$ can be approximated with a PWA map,

$$f(x) = \begin{cases} [x, 1]\ \alpha^1, & \text{if } x \in \chi^1 \\ [x, 1]\ \alpha^2, & \text{if } x \in \chi^2 \\ \quad\vdots & \\ [x, 1]\ \alpha^s, & \text{if } x \in \chi^s \end{cases}, \tag{6.17}$$

where x is the input of the static function, χ is a bounded polyhedron, termed the "regressor set", which can be partitioned into s regions, and α^i are the parameter vectors (PVs) of the affine function.

As the main task of the identification, the map f will be reconstructed based on the dataset $(f(x_k), x_k)_{k=1}^n$. The clustering-based identification method consists of the following steps.

Step 1: Build local datasets A local dataset (LD) C_k collects a point $(f(x_k), x_k)$ and its nearest $c-1$ points. The distance is measured in the Euclidean metric. A pure C_k refers to a LD whose outputs are generated by only one model. Otherwise, the LD is mixed. The cardinality c is a tuning parameter. The ratio between pure and mixed LDs should be high, in order to achieve good approximation results. To be noticed, proper tuning of c is important. A small c leads to a small number of mixed points. However, a bigger c is required to counteract the effect of noise on the accuracy of the local model. The local parameter vector α^{LS_k} is computed for each C_k through least squares. The feature vector ξ_k is built by the parameter values α^{LS_k} and the mean value of the inputs m_k: $\xi_k = [(\alpha^{LS_k})', m'_k]'$, which holds both information about localization of C_k and its corresponding affine model parameters. Furthermore, the variance R_k of the feature vectors is computed as a diagonal block matrix, including the co-variance V_k of α^{LS_k} and the scatter matrix Q_k of m_k. The variance R_k is used to compute a confidence measure of ξ_k.

Step 2: Partition the feature vectors into clusters In this step, the derived feature vectors are partitioned into s clusters $\{F_i\}_{i=1}^s$. The partitioning can be done either by a supervised clustering method (K-means) or by an unsupervised clustering method (single-linkage). In this study, the K-means method is used so as to reduce the complexity and s is pre-fixed. It exploits pre-computed confidence measures on feature vectors, which are used to reduce the negative effects of LDs suspected to be mixed.

Step 3: Estimate the submodel parameters In this step, the local models with similar features are collected into one cluster. The data points are classified into one of the data subsets D_i using the mapping,

$$(f(u_k), u_k) \in D_i, \text{if } \xi_k \in F_i. \tag{6.18}$$

The submodel parameter vector α_i is estimated by weighted least squares over the data subset D_i.

Step 4: Estimate the region In this step, the complete polyhedral partitions of the regressor set are derived. Polytopes χ_i $(i = 1, ..., s)$ are decided by solving a multicategory classification problem. Different classification methods have been introduced and compared by Ferrari-Trecate [10], including multi-category robust linear programming, support vector classification and proximal support vector classification.

6.5 Discrete-time PWA Modeling of Wind Turbines

In this section, the discrete-time PWA model of a wind turbine is specified using the clustering-based method introduced in Section 6.4. Specifically, the non-linearities of the simplified NREL wind turbine model, T_a, T_g, F_t and K_c, are identified.

6.5.1 Identification of Aerodynamic Torque T_a

According to (6.1), T_a is a non-linear function of ω_r, v_w, and θ. In the three-dimensional space $[\omega_r, v_w, \theta]'$, the locations of the regressors should be close to the operation regions, as shown in Figure 6.2. Due to the practical constraints, the ranges of ω_r, v_w, and θ are limited. To get a good approximation of the non-linear function, sufficient regressors should be generated. In this study, 500 regressors were randomly generated with a Gaussian distribution characterized by the mean value on both operation regions and dispersion σ^2 around them, as shown in Figure 6.4.

C_p is derived by interpolation of the look-up table published by Jonkman et al. [11]. T_a is computed according to (6.1) for each regressor. The region number s and the number of the points in LD c should be predefined. Although a larger s can give a better approximation, it leads to an increase in computational complexity for searching, and more storage spaces are required. To achieve the best trade-off, $s = 4$ and $c = 20$. The clustering-based identification is performed based on the output–input pairs $(T_a, [\omega_r, v_w, \theta]')$. The partitioned regions for identified T_a are depicted in Figure 6.5.

With the parameter vector $\alpha^i_{T_a} = [\alpha^i_1, \alpha^i_2, \alpha^i_3, \alpha^i_4]'$ and $[\omega_r(k), \theta(k), v_w(k)] \in \chi^i_{T_a}$, T_a can be calculated from:

$$T_a(k) = \alpha^i_1\omega_r(k) + \alpha^i_2\theta(k) + \alpha^i_3 v_w(k) + \alpha^i_4, \tag{6.19}$$

where i and k indicate the region index and time step index, respectively.

6.5.2 Identification of Generator Torque T_g

According to (6.7), T_g is a non-linear function of $P^{\text{ref}}_{\text{wt}}$ and ω_g (that is, ω_r). The locations of the regressors should be distributed in the two-dimensional space $[\omega_g, P^{\text{ref}}_{\text{wt}}]'$, bounded

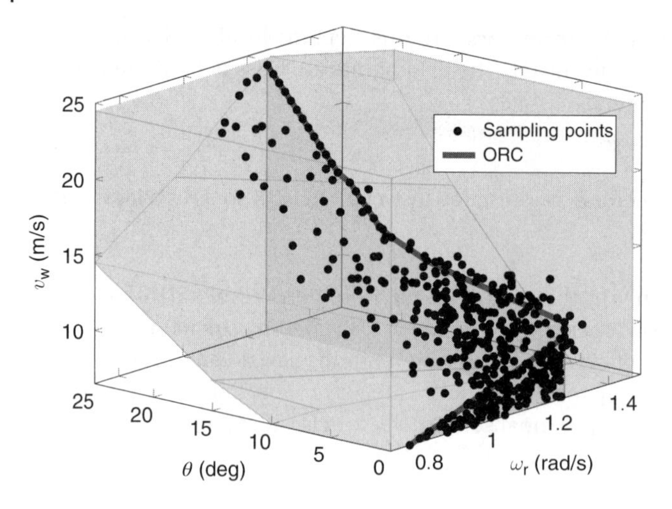

Figure 6.4 Regressors for identification of T_a.

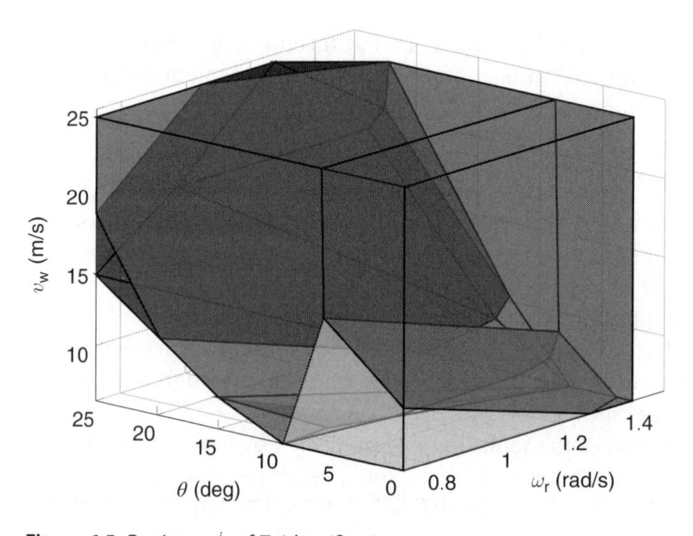

Figure 6.5 Regions $\chi^i_{T_a}$ of T_a identification.

by the practical constraints. In this study, 500 regressors are uniformly generated. With $s = 4$ and $c = 10$, clustering-based identification is performed. The partitioned regions for identified T_g are shown in Figure 6.6.

With the parameter vector $\alpha^i_{T_g} = [\alpha^i_5, \alpha^i_6, \alpha^i_7]'$ and $[\omega_r(k), P^{ref}_{wt}(k)] \in \chi^i_{T_g}$, T_g can be calculated as:

$$T_g(k) = \alpha^i_5 \omega_r(k) + \alpha^i_6 P^{ref}_{wt}(k) + \alpha^i_7. \tag{6.20}$$

6.5.3 Identification of Thrust Force F_t

The identification of F_t is similar to that of T_a. C_t is derived by interpolation of the look-up table. With the same regressor generation, F_t to each regressor can be calculated

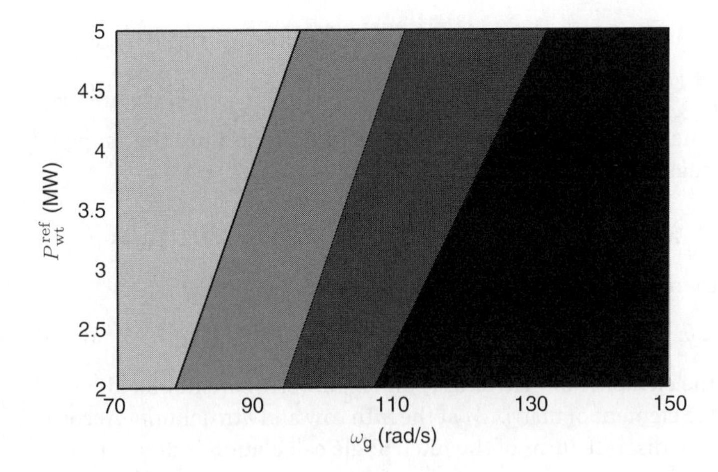

Figure 6.6 Regions $\chi^i_{T_g}$ of T_g identification.

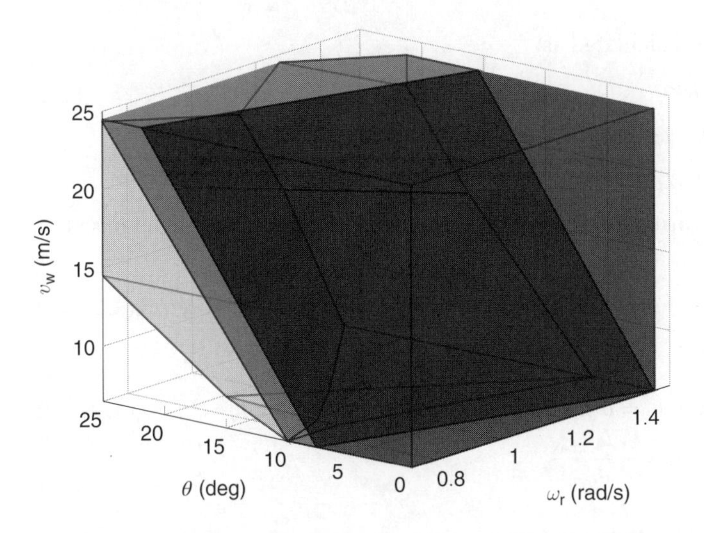

Figure 6.7 Regions $\chi^i_{F_t}$ of F_t identification.

according to (6.2). In this study, $s = 3$ and $c = 20$. The partitioned regions for identified F_t are illustrated in Figure 6.7.

With the parameter vector $\alpha^i_{F_t} = [\alpha^i_8, \alpha^i_9, \alpha^i_{10}, \alpha^i_{11}]'$ and $[\omega_r(k), \theta(k), v_w(k)] \in \chi^i_{F_t}$, F_t can be calculated as:

$$F_t(k) = \alpha^i_8\omega_r(k) + \alpha^i_9\theta(k) + \alpha^i_{10}v_w(k) + \alpha^i_{11}. \tag{6.21}$$

6.5.4 Identification of Correction Factor K_c

According to (6.14), in gain-scheduled pitch control, K_c is a function of P_{wt}^{ref} and θ. In this study, K_c is regarded as a fixed value for each region, which avoids creating new non-linearities. The parameter vector $\alpha^i_{K_c} = \alpha^i_{12}$, if $[P^{\text{ref}}_{\text{wt}}(k), \theta(k)] \in \chi^i_{K_c}$. Accordingly,

$$K_c = \alpha^i_{12}. \tag{6.22}$$

In the NREL model, the proposed K_c is only related to θ. Therefore, the range of θ ($[\theta_{\min}, \theta_{\max}]$) can be equally divided into s segments. In this study, $s = 4$.

6.5.5 Formulation of A'_d and B'_d

In practical operation, there is ramp-rate constraint of the pitch control:

$$-\Delta\theta_{\max} \leq \Delta\theta \leq \Delta\theta_{\max}, \tag{6.23}$$

Due to this constraint, the elements of matrices A'_d and B'_d in (6.16) are not fixed.

Let $M(m, n)$ indicate the element of matrix M at the mth row and nth column. According to (6.12) and (6.16), the discrete form of the pitch angle calculation is derived as:

$$\theta(k+1) = A'_d(3,1) \cdot \omega_r(k) + A'_d(3,2) \cdot \omega_f(k) + A'_d(3,3) \cdot \theta(k) + B'_d(3,4) \cdot \omega_{\text{rated}}. \tag{6.24}$$

Accordingly, $\Delta\theta$ can be calculated as:

$$\begin{aligned}\Delta\theta &= \theta(k+1) - \theta(k) \\ &= A'_d(3,1) \cdot \omega_r(k) + A'_d(3,2) \cdot \omega_f(k) + (A'_d(3,3) - 1) \cdot \theta(k) \\ &\quad + B'_d(3,4) \cdot \omega_{\text{rated}}.\end{aligned} \tag{6.25}$$

Once $\Delta\theta$ violates its constraint, the matrices A'_d and B'_d should be updated based on the following rules:

- If $\Delta\theta > \Delta\theta_{\max}$, $\theta(k+1) = \theta(k) + \Delta\theta_{\max}$; accordingly, $A'_d(3,1) = 0$, $A'_d(3,2) = 0$, $A'_d(3,3) = 1$, and $B'_d(3,4) = \dfrac{\Delta\theta_{\max}}{\omega_{\text{rated}}}$.
- If $\Delta\theta < -\Delta\theta_{\max}$, $\theta(k+1) = \theta(k) - \Delta\theta_{\max}$; accordingly, $A'_d(3,1) = 0$, $A'_d(3,2) = 0$, $A'_d(3,3) = 1$, and $B'_d(3,4) = -\dfrac{\Delta\theta_{\max}}{\omega_{\text{rated}}}$.

6.5.6 Region Construction through Intersection

In this study, a five-dimensional space $[\omega_r, \omega_f, v_w, \theta, P^{\text{ref}}_{\text{wt}}]'$ is introduced to include all the polytopes identified above. When the wind turbine operates in Region II, pitch angle control is deactivated: $\theta = 0$. Therefore, the space is only three-dimensional: $[\omega_r, v_w, P^{\text{WT}}_{\text{ref}}]'$. Each polytope constructed by the identification has to be intersected with others.

6.5.7 PWA Model of a Wind Turbine

For each polytope χ^i, the constant K_c is determined and brought into A in (6.15). Subsequently, according to the rules in Section 6.5.5, A'_d and B'_d are updated. Based on (6.19)–(6.22), $g(k)$ in (6.16) can be calculated by the identified parameters:

$$g(k) = A_p x(k) + B_p u(k) + E_p d(k) + F_p, \tag{6.26}$$

where $u = P_{\mathrm{wt}}^{\mathrm{ref}}$, $d = v_{\mathrm{w}}$,

$$A_{\mathrm{p}} = \begin{bmatrix} \alpha_1^i & 0 & \alpha_2 \\ \alpha_5^i & 0 & 0 \\ \alpha_8^i & 0 & \alpha_9^i \\ 0 & 0 & 0 \end{bmatrix}, \quad B_{\mathrm{p}} = \begin{bmatrix} \alpha_3^i \\ 0 \\ \alpha_{10}^i \\ 0 \end{bmatrix}, \quad E_{\mathrm{p}} = \begin{bmatrix} 0 \\ \alpha_6^i \\ 0 \\ 0 \end{bmatrix}, \quad F_{\mathrm{p}} = \begin{bmatrix} \alpha_4^i \\ \alpha_7^i \\ \alpha_{11}^i \\ \omega_{\mathrm{rated}} \end{bmatrix}.$$

Substituting (6.26) into (6.16), a standard PWA format can be obtained,

$$\begin{aligned} x(k+1) &= A_{\mathrm{d}}x(k) + B_{\mathrm{d}}u(k) + E_{\mathrm{d}}d(k) + F_{\mathrm{d}}, \\ y(k) &= C_{\mathrm{d}}x(k) + D_{\mathrm{d}}u(k) + G_{\mathrm{d}}d(k) + H_{\mathrm{d}}, \\ &\text{if } \begin{bmatrix} x(k) \\ u(k) \\ d(k) \end{bmatrix} \in \chi_i. \end{aligned} \tag{6.27}$$

with

$$\begin{aligned} &A_{\mathrm{d}} = A'_{\mathrm{d}} + B'_{\mathrm{d}}A_{\mathrm{p}},\ B_{\mathrm{d}} = B'_{\mathrm{d}}B_{\mathrm{p}},\ E_{\mathrm{d}} = B'_{\mathrm{d}}E_{\mathrm{p}},\ F_{\mathrm{d}} = B'_{\mathrm{d}}F_{\mathrm{p}}, \\ &C_{\mathrm{d}} = C'_{\mathrm{d}} + D'_{\mathrm{d}}A_{\mathrm{p}},\ D_{\mathrm{d}} = D'_{\mathrm{d}}B_{\mathrm{p}},\ G_{\mathrm{d}} = D'_{\mathrm{d}}E_{\mathrm{p}},\ H_{\mathrm{d}} = D'_{\mathrm{d}}F_{\mathrm{p}}. \end{aligned}$$

6.6 Case Study

In this section, the PWA model developed was verified by the comparison with the 5-MW NREL non-linear wind turbine model. The PWA modeling and simulation were executed in Multi-Parametric Toolbox – a Matlab-based toolbox for parametric optimization, computational geometry, and model predictive control [12].

Two scenarios were designed to test the developed PWA wind turbine model under both low- and high-wind conditions. The simulation time for both cases is 300 s. The wind speed variations for both scenarios are shown in Figure 6.8. In addition, the power reference also varies during the simulation, as shown in Figure 6.9; there are two step changes at $t = 100$ s and $t = 200$ s, respectively.

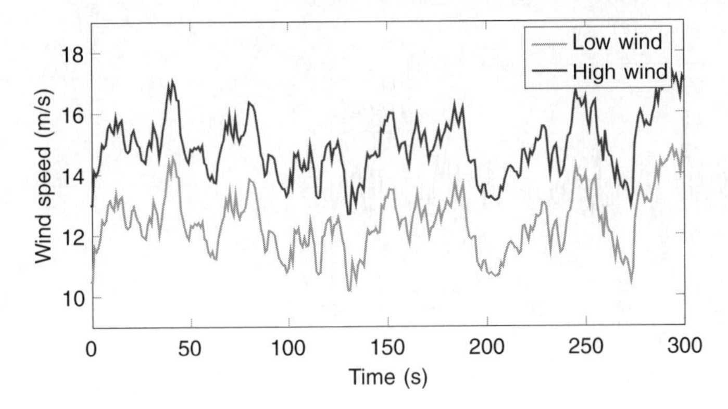

Figure 6.8 Wind speed variation.

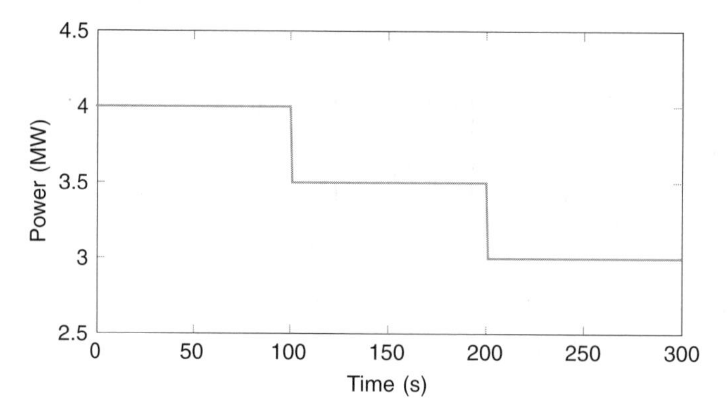

Figure 6.9 Power reference during simulation.

6.6.1 Low Wind Speed Case

As shown in Figure 6.8, the wind speed variation of this scenario covers the range between 10 m/s and 15 m/s. The simulation results for the non-linear NREL model (NL) and the PWA model (PWA) are shown in Figures 6.10 and 6.11, the former covering the state variables (θ and ω_r) and the latter the output variables (T_s, F_t). The percentage root mean square error (%RMSE) is used for model mismatch evaluation. Lower %RMSE means less model mismatch and better model estimation. The commonly applied threshold value for %RMSE is 10.

Following the wind variation, the pitch angle θ is regulated to track the power production. Once θ is reduced to 0, pitch angle control is deactivated and the system gets

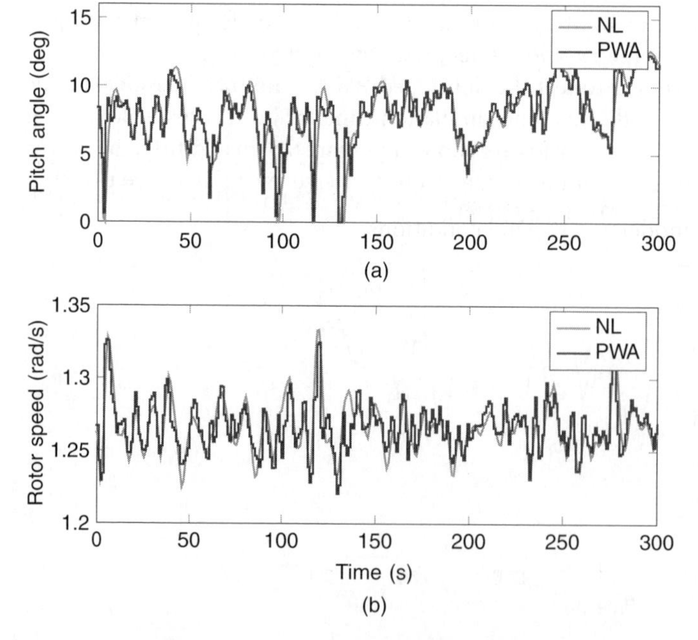

Figure 6.10 State variable comparison under low-wind conditions.

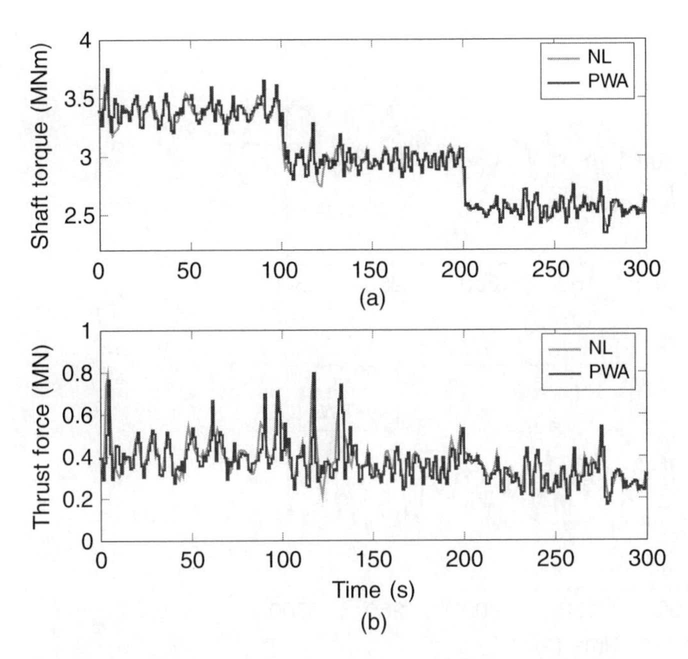

Figure 6.11 Output variable comparison under low-wind conditions.

Table 6.1 %RMSE under low-wind conditions.

	θ	ω_r	T_s	F_t
%RMSE	8.1028	0.8865	1.2365	9.7556

into Region II. Otherwise, the rotor speed ω_r is regulated to prevent the rotor from over-speeding ($\omega_{rated} = 1.2671$ rad/s). The shaft torque T_s follows the variation of the power reference. The %RMSE values of θ, ω_r, T_s, and F_t are listed in Table 6.1. The maximum %RMSE value is 9.7556. Therefore the agreement between the two models is good, thus verifying the PWA model in low-wind conditions.

6.6.2 High Wind Speed Case

The procedure is similar for the high wind speed case. The wind-speed variation is shown in Figure 6.8, and covers the range between 13 m/s and 18 m/s. The simulation results of the non-linear NREL model (NL) and the PWA model (PWA) are again illustrated and compared, in Figure 6.12 for the system states (θ and ω_r) and in Figure 6.13 for the outputs (T_s, F_t).

Following the wind variation, θ is regulated to track the power production. The pitch angle control is always in the active status (Region I). ω_r is regulated around its rated value of 1.2671. In essence, T_s follows the variation of the power reference. The %RMSE values in the high-wind case are listed in Table 6.2, ranging from 5.7360 to 8.5764. There is good agreement between the two models, so the PWA model developed is verified in the high-wind conditions.

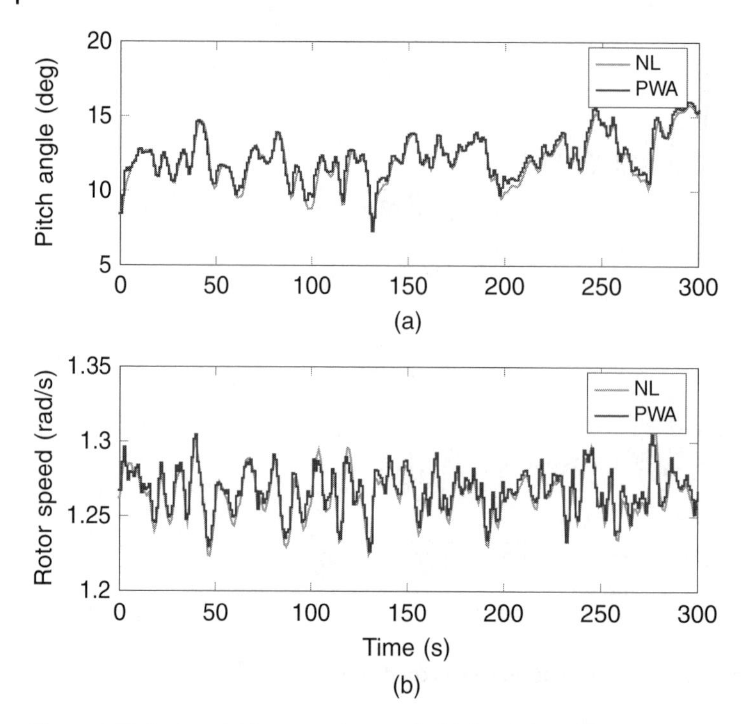

Figure 6.12 State variable comparison under high-wind conditions.

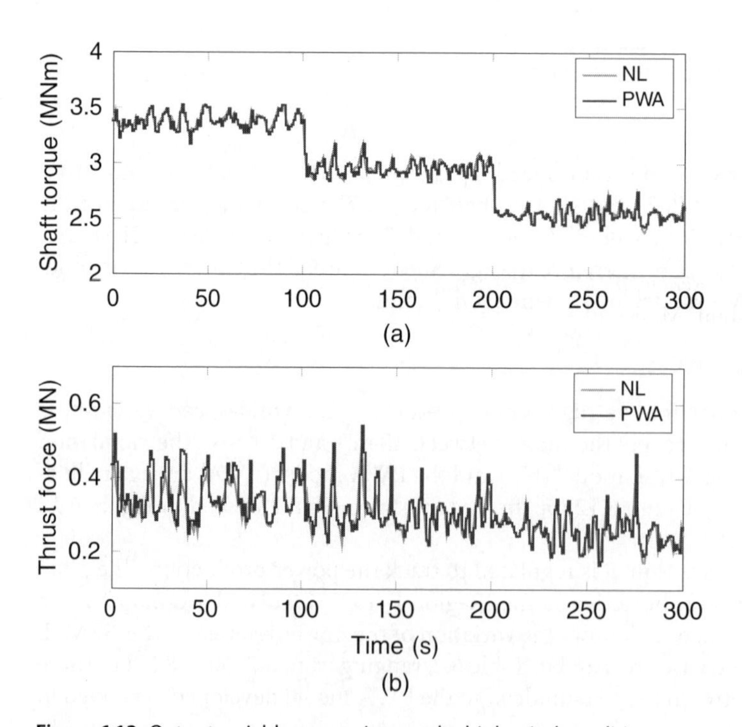

Figure 6.13 Output variable comparison under high-wind conditions.

Table 6.2 %RMSE under high-wind conditions.

	θ	ω_r	T_s	F_t
%RMSE	5.7360	0.5593	0.9441	8.5764

The reason for the error is twofold. Firstly, the model has been simplified and some dynamics are ignored, as described in Section 6.3. Secondly, the number of affine sub-models is limited, so as to reduce the complexity of the PWA model. Thus the approximation accuracy is reduced to some extent. However, this error will not significantly affect wind farm control performance. In this study, the simulation time is set at 300 s. For a real-time MPC, the prediction horizon is much shorter than this range, normally around several seconds. In addition, the prediction error can be compensated with a feedback mechanism.

6.7 Conclusion

In this chapter, a dynamic discrete-time PWA model of a power-controlled wind turbine is developed for optimal active power control of a wind farm. The non-linearities are identified by a clustering-based identification method.

Compared to the Taylor approximation at selected operating points, it has the advantage of estimating the linear submodels by classifying the multidimensional operating points and reconstructing the regions optimally.

The developed model is verified by comparison with a non-linear model under both high- and low-wind conditions. The developed PWA model is suitable for advanced optimal control at wind farm level, including MPC and the linear-quadratic regulator.

References

1 Tsili, M. and Papathanassiou, S. (2009) A review of grid code technical requirements for wind farms. *IET Renewable Power Generation*, **3** (3), 308–332.

2 Grunnet, J.D., Soltani, M., Knudsen, T., Kragelund, M.N., and Bak, T. (2010) Aeolus toolbox for dynamics wind farm model, simulation and control, in *The European Wind Energy Conference & Exhibition, EWEC 2010*.

3 Spudić, V., Jelavić, M., and Baotić, M. (2011) Wind turbine power references in coordinated control of wind farms. *Automatika – Journal for Control, Measurement, Electronics, Computing and Communications*, **52** (2), 82–94.

4 Shi, Y.T., Kou, Q., Sun, D.H., Li, Z.X., Qiao, S.J., and Hou, Y.J. (2014) H_∞ fault tolerant control of WECS based on the PWA model. *Energies*, **7** (3), 1750–1769.

5 Soliman, M., Malik, O., and Westwick, D.T. (2011) Multiple model predictive control for wind turbines with doubly fed induction generators. *IEEE Transactions on Sustainable Energy*, **2** (3), 215–225.

6 Ferrari-Trecate, G., Muselli, M., Liberati, D., and Morari, M. (2003) A clustering technique for the identification of piecewise affine systems. *Automatica*, **39** (2), 205–217.

7 Zhao, H., Wu, Q., Rasmussen, C.N., and Blanke, M. (2014) Adaptive speed control of a small wind energy conversion system for maximum power point tracking. *IEEE Transactions on Energy Conversion*, **29** (3), 576–584.
8 Spudic, V., Jelavic, M., Baotic, M., and Peric, N. (2010), Hierarchical wind farm control for power/load optimization, *The Science of making Torque from Wind* (Torque2010).
9 Maciejowski, J.M. (2002) *Predictive Control: With constraints*, Pearson Education.
10 Ferrari-Trecate, G. (2005), Hybrid identification toolbox (HIT).
11 Jonkman, J., Butterfield, S., Musial, W., and Scott, G. (2009) Definition of a 5-MW reference wind turbine for offshore system development. *National Renewable Energy Laboratory, Golden, CO, Technical Report No. NREL/TP-500-38060.*
12 Herceg, M., Kvasnica, M., Jones, C., and Morari, M. (2013) Multi-parametric toolbox 3.0, in *Proceedings of the European Control Conference*, EPFL-CONF-186265.

7

Adaptive Control of Wind Turbines for Maximum Power Point Tracking

Haoran Zhao and Qiuwei Wu

Technical University of Denmark

The control objectives for a variable-speed wind turbine normally include limiting the captured wind power in the full-load regime and maximizing the harvested wind power in the partial-load regime. In this chapter, an $\mathcal{L}_1$ adaptive controller for maximum power point tracking (MPPT) of a small variable-speed wind energy conversion system (WECS) is presented. The proposed control scheme has a generic structure and can be applied for different generator types. The simulation results show that the designed $\mathcal{L}_1$ adaptive controller has good tracking performance, even with unmodeled dynamics and in the presence of parameter uncertainties and unknown disturbances.

7.1 Introduction

According to the speed control criterion, WECSs can be classified into two types: fixed speed and variable speed. Since the rotor speed can be regulated within a large range, variable-speed wind turbines have a higher aerodynamic efficiency and are more widely used [1]. Generally, the control system of a variable-speed WECS consists of three main sub-controllers: for pitch control, generator control and power transfer control, as illustrated in Figure 7.1.

The objectives of the sub-controllers are dependent on the wind turbine operation. Depending on the wind speed, the wind turbine operation is divided into two regimes, as shown in Figure 7.2:

- partial-load regime, where the wind speed v_w is between the cut-in and the rated speed
- full-load regime, where the wind speed v_w exceeds the rated wind speed.

In the partial load regime, the pitch control system is typically deactivated. There are two exceptions:

- assistance for the start-up process, as the starting torque of wind turbines is relatively low
- limitation of the rotation speed, as the wind speed approaches the rated value.

Modeling and Modern Control of Wind Power, First Edition. Edited by Qiuwei Wu and Yuanzhang Sun.

Companion website: www.wiley.com/go/wu/modeling

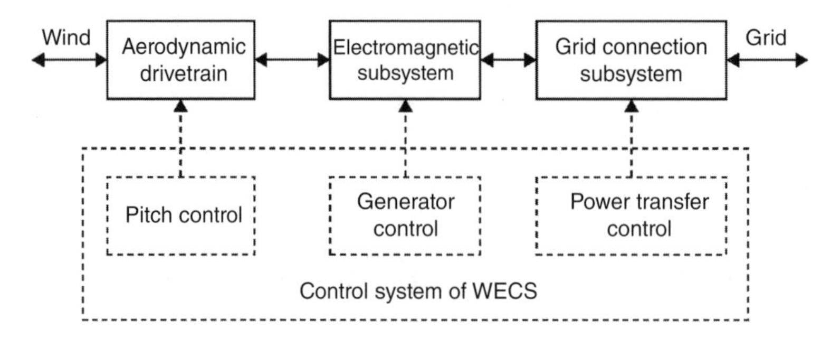

Figure 7.1 Control system of variable speed WECS [2].

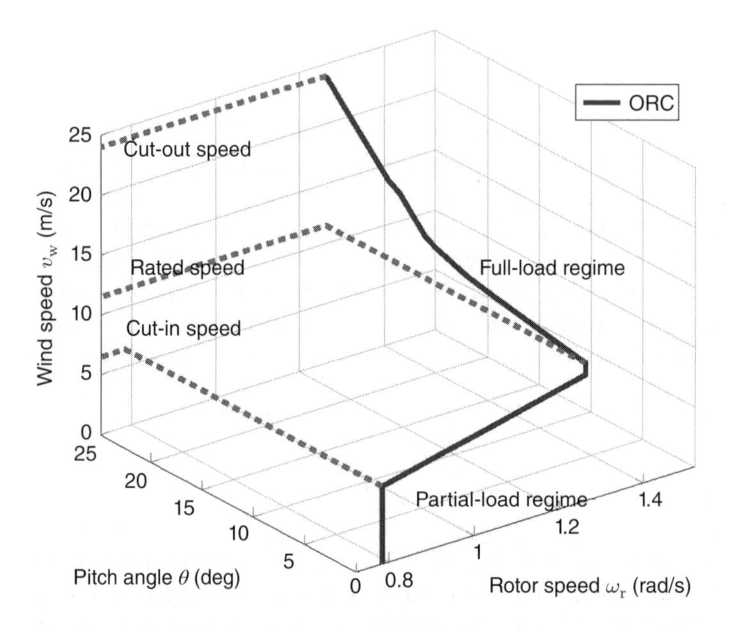

Figure 7.2 Optimal regimes characteristic (ORC) of variable speed WECS [2].

Generator control aims to maximize the energy captured from the wind. By regulating the generator torque, the generator speed can be controlled to track the optimum tip-speed ratio (TSR) [3].

During the full-load regime, pitch control is active, so as to limit the aerodynamic power to the rated value. Generator control is used to prevent the generator from over-speeding.

This chapter focuses on the generator control of wind turbines for maximum power point tracking in the partial-load regime. The MPPT control methods proposed in the literature can be classified into the following three categories [4, 5]:

7.1.1 Hill-climbing Search Control

The hill climbing search control algorithm is simple, robust and has the advantage of having very few parameters and little feedback information [6]. Normally, the rotor speed and the turbine power variation are required. Recently, other variables, such as the DC-link voltage and the duty cycle have been used as control inputs [7]. However, the approach is more suitable for slow-varying systems, such as photovoltaics, rather than WECS, which are liable to fast wind-speed Changes [8].

7.1.2 Power Signal Feedback Control

Power signal feedback control is based on the maximum power curve of the wind turbine and requires rotor speed information for calculating the power reference [9]. The tracking speed is dependent on the rotor inertia. Larger rotor inertia leads to slower tracking speeds, which further affects the efficiency, especially in the low wind-speed conditions [10].

7.1.3 Tip-speed Ratio Control

This approach regulates the rotational speed to maintain the optimal tip-speed ratio (TSR). In contrast to the other two control methods, wind-speed information is required. Since the external anemometer may increase the capital cost and its measurement accuracy is limited due to locational constraints [9, 11], the effective wind speed is sometimes estimated [12].

Many control schemes have been proposed based on TSR control. The proportional integration (PI) controller is the simplest and most robust. Its parameters are designed based on a linearized WECS model at a steady-state operating point. However, with a shift of the operation point, its control performance cannot been guaranteed. The sliding mode controller (SMC) is effective and intrinsically robust, requiring little information and being insensitive to parameter variations [13, 14]. However, the SMC can lead to chattering, which may result in unmodeled dynamics and produce destructive oscillations [3]. The feedback linearization is effective for non-linear WECSs [15]. However, its formulation is based on the specific system model, which may not be available.

Recently, $\mathcal{L}_1$ adaptive control has been developed by Hovakimyan and Cao. Compared with conventional model reference adaptive control, it has guaranteed transient performance and robustness in the presence of fast adaptions to uncertainties [16–18]. It is very suitable for the design of MPPT controllers for WECSs with non-linearities, uncertainties and disturbances.

7.2 Generator Control System for WECSs

A full-converter wind turbine with an squirrel cage induction generator (SCIG) is used to illustrate the generator control system for a variable-speed WECS. The generator control system based on the TSR concept is illustrated in Figure 7.3. With knowledge of the generator speed ω_g and power P_g, the optimal speed reference ω_g^{ref} can be calculated and fed to the speed control block. The speed controller generates the torque reference T_g^{ref} as the input for the torque controller. By using the vector control scheme, the control signal $u_{(a,b,c)}$ for pulse-width modulation (PWM) is generated by the torque controller and fed to the AC/DC converter. More details about these control blocks are described in the following.

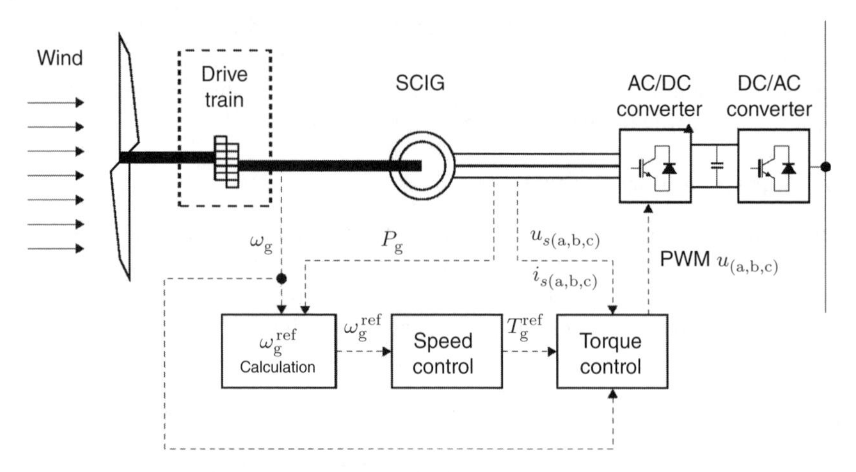

Figure 7.3 Generator control of WECS based on TSR.

7.2.1 Speed Reference Calculation

The power P_{wt} captured by a wind turbine can be expressed as,

$$P_{wt} = \frac{1}{2} \cdot \rho \pi R^2 {v_w}^3 C_p(\lambda, \theta) = \frac{1}{2} \cdot \frac{C_p(\lambda)}{\lambda^3} \rho \pi R^5 \omega_r^3, \lambda = \frac{\omega_r R}{v_w}, \tag{7.1}$$

where ρ is the air density, R is the blade length, v_w is the wind speed, ω_r is the rotor speed, and C_p represents the power efficiency. C_p is a function of the TSR λ and pitch angle θ in a pitch-controlled wind turbine. As mentioned in Section 7.1, the pitch control system is deactivated: $\theta = 0$ [19]. Conventionally, C_p can be expressed as a polynomial in λ,

$$C_p = \alpha_6 \lambda^6 + \alpha_5 \lambda^5 + \alpha_4 \lambda^4 + \alpha_3 \lambda^3 + \alpha_2 \lambda^2 + \alpha_1 \lambda + \alpha_0, \tag{7.2}$$

where $\alpha_0 \ldots \alpha_6$ are polynomial coefficients. There exists an optimal TSR λ_{opt}, at which the power efficiency C_p has its maximum value: $C_p^{max} = C_p(\lambda_{opt})$. Accordingly, the maximum power can be captured. In order to drive the WECS working at λ_{opt}, ω_r should be regulated, so as to follow the wind variation. It is assumed that the drivetrain is stiff, and the ratio between ω_r and ω_g is fixed and equal to the gearbox ratio η. The generator speed reference ω_g^{ref} can be calculated by:

$$\omega_g^{ref} = \eta \frac{\lambda_{opt} v_w}{R}. \tag{7.3}$$

Since λ_{opt}, R and η are constants, it is essential to get v_w to determine ω_g^{ref}.

As described in Section 7.1, v_w can either be measured or estimated. In this study, the power balance estimator is used, an approach which is widely applied in the wind turbine industry as well as wind research community for wind speed estimation [20]. The estimated variable is noted with hat symbol {. The calculation procedure is as follows:

Step 1: The smoothed aerodynamic torque $\hat{T}_a$ and rotor speed $\hat{\omega}_r$ are estimated from:

$$\hat{T}_a(s) = \frac{1}{1 + s\tau}(J_{eq} s \hat{\omega}_r + \eta T_g(s) + T_{loss}(s)), \tag{7.4}$$

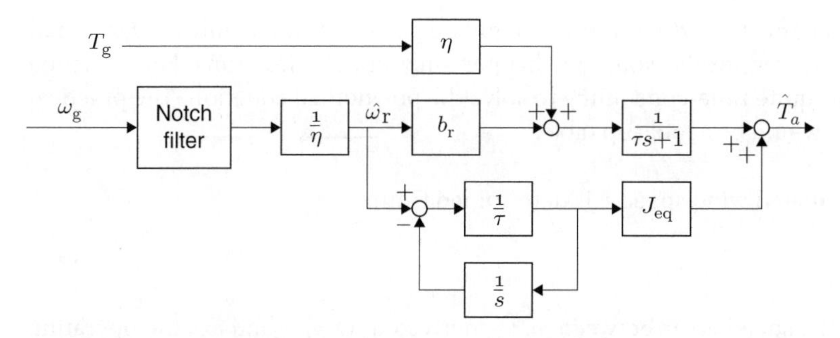

Figure 7.4 Estimation of $\hat{T}_a$.

where J_{eq} is the equivalent inertia: $J_{eq} = J_{wt} + \eta^2 J_g$. J_{wt} and J_g are the wind turbine rotor inertia and generator inertia, respectively, and $\hat{\omega}_r$ is equivalent to ω_g, with a notch filter used to remove the drivetrain eigen-frequency component. A low-pass filter with time constant τ is applied to smooth out $\hat{T}_a$. The loss term T_{loss} represents viscous friction: $T_{loss} = b_r \omega_r$, where b_r is the viscous coefficient. Figure 7.4 shows the implementation of the estimator.

Step 2: Equation 7.1 can be rewritten as:

$$C_p(\lambda)\lambda^{-3} = \frac{2P_{wt}}{\rho \pi R^5 \omega_r^3} = \frac{2T_a}{\rho \pi R^5 \omega_r^2}. \tag{7.5}$$

By replacing ω_r and T_a with the derived estimation values $\hat{\omega}_r$ and $\hat{T}_a$ from Step 1, $C_p(\lambda)\lambda^{-3}$ as a whole can be calculated as,

$$C_p(\hat{\lambda})\hat{\lambda}^{-3} = \frac{2\hat{T}_a}{\rho \pi R^5 \hat{\omega}_r^{\,2}}. \tag{7.6}$$

By substituting (7.2) into (7.6), $C_p(\hat{\lambda})\hat{\lambda}^{-3}$ is transformed into a polynomial equation and $\hat{\lambda}$ is the root. The curve is shown in Figure 7.5, and is monotonic for the case $\theta = 0$. As

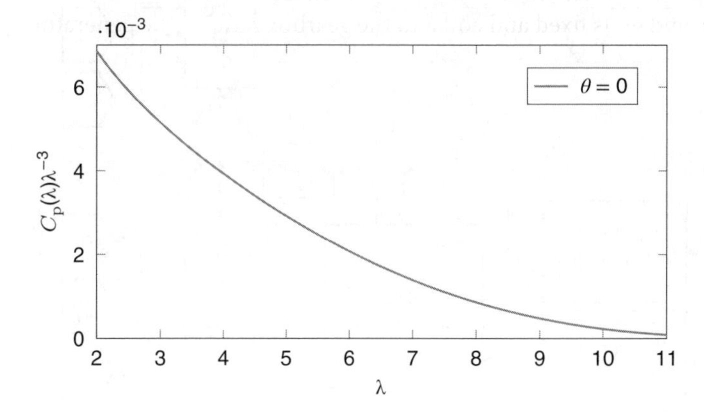

Figure 7.5 $C_p(\lambda)\lambda^{-3}$ curve.

mentioned by Soltani et al. [20], the curve might be non-monotonic only for very small pitch angles. However, invalid solutions happen only in rare situations. For real-time control, it will be quite time-consuming to solve the polynomial equation. The practical implementation would be a look-up table.

Step 3: The estimated wind speed $\hat{v}$ is determined from:

$$\hat{v} = \frac{\hat{\omega}_r R}{\lambda}. \tag{7.7}$$

By eliminating the speed error between ω_g^{ref} – derived in (7.3) – and ω_g, the operating point moves to the maximum power point.

7.2.2 Generator Torque Control

The wind turbine rotor is coupled to the generator via the gearbox, as shown in Figure 7.3. To drive the rotor following the speed reference $\omega_r^{ref} = \frac{\omega_g^{ref}}{\eta}$, the generator torque T_g should be regulated by torque control.

The inputs of the torque controller include the torque reference T_g^{ref}, derived by the speed controller, the measured generator speed ω_g, and the measured stator voltages $u_{s(a,b,c)}$ and currents $i_{s(a,b,c)}$. As the outputs of the torque controller, the references of the generator stator voltages $u_{s(a,b,c)}^{ref}$ are sent to the AC/DC converter in order to generate the switching signals through PWM [3].

The rotor flux orientation based vector control structure for the torque control of an SCIG is illustrated in Figure 7.6. It comprises two decoupled loops. The first is the rotor flux loop, which ensures the field orientation of the induction machine in order to control i_{sd}. The second is the torque control loop, which imposes an electromagnetic torque in order to control i_{sq}.

Vector control can provide fast and accurate responses. Thus the torque control can be described by a first-order function with a time constant τ_g, which is normally of the

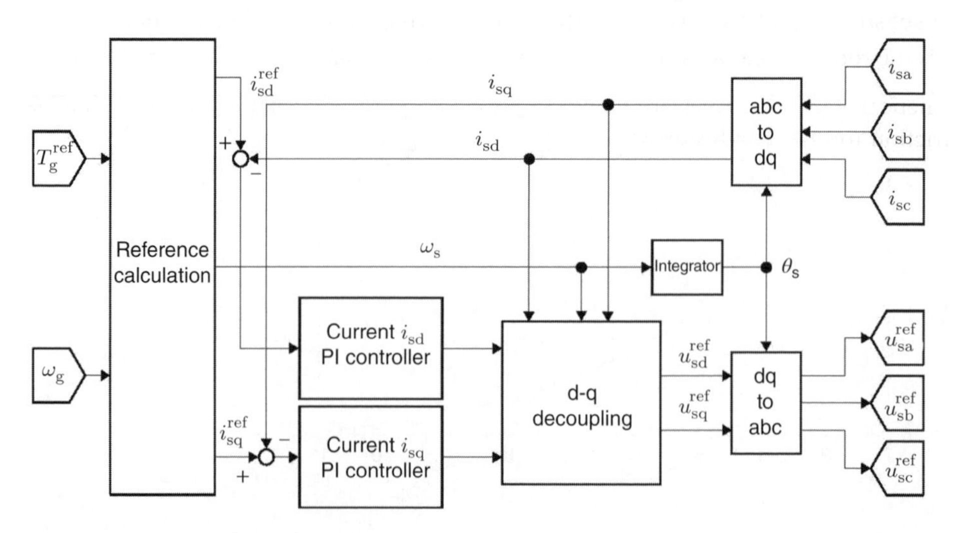

Figure 7.6 Vector control of the SCIG.

order of milliseconds:

$$T_g = \frac{1}{1 + s\tau_g} T_g^{ref}. \tag{7.8}$$

7.2.3 Speed Control

This controller aims to generate the torque reference T_g^{ref} in order to regulate the generator speed ω_g by tracking the reference ω_g^{ref}. The wind turbine can be described as the following fourth-order state-space model, which covers both the drivetrain and the generator dynamics,

$$\begin{aligned} \dot{x}(t) &= Ax(t) + B(u(t) + u_d(t)) + E\xi(t) \\ y(t) &= Cx(t) \end{aligned} \tag{7.9}$$

with

$$A = \begin{bmatrix} -\frac{b_r}{J_{wt}} & 0 & -\frac{\eta}{J_{wt}} & 0 \\ 0 & -\frac{b_g}{J_g} & \frac{1}{J_g} & -\frac{1}{J_g} \\ \eta k_s - \eta\frac{b_r d_s}{J_{wt}} & \frac{b_g d_s}{J_g} - k_s & -\frac{d_s \eta^2}{J_{wt}} - \frac{d_s}{J_g} & \frac{d_s}{J_g} \\ 0 & 0 & 0 & -\frac{1}{\tau_g} \end{bmatrix}, B = \begin{bmatrix} 0 \\ 0 \\ 0 \\ \frac{1}{\tau_g} \end{bmatrix},$$

$$C = \begin{bmatrix} 0 & 1 & 0 & 0 \end{bmatrix}, E = \begin{bmatrix} \frac{1}{J_{wt}} \\ 0 \\ \frac{\eta d_s}{J_{wt}} \\ 0 \end{bmatrix}.$$

where the state variables are defined as: $x = [\omega_r, \omega_g, T_s, T_g]'$, the input u is T_g^{ref}, u_d is friction torque, ξ indicates the aerodynamic torque T_a, the output y is ω_g, k_s and d_s are the stiffness and the damping coefficients of the drivetrain, respectively, b_r and b_g are the viscous friction coefficients for wind turbine rotor and generator, which are time-varying with known upper and lower bounds, and T_s is the shaft torque for the input. The transfer function in the s domain is expressed as,

$$y(s) = C(sI - A)^{-1}B(u(s) + u_d(s) + B^{-1}E\xi(s)). \tag{7.10}$$

Considering u_d and ξ as unmodeled disturbances, (7.10) can be changed into

$$y(s) = C(sI - A)^{-1}B(u(s) + d(s)) \tag{7.11}$$

with $d(s) = u_d(s) + B^{-1}E\xi(s)$.

This is a single-input, single-output (SISO) system with unmodeled disturbances and parameter uncertainties. Conventionally, a PI controller can be designed considering the closed-loop stability and dynamic performance around the operating point. However, due to the movement of the operating points and disturbances, the transient performance cannot be guaranteed. To overcome these problems, an $\mathcal{L}_1$ adaptive controller is designed in this study.

7.3 Design of $\mathcal{L}_1$ Adaptive Controller

7.3.1 Problem Formulation

In this section, an $\mathcal{L}_1$ adaptive output feedback controller is designed for the SISO WECS system described in Section 7.2.3, following the scheme proposed by Zhao et al. [1]. The control object is to enable the system output ω_g to follow the reference ω_g^{ref}. As the first-order system has good dynamic performance, the reference transfer function model $M(s)$ with time constant τ_m is,

$$M(s) = \frac{1}{1+\tau_m s} = \frac{m}{s+m}, \tau_m = \frac{1}{m}, m > 0. \tag{7.12}$$

Consider the system:

$$y(s) = A(s)(u(s) + d(s)) \tag{7.13}$$

where $u(s)$ and $y(s)$ are $T_g^{ref}(s)$ and $\omega_g(s)$, respectively. $A(s)$ represents the strictly proper unknown transfer function of WECS, and $d(s)$ represents the time-varying non-linear uncertainties and disturbance, denoted by $d(t) = f(t, y(t))$ in the time domain. Two assumptions are made for $f(t, y(t))$.

Assumption 1 ***(Lipschitz continuity)*** There exist constants $L > 0$ (Lipschnitz gain) and $L_0 > 0$, possibly arbitrarily large, such that the following inequalities hold uniform,

$$\begin{cases} |f(t,y_1) - f(t,y_2)| \leq L|y_1 - y_2| \\ |f(t,y)| \leq L|y| + L_0, \qquad L > 0, L_0 > 0. \end{cases} \tag{7.14}$$

Assumption 2 ***(Uniform boundedness of the variation rate of uncertainties)*** The variation rate of uncertainties is uniform bounded: there exist constants $L_1, L_2, L_3 > 0$, such that, for any $t \geq 0$,

$$|\dot{d}(t)| \leq L|y_1 - y_2| \leq L_1|\dot{y}(t)| + L_2|y(t)| + L_3 \tag{7.15}$$

Based on (7.12) and (7.13), the system can be rewritten in terms of the reference model as,

$$y(s) = M(s)(u(s) + \sigma(s)), \tag{7.16}$$

where uncertainties due to $A(s)$ and $d(s)$ are lumped into the signal $\sigma(s)$,

$$\sigma(s) = \frac{(A(s) - M(s))u(s) + A(s)d(s)}{M(s)}. \tag{7.17}$$

7.3.2 Architecture of the $\mathcal{L}_1$ Adaptive Controller

Figure 7.7 shows the closed loop $\mathcal{L}_1$ adaptive controller, which consists of three parts: state predictor, adaption law, and control law.

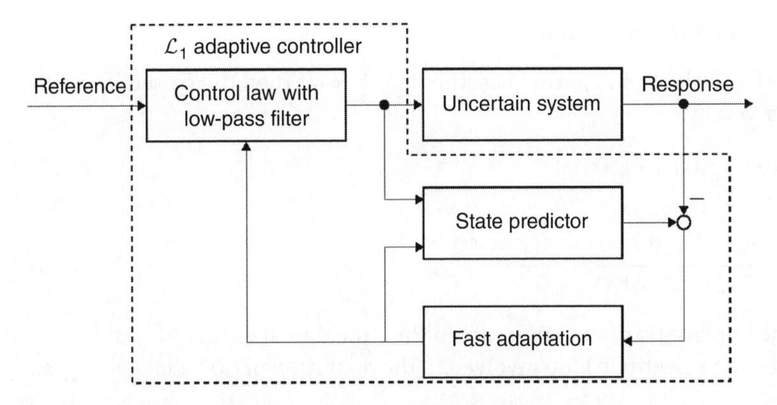

Figure 7.7 The closed-loop $\mathcal{L}_1$ adaptive controller [16].

State (Output) Predictor

$$\dot{\hat{y}}(t) = -m\hat{y}(t) + m(u(t) + \hat{\sigma}(t)), \quad \hat{y}(0) = 0, \tag{7.18}$$

where $\hat{\sigma}(t)$ is the adaptive estimate.

Adaption Law

The adaption of $\hat{\sigma}(t)$ is defined as,

$$\dot{\hat{\sigma}}(t) = \Gamma\text{Proj}(\hat{\sigma}(t), -\tilde{y}(t)), \quad \hat{\sigma}(0) = 0, \tag{7.19}$$

where $\tilde{y}(t) \triangleq \hat{y}(t) - y(t)$ is the error signal between the output and the predictor, and Γ is the adaptive gain. Projection methods are introduced to ensure the boundary of the parameter estimates. More details about projection operators are described by Hovakimyan and Cao [16]. The projection is bounded by,

$$\hat{\sigma}(t) \leq \Delta \tag{7.20}$$

where Δ is the boundary.

The adaptive system tracks the state predictor, while the state predictor matches the expected closed-loop reference system output very closely. As adaptive gain $\Gamma \to \infty$, the tracking errors of these signals go to zero.

Control Law with Low-pass Filter

This block is used to limit the compensation of the uncertainties within the bandwidth,

$$u(s) = C(s)(r(s) - \hat{\sigma}(s)), \tag{7.21}$$

where $r(s)$ is the reference signal that $y(s)$ should follow, $C(s)$ is the low-pass filter with cut-off frequency $f_{\text{cut}} = \frac{w}{2\pi}$, with $C(0) = 1$. The DC gain of one ensures reference tracking. Because of the approximation of $\hat{\sigma}$, the low-pass filter $C(s)$ is used to prevent any high-frequency chatter from passing into the system,

$$C(s) = \frac{w}{s + w}. \tag{7.22}$$

7.3.3 Closed-loop Reference System

Assume the adaptive variable $\hat{\sigma}$ is perfectly estimated: $\hat{\sigma} = \sigma$. Then the closed loop reference system is given by:

$$y_{\text{ref}}(s) = M(s)(u_{\text{ref}}(s) + \sigma_{\text{ref}}(s)), \tag{7.23}$$

$$u_{\text{ref}}(s) = C(s)(r(s) - \sigma_{\text{ref}}(s)), \tag{7.24}$$

$$\sigma_{\text{ref}}(s) = \frac{(A(s) - M(s))u_{\text{ref}}(s) + A(s)d_{\text{ref}}(s)}{M(s)}, \tag{7.25}$$

where $d_{\text{ref}}(s)$ is the Laplace transform of $d_{\text{ref}}(t)$ in the time domain, $d_{\text{ref}}(t) \triangleq f(t, y_{\text{ref}}(t))$. It is seen that there is no algebraic loop involved in the definition of $\sigma(s)$, $u(s)$ and $\sigma_{\text{ref}}(s)$, $u_{\text{ref}}(s)$.

The closed-loop reference to output transfer function $H(s)$ is,

$$H(s) = \frac{A(s)M(s)}{C(s)A(s) + (1 - C(s))M(s)}. \tag{7.26}$$

Therefore, $C(s)$ and $M(s)$ should be selected to ensure that $H(s)$ is stable. Then the lemma of Cao and Hovakimyan is used [17].

Lemma Let $C(s)$ and $M(s)$ verify the $\mathcal{L}_1$-norm condition,

$$||G(s)||_{\mathcal{L}_1} L < 1, \text{ with } G(s) \triangleq H(s)(1 - C(s)). \tag{7.27}$$

Then the closed-loop reference functions (7.23)–(7.25) are bounded input and bounded output (BIBO) stable. Similarly, by substituting (7.24) into (7.25), the following closed-loop function can be derived,

$$\sigma_{\text{ref}}(s) = \frac{C(s)(A(s) - M(s))r(s) + A(s)d_{\text{ref}}(s)}{C(s)A(s) + (1 - C(s))M(s)}. \tag{7.28}$$

The denominator is the same as $H(s)$. Accordingly, σ_{ref} is stable if $H(s)$ is stable.

7.3.4 Design of $\mathcal{L}_1$ Adaptive Controller Parameters

It is expected that small variations in $A(s)$ will not change the derived properties of the control method. As an approximation of the system, in this study, $A(s)$ is defined as the transfer function of a 6-kW variable speed WECS [3], based on the matrices A, B, and C derived in (7.9):

$$A(s) = \frac{A_{\text{n}}(s)}{A_{\text{d}}(s)} = C(sI - A)^{-1}B \tag{7.29}$$

where $A_{\text{n}}(s)$ and $A_{\text{d}}(s)$ are the numerator and denominator of $A(s)$, respectively. Accordingly, the closed-loop reference transfer function is,

$$H(s) = \frac{H_{\text{n}}(s)}{H_{\text{d}}(s)} = \frac{m(s + w)A_{\text{n}}(s)}{w(s + m)A_{\text{n}}(s) + msA_{\text{d}}(s)} \tag{7.30}$$

where $H_{\text{n}}(s)$ and $H_{\text{d}}(s)$ are the numerator and denominator of $H(s)$, respectively.

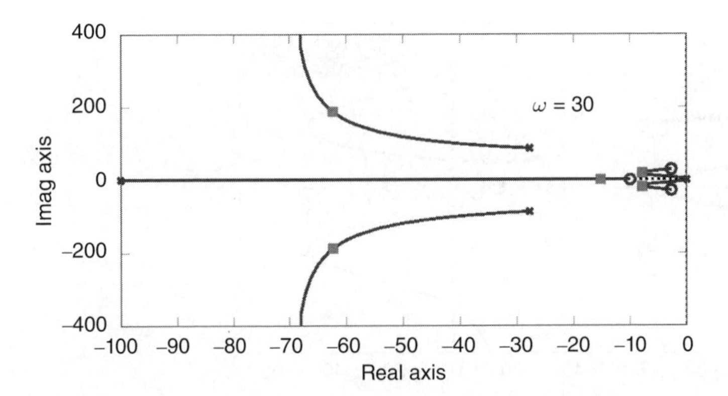

Figure 7.8 Root locus for the choice of w.

Determination of w

To get a faster response, the closed-loop reference time constant τ_m is set as $\tau_m = 0.1$ s; that is, $m = \frac{1}{\tau_m} = 10$. To ensure the stability of $H(s)$, the poles of $H(s)$ must be in the left-hand side of the complex plane, which restricts the choices of $M(s)$ and $C(s)$. The root locus method via loop transfer function $L(s)$ is used to check the stability of $H(s)$. $L(s)$ is defined by,

$$L(s) = \frac{\omega(s+m)}{ms}A(s). \tag{7.31}$$

As shown in Figure 7.8, the poles are always located in the left-hand half of the complex plane if $w > 0$. If the cut-off frequency $f_{cut} = \frac{w}{2\pi}$ of the low-pass filter $C(s)$ is too large, $C(s)$ is like an all-pass filter and $C(s) \approx 1$. Accordingly, it draws $H(s) \approx M(s)$ from (7.26) and the system responds as the reference model. Therefore, a large cut-off frequency can lead to a better match. However, a low-pass filter with a large cut-off frequency may result in a high gain feedback and thus lead to closed-loop systems with small robustness margins and too much amplification of measured noise. To keep the balance, $w = 30$ in this study.

Determination of Γ

Assume that the estimator $\hat{\sigma}(s)$ in (7.19) is within the boundary

$$\dot{\hat{\sigma}}(t) = \Gamma(y(t) - \hat{y}(t)). \tag{7.32}$$

The following closed-loop system can be derived:

$$\hat{\sigma}(s) = \frac{C(s)(A(s) - M(s))r(s) + A(s)d(s)}{\frac{1}{\Gamma}s + C(s)A(s) + (1 - C(s))M(s)}. \tag{7.33}$$

Obviously, the stability of estimator $\hat{\sigma}(s)$ is dependent on Γ, the feasible range of which can be derived by performing a root locus via loop function $L(s)$, where

$$L(s) = \Gamma\frac{C(s)A(s) + (1 - C(s))M(s)}{s}. \tag{7.34}$$

According to the root locus method, the adaptive gain can be limited. As depicted in Figure 7.9, Γ should fulfill $\Gamma < 160$ or $\Gamma > 1.64 \times 10^4$. In this study, $\Gamma = 115$.

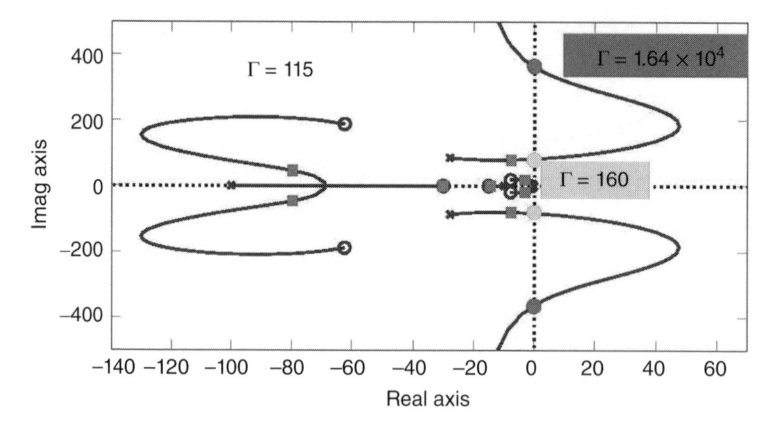

Figure 7.9 Root locus for the choice of Γ.

7.4 Case Study

In this section, several case studies have been carried out to verify the developed $\mathcal{L}_1$ adaptive controller for WECSs.

7.4.1 Wind Speed Estimation

In this case, the wind speed estimation accuracy is tested. The wind profile has been derived using the von Karman spectrum in the IEC standard, which covers most partial-load regimes: from 6.5 m/s to 9.8 m/s. Following the method in Section 7.2.1, the wind speed is estimated and compared with the actual wind speed. The comparison is depicted in Figure 7.10. A good match between both curves can be observed.

7.4.2 MPPT Performance

Based on the estimated wind speed, the tracking performance of the speed reference ω_g^{ref} is the key factor that determines the power efficiency and energy production. According

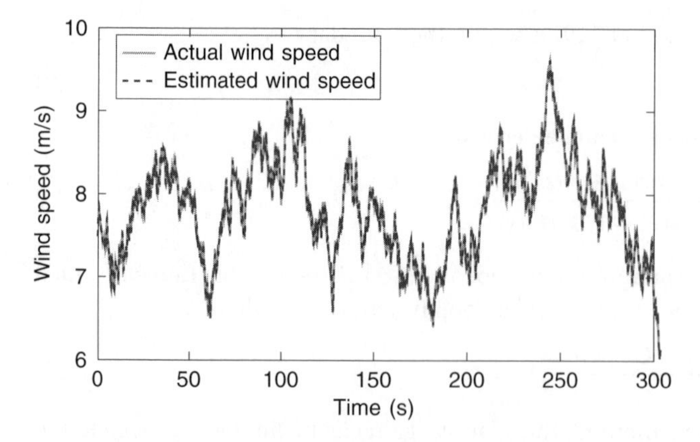

Figure 7.10 Comparison of actual and estimated wind speed.

to different turbulence intensities of the wind profiles, this case study is divided into two scenarios:

- Scenario 1: small turbulence intensity
- Scenario 2: large turbulence intensity.

The time-varying viscous coefficients (b_r, b_g) and the other disturbances except viscous friction are randomly generated within bounds. The simulation time is 300 s. The optimal values of the TSR and power coefficient are: $\lambda_{opt} = 7, C_p^{max} = 0.4763$.

Conventionally, a PI controller is designed using the same SISO WECS, considering the closed-loop system stability and transient performances. To some extent, it can also stabilize the system in the presence of unmodeled dynamics and disturbances. In this study, the tracking performances of the $\mathcal{L}_1$ adaptive controller and the time-invariant PI controller are compared (see Figures 7.11–7.14).

For Scenario 1, due to the small turbulence intensity, both controllers are good at keeping the TSR λ around λ_{opt} during the simulation, as shown in Figure 7.11a. The range is between 6.88 and 7.13. From the density point of view (Figure 7.11b), it can be seen clearly that both controllers have similar tracking performance. Accordingly, the power efficiency C_p is kept at a high level ($0.475 \sim 0.4763$) (Figure 7.12).

For Scenario 2, due to the large turbulence intensity, the tracking performances of both controllers are affected and become worse than in Scenario 1. As illustrated in Figure 7.13, the $\mathcal{L}_1$ adaptive controller is better at keeping λ at its optimal value λ_{opt} than

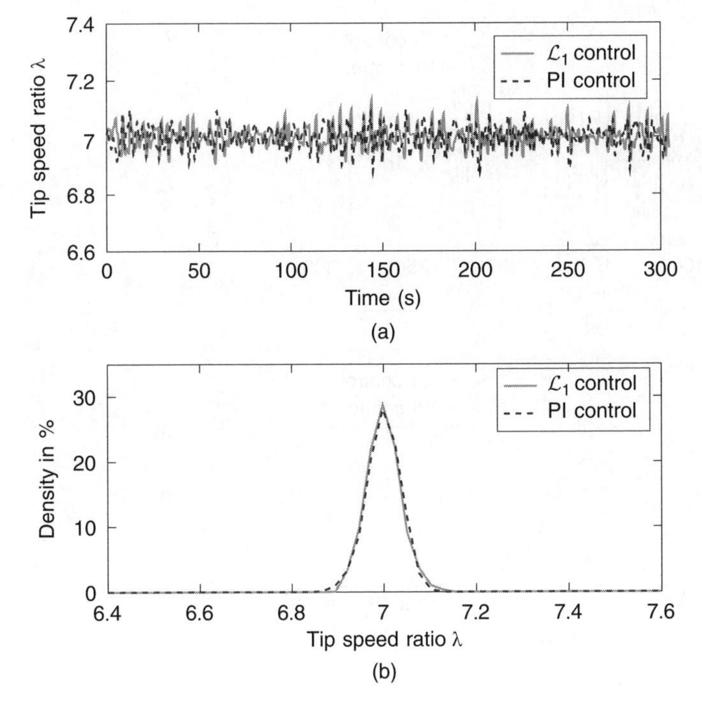

Figure 7.11 Comparison of λ in Scenario 1.

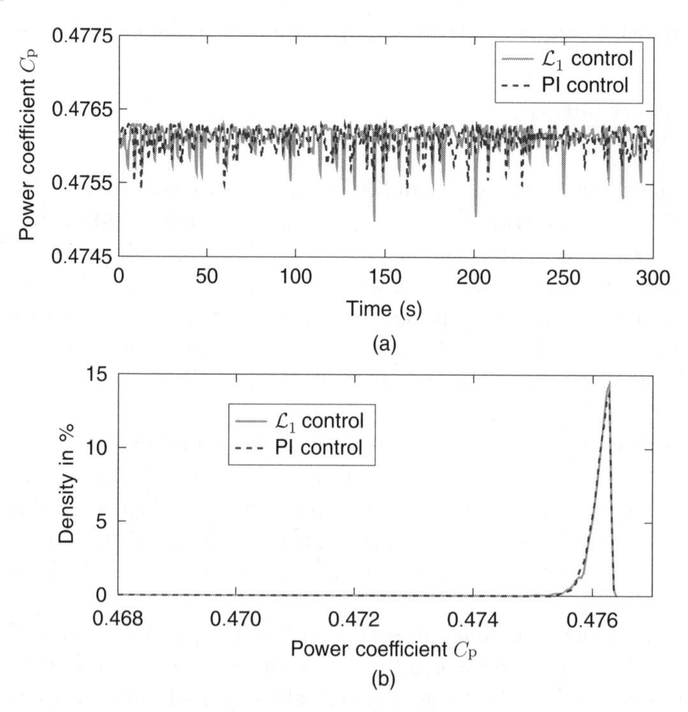

Figure 7.12 Comparison of C_p in Scenario 1.

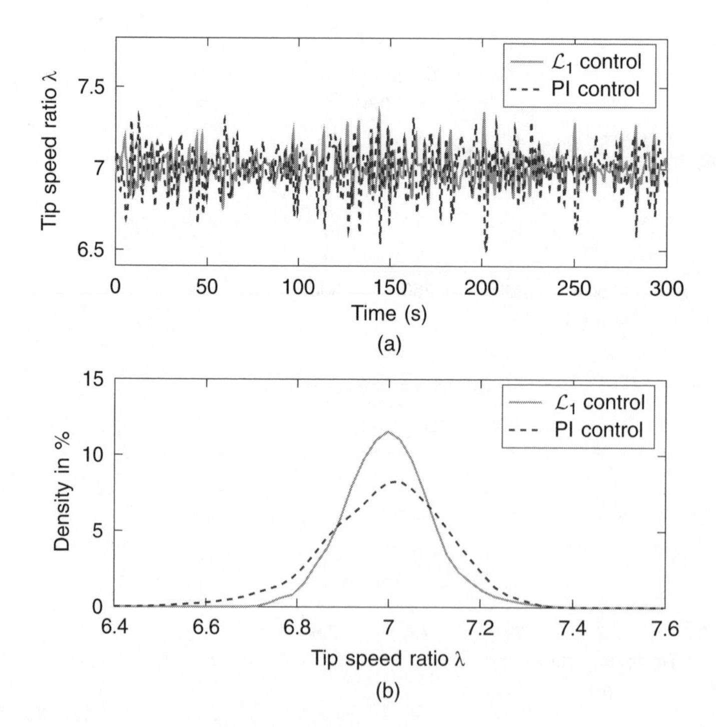

Figure 7.13 Comparison of λ in Scenario 2.

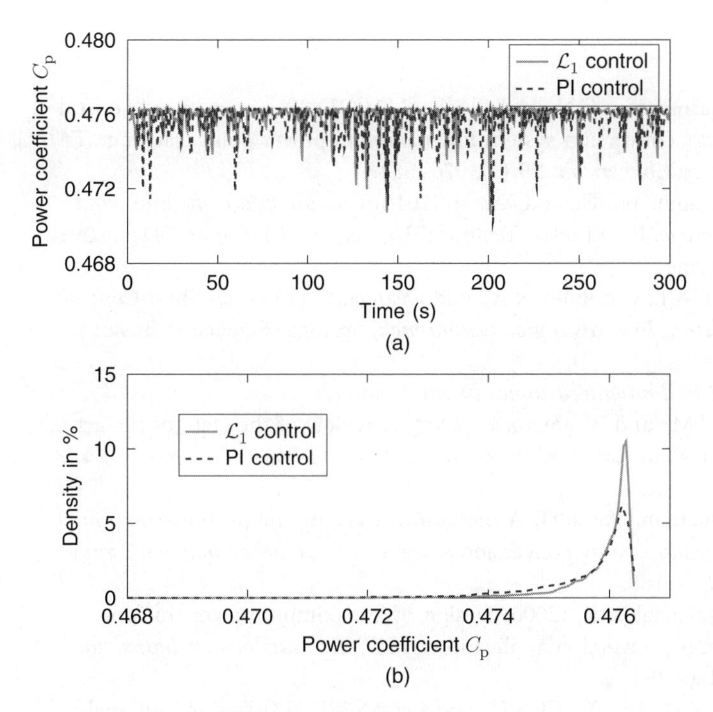

Figure 7.14 Comparison of C_p in Scenario 2.

the PI controller. The TSRs with the $\mathcal{L}_1$ controller are distributed mainly between 6.8 and 7.2, while those with the PI controller are distributed between 6.6 and 7.4. Accordingly, without precise tracking, the power efficiency C_p using the PI control is lower, as depicted in Figure 7.14.

7.5 Conclusion

WECSs are non-linear systems with parameter uncertainties and which are subject to disturbances, in the form of non-linear and unmodeled aerodynamics. The proposed $\mathcal{L}_1$ adaptive speed control approach avoids accurate WECS modeling. Instead, the WECS is modeled as a transfer function with the generator torque as the input and the generation speed as the output. Based on this simplified model, an $\mathcal{L}_1$ adaptive speed controller was designed to achieve a trade-off between transient performance and the stability of MPPT under disturbance.

As shown in detailed simulation cases, the proposed control scheme has good tracking performance for optimum TSR and robustness, with fast adaptation to uncertainties and disturbances. Compared to the conventional PI controller, the proposed $\mathcal{L}_1$ adaptive speed control for MPPT has better power efficiency during operation, especially in highly turbulent wind conditions. Therefore, it has potential for use in MPPT of modern WECSs.

References

1 Zhao, H., Wu, Q., Rasmussen, C.N., and Blanke, M. (2014) Adaptive speed control of a small wind energy conversion system for maximum power point tracking. *IEEE Transactions on Energy Conversion*, **29** (3), 576–584.

2 Zhao, H., Wu, Q., Rasmussen, C., and Xu, H. (2014) *Coordinated control of wind power and energy storage*, PhD thesis, Technical University of Denmark, Department of Electrical Engineering.

3 Munteanu, I., Bratcu, A.I., Cutululis, N.A., and Ceanga, E. (2008) *Optimal Control of Wind Energy Systems: Towards a global approach*, Springer Science & Business Media.

4 (2011) *Fundamental and advanced topics in wind power.*

5 Chen, Z., Guerrero, J.M., and Blaabjerg, F. (2009) A review of the state of the art of power electronics for wind turbines. *IEEE Transactions on Power Electronics*, **24** (8), 1859–1875.

6 Datta, R. and Ranganathan, V. (2003) A method of tracking the peak power points for a variable speed wind energy conversion system. *IEEE Transactions on Energy Conversion*, **18** (1), 163–168.

7 Koutroulis, E. and Kalaitzakis, K. (2006) Design of a maximum power tracking system for wind-energy-conversion applications. *IEEE Transactions on Industrial Electronics*, **53** (2), 486–494.

8 Dalala, Z.M., Zahid, Z.U., Yu, W., Cho, Y., and Lai, J.S.J. (2013) Design and analysis of an MPPT technique for small-scale wind energy conversion systems. *IEEE Transactions on Energy Conversion*, **28** (3), 756–767.

9 Barakati, S.M., Kazerani, M., and Aplevich, J.D. (2009) Maximum power tracking control for a wind turbine system including a matrix converter. *IEEE Transactions on Energy Conversion*, **24** (3), 705–713.

10 Kim, K.H., Van, T.L., Lee, D.C., Song, S.H., and Kim, E.H. (2013) Maximum output power tracking control in variable-speed wind turbine systems considering rotor inertial power. *IEEE Transactions on Industrial Electronics*, **60** (8), 3207–3217.

11 Pan, C.T. and Juan, Y.L. (2010) A novel sensorless MPPT controller for a high-efficiency microscale wind power generation system. *IEEE Transactions on Energy Conversion*, **25** (1), 207–216.

12 Abo-Khalil, A.G. and Lee, D.C. (2008) MPPT control of wind generation systems based on estimated wind speed using SVR. *IEEE Transactions on Industrial Electronics*, **55** (3), 1489–1490.

13 Munteanu, I., Bacha, S., Bratcu, A.I., Guiraud, J., and Roye, D. (2008) Energy-reliability optimization of wind energy conversion systems by sliding mode control. *IEEE Transactions on Energy Conversion*, **23** (3), 975–985.

14 Beltran, B., Benbouzid, M.E.H., and Ahmed-Ali, T. (2012) Second-order sliding mode control of a doubly fed induction generator driven wind turbine. *IEEE Transactions on Energy Conversion*, **27** (2), 261–269.

15 Mauricio, J.M., Leon, A.E., Gomez-Exposito, A., and Solsona, J.A. (2008) An adaptive nonlinear controller for DFIM-based wind energy conversion systems. *IEEE Transactions on Energy Conversion*, **23** (4), 1025–1035.

16 Hovakimyan, N. and Cao, C. (2010) $\mathcal{L}_1$ Adaptive Control Theory: Guaranteed robustness with fast adaptation, Siam.

17 Cao, C. and Hovakimyan, N. (2008) Design and analysis of a novel adaptive control architecture with guaranteed transient performance. *IEEE Transactions on Automatic Control*, **53** (2), 586–591.

18 Cao, C. and Hovakimyan, N. (2010) Stability margins of adaptive control architecture. *IEEE Transactions on Automatic Control*, **55** (2), 480–487.

19 Geng, H., Yang, G., Xu, D., and Wu, B. (2011) Unified power control for PMSG-based WECS operating under different grid conditions. *IEEE Transactions on Energy Conversion*, **26** (3), 822–830.

20 Soltani, M.N., Knudsen, T., Svenstrup, M., Wisniewski, R., Brath, P., Ortega, R., and Johnson, K. (2013) Estimation of rotor effective wind speed: A comparison. *IEEE Transactions on Control Systems Technology*, **21** (4), 1155–1167.

8

Distributed Model Predictive Active Power Control of Wind Farms

Haoran Zhao and Qiuwei Wu

Technical University of Denmark

In this chapter, a distributed model predictive control (D-MPC) approach for optising active power of a wind farm is proposed. The control scheme is based on the fast gradient method via dual decomposition. The developed D-MPC approach is implemented using the clustering-based piecewise affine (PWA) wind turbine model, as described in Chapter 6.

8.1 Introduction

The modern wind farm is required to operate much more like a conventional power plant and should ultimately replace conventional power plants. Specifically, for active power control, the wind farm must be capable of tracking the power references issued by the system operator.

Wind farm control can be implemented either by the utilization of a separate energy storage device or through derated operation of the wind turbines [1]. In the latter case, the additional capital investment and maintenance cost of the energy storage system would be saved. Since the wind farm is required to produce less than the maximum possible power, the wind turbines must limit their power production. As long as the total power production of the farm meets demand, the wind turbines can vary their power production in response to wind speed fluctuations.

Conventionally, the dispatch function of each turbine is based on either the available power [2] or the actual output power [1]. The conventional method focuses only on tracking the power reference. The mechanical load ("load" for short hereinafter), which can significantly shorten the service lifetime of wind turbines [3], is not considered. In this study, load mainly refers to the forces and moments experienced by the wind turbine structure. Recently, wind farm control with load optimization has been much studied [4–6]. By dynamical power distribution, the loads experienced by the turbines are minimized while the desired power production is maintained at all times.

Model predictive control (MPC) is an effective scheme for multi-objective wind farm control. Using the receding horizon principle, a finite-horizon optimal control problem is solved over a fixed interval of time [7]. The main advantage of MPC is

Modeling and Modern Control of Wind Power, First Edition. Edited by Qiuwei Wu and Yuanzhang Sun.

Companion website: www.wiley.com/go/wu/modeling

that it can realize control within input and output constraints. A centralized MPC (C-MPC) algorithm has been developed for wind farm control by Spudić et al. [6]. The optimal power references are explicitly computed offline by multi-parametric programming for each turbine. These references are based on other auxiliary power variables. They are then used for online coordination of the turbines to meet the total power demand. However, the wind farm model is described as a coupled, constrained multiple-input, multiple-output (MIMO) system. With an increasing number of wind turbines, the order of the wind farm model grows dramatically. The computational burden of C-MPC makes it impractical for real-time wind farm applications.

Distributed MPC (D-MPC) was developed to solve the optimization problems with a much reduced computational burden compared to C-MPC. Each wind turbine can be regarded as a single distributed unit. These units are coupled by the shared wind field and the power demand to the wind farm. Among current distributed algorithms, some are based on the property that the (sub)gradient to the dual of optimization problems can be handled in a distributed manner [8]; this is referred to as the "dual decomposition". In recent years, the fast gradient method used in dual decomposition has attracted more and more attention for the D-MPC approach [9–12]. The convergence rate may be significantly better than with standard gradient methods. In this study, a D-MPC approach is used for wind farm control, based on the parallel generalized fast dual gradient method. This control structure is independent from the scale of the wind farm, and the communication burden between the local D-MPC and the central unit is greatly reduced.

8.2 Wind Farm without Energy Storage

8.2.1 Wind Farm Control Structure

Since the wind speed can be considered as a mean value with turbulent fluctuations superimposed on it, the wind field dynamics can be represented by two decoupled time scales [13]. The slow time-scale dynamic is related to the mean wind speed, which represents the propagation of the wind stream travelling through the wind farm. There is coherence between wind turbines due to wake effects. According to the wind field model and measurements of the wind farm, the mean wind speed of individual wind turbine can be estimated. The fast time-scale dynamic is related to wind turbulence and gusts. The turbulence of different wind turbines is considered uncoupled, and leads to a load increase.

The hierarchical structure of D-MPC-based wind farm control is shown in Figure 8.1.

High-level Control

High-level control operates at the slow time scale. To be specific, the wind farm power reference P_{wf}^{ref} is generated according to the requirements from the system operator and the available wind farm power. With the wind field model and the measurements, the mean wind speed of a certain period (several minutes) can be estimated. As reviewed by Spudić et al. [6], several approaches have been developed to distribute the mean power references for individual wind turbines ($\overline{P_{wt\_1}^{ref}}$, $\overline{P_{wt\_2}^{ref}}$, …,

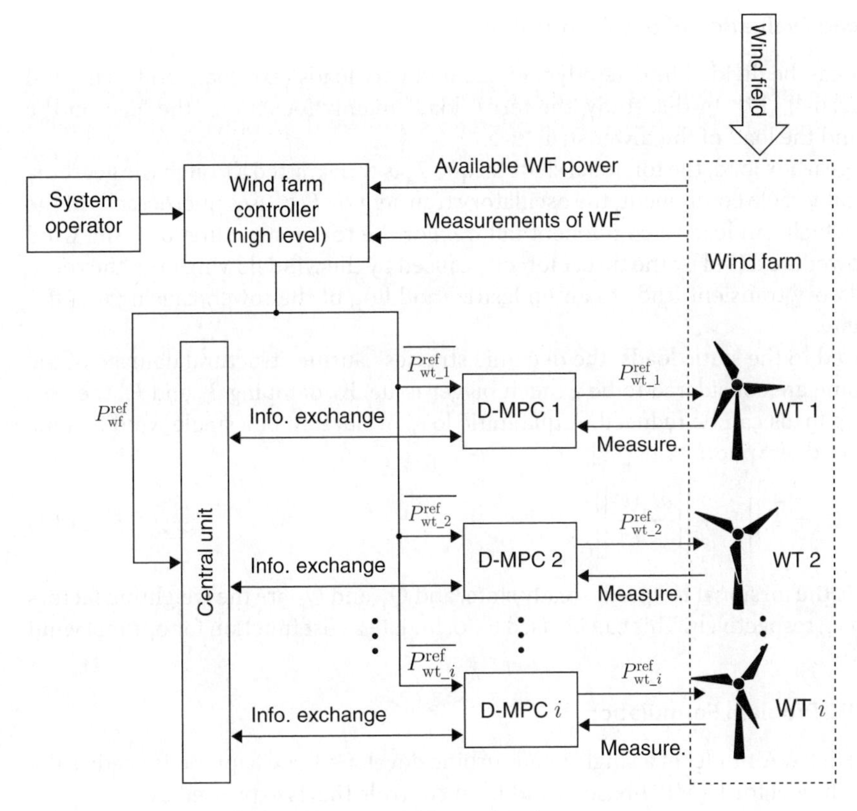

Figure 8.1 Structure of D-MPC-based wind farm control.

$\overline{P_{\mathrm{wt\_}i}^{\mathrm{ref}}}$). In this study, the proportional distribution algorithm proposed by Sørensen is adopted [2].

Low-level Control

Conventionally, these mean power references are directly assigned to the individual wind turbines without consideration of turbulence effects. In this study, these references are delivered to the D-MPC controller in each wind turbine. Local D-MPC is regarded as low-level control of the short time-scale dynamics.

With the PWA model developed in Chapter 6 and the measurements, the operating region in which the wind turbine is currently located can be determined. Consequently, the corresponding parameters for the linearized state-space model can be decided. This model is used as the prediction model to formulate the local optimization problem. Through communication with the central unit, iterations are performed to reach the global constraints. To be clear, the central unit is used to update the dual variables by collecting the matrices from the wind turbines, with the computations then performed offline. Thus, the central unit does not undertake much computation. More details are set out in the following sections. As the outputs of the D-MPC, the modified power references ($P_{\mathrm{wt\_1}}^{\mathrm{ref}}, P_{\mathrm{wt\_2}}^{\mathrm{ref}}, \cdots, P_{\mathrm{wt\_}i}^{\mathrm{ref}}$) are distributed to the individual wind turbine controllers.

8.2.2 Load Evaluation of the Wind Turbine

The loads can be divided into aerodynamic and gravity loads (external), and structural loads (internal) [14]. In this study, the term "load" mainly focuses on the load on the gearbox and the load of the tower structure.

For the gearbox load, the torsional shaft torque T_s is transmitted through the gearbox. As it is a vulnerable component, the oscillatory transient of T_s causes microcracks in the material, which can lead to component failure. For the tower structure load, the wind turbine tower is excited by the thrust force F_t caused by the wind flowing onto the rotor. The oscillatory transient leads to an undesired nodding of the tower, causing it suffer from fatigue.

Compared to the static loads, the dynamic stresses causing structural damage of the wind turbine are considered to be a much bigger issue. By damping T_s and F_t, the corresponding loads can be reduced. A quadratic load evaluation of a single wind turbine was proposed by Spudić et al. [13]:

$$\left\|T_s(t) - \overline{T_s}\right\|_{Q_T}^2 + \left\|\frac{\partial F_t(t)}{\partial t}\right\|_{Q_F}^2, \tag{8.1}$$

where $\overline{T_s}$ is the torsional torque at steady state, and Q_T and Q_F are the weighting factors for T_s and F_t, respectively. This can be used to formulate a cost function for optimal wind farm control.

8.2.3 MPC Problem Formulation

The discrete PWA model of a single wind turbine developed in Chapter 6 is used as the prediction model for D-MPC-based wind farm control. This is expressed by:

$$x(k+1) = A_d x(k) + B_d u(k) + E_d d(k) + F_d, \tag{8.2}$$

$$y(k) = C_d x(k) + D_d u(k) + G_d d(k) + H_d, \tag{8.3}$$

$$x \in \mathcal{X}, u \in \mathcal{U}, d \in \mathcal{D},$$

where x, u, d, and y are state, input, disturbance, and output vectors, respectively. $\mathcal{X}$, $\mathcal{U}$ and $\mathcal{D}$ are the feasible regions for x, u, and d, respectively, defined by,

$$x = [\omega_r, \omega_f, \theta]', \mathcal{X} = \{[\underline{\omega_r}, \overline{\omega_r}], [\eta_g \underline{\omega_r}, \eta_g \overline{\omega_r}], [\underline{\theta}, \overline{\theta}]\},$$

$$u = P_{wt}^{ref}, \mathcal{U} = [\underline{P_{wt}}, \overline{P_{wt}}],$$

$$d = v_w, \mathcal{D} = [\underline{v_w}, \overline{v_w}],$$

$$y = [T_s, F_t]',$$

where θ is the pitch angle, $\overline{\theta}$ and $\underline{\theta}$ are the upper and lower limits of θ, ω_r is the rotor speed, $\overline{\omega_r}$ and $\underline{\omega_r}$ are the upper and lower limits of ω_r, ω_f is the filtered generator speed, η_g is the gearbox ratio, v_w is the wind speed, $\overline{v_w}$ and $\underline{v_w}$ are the cut-in and cut-out speeds, T_s is the shaft torque, F_t is the thrust force, and P_{wt}^{ref} is the power reference derived from the wind farm. More details regarding the formulation of model parameters A_d, B_d, C_d, D_d, E_d, F_d, G_d, and H_d are described in Chapter 6.

With the state measurement and the estimated wind speed, the parameters of the prediction model can be updated by searching the current operating region for each time step. It is assumed that the prediction model obtained is kept invariant during the prediction horizon. In this study, the wind speed v_{w} refers to "effective wind speed", which is used to describe the wind speed affecting the entire rotor.

Both the tracking performance and the minimization of the wind turbine load are considered in the cost function of the D-MPC design. It is assumed that the mean wind speed $\overline{v_{\mathrm{w}}}$ of a certain period (10 min in the paper by Spudić et al. [13]) can be estimated during the wind farm operation. Moreover, an initial distribution of individual wind turbine power references for this period is known. Therefore, the mean power reference for the ith wind turbine $\overline{P^{\mathrm{ref}}_{\mathrm{wt\_}i}}$ can be calculated by a simple proportional algorithm according to the available power:

$$\overline{P^{\mathrm{ref}}_{\mathrm{wt\_}i}} = \alpha_i P^{\mathrm{ref}}_{\mathrm{wf}}, \quad \sum_{i=1}^{n_{\mathrm{t}}} \alpha_i = 1, \tag{8.4}$$

where n_{t} is the wind turbine number in the wind farm, $P^{\mathrm{ref}}_{\mathrm{wf}}$ is the power reference for the wind farm, and α_i is the distribution factor for the ith wind turbine. Accordingly, other steady state variables, such as the shaft torque of the ith wind turbine $\overline{T_{\mathrm{s\_}i}}$, can be decided.

By defining $S_1 = [1, 0]$, $S_2 = [0, 1]$,

$$\begin{aligned}
&P^{\mathrm{ref}}_{\mathrm{wt_}i}(k) = u_i(k),\\
&T_{\mathrm{s_}i}(k) = S_1 \cdot y_i(k),\\
&F_{\mathrm{t}}^{\mathrm{WT}_i}(k) = S_2 \cdot y_i(k),\\
&u_i = [u_i(0), \ldots, u_i(n_{\mathrm{p}} - 1)]', u_i \in \mathbb{R}^{n_{\mathrm{p}} \times 1},\\
&u = [u_1', \ldots, u_{n_{\mathrm{t}}}']', u \in \mathbb{R}^{(n_{\mathrm{p}} n_{\mathrm{t}}) \times 1},
\end{aligned}$$

where n_{p} indicates the prediction horizon, and k indicates the prediction step index. The MPC problem at time t can be formulated as follows,

$$\min_u \sum_{i=1}^{n_{\mathrm{t}}} \left(\underbrace{\sum_{k=0}^{n_{\mathrm{p}}-1} \| u_i(k) - \overline{P^{\mathrm{ref}}_{\mathrm{wt\_}i}} \|^2_{Q_P}}_{\text{Term1}} + \underbrace{\sum_{k=0}^{n_{\mathrm{p}}-1} \| S_1 \cdot y_i(k) - \overline{T_{\mathrm{s\_}i}} \|^2_{Q_T}}_{\text{Term2}} + \underbrace{\sum_{k=0}^{n_{\mathrm{p}}-1} \| \Delta(S_2 \cdot y_i(k)) \|^2_{Q_F}}_{\text{Term3}} \right), \tag{8.5}$$

subject to:

$$x_i(k+1) = A_{\mathrm{d}} x_i(k) + B_{\mathrm{d}} u_k(t) + E_{\mathrm{d}} d_i(k) + F_{\mathrm{d}}, \tag{8.6}$$

$$y_i(k) = C_{\mathrm{d}} x_i(k) + D_{\mathrm{d}} u_i(k) + G_{\mathrm{d}} d_i(k) + H_{\mathrm{d}}, i \in [1, n_{\mathrm{t}}], k \in [0, n_{\mathrm{p}} - 1],$$

$$x_i \in \mathcal{X}_i, \tag{8.7}$$

$$u_i \in \mathcal{U}_i, \tag{8.8}$$

$$x_i(0) = x_i(t), \tag{8.9}$$

$$\sum_{i=1}^{n_t} u_i(0) = P_{\text{ref}}^{\text{wfc}}, \tag{8.10}$$

where Q_P, Q_T and Q_F are the weighting factors. Term 2 and Term 3 in the cost function are used to penalize the deviation of the shaft torque from the steady state and the derivative of the thrust force in order to reduce the wind turbine load. The initial value of the optimization variable, $u_i(0)$, is taken as the control input for each turbine. The summary of the coupled control inputs equals the power reference of the wind farm $P_{\text{wf}}^{\text{ref}}$.

8.2.4 Standard QP Problem

The formulated MPC problem (8.5) can be reformulated as a standard quadratic programming (QP) problem.

Single Wind Turbine

By substituting (8.3) into (8.2) recursively,

$$y = \underbrace{\begin{bmatrix} C_d \\ C_d A_d \\ \vdots \\ C_d A_d^{n_p-1} \end{bmatrix}}_{\Phi} x(0) + \underbrace{\begin{bmatrix} D_d & 0 & \cdots & 0 \\ C_d B_d & D_d & \cdots & 0 \\ \vdots & \vdots & \ddots & \vdots \\ C_d A_d^{n_p-2} B_d & C_d A_d^{n_p-3} B_d & \cdots & D_d \end{bmatrix}}_{\Gamma_u} u$$
$$+ \underbrace{\begin{bmatrix} G_d & 0 & \cdots & 0 \\ C_d E_d & G_d & \cdots & 0 \\ \vdots & \vdots & \ddots & \vdots \\ C_d A_d^{n_p-2} E_d & C_d A_d^{n_p-3} E_d & \cdots & G_d \end{bmatrix}}_{\Gamma_d} d$$
$$+ \underbrace{\begin{bmatrix} H_d & 0 & \cdots & 0 \\ C_d F_d & H_d & \cdots & 0 \\ \vdots & \vdots & \ddots & 0 \\ C_d A_d^{n_p-2} F_d & C_d A_d^{n_p-3} F_d & \cdots & H_d \end{bmatrix}}_{\Gamma_f}, \tag{8.11}$$

where Φ, Γ_u, Γ_d, Γ_f are the coefficient matrices of u, d, and the constant matrix, respectively. Since $y = [T_s, F_t]'$, let y_1, y_2 indicate T_s and F_t. Accordingly, (8.11) can be rewritten as,

$$y_1 = \Phi_1 x(0) + \Gamma_{1u} u + \Gamma_{1d} d + \Gamma_{1f}, \tag{8.12}$$

$$y_2 = \Phi_2 x(0) + \Gamma_{2u} u + \Gamma_{2d} d + \Gamma_{2f}. \tag{8.13}$$

Similar to 8.5, the cost function for a single wind turbine can be divided into three terms,

$$\underbrace{\sum_{k=0}^{n_p-1} \frac{1}{2} \| u(k) - r_0 \|_{Q_0}^2}_{\text{Part1}} + \underbrace{\sum_{k=0}^{n_p-1} \frac{1}{2} \| y_1(k) - r_1 \|_{Q_1}^2}_{\text{Part2}} + \underbrace{\sum_{k=1}^{n_p-1} \frac{1}{2} \| \Delta y_2(k) \|_{Q_2}^2}_{\text{Part3}}, \tag{8.14}$$

where r_0, r_1 represent the reference values $\overline{P_{\mathrm{wt}}^{\mathrm{ref}}}$, $\overline{T_{\mathrm{s}}}$; Q_0, Q_1, and Q_2 correspond to Q_P, Q_T, and Q_F.

Term 1 By defining $\Lambda = [1, \cdots, 1]', \Lambda \in \mathbb{R}^{n_{\mathrm{p}} \times 1}$ and identity matrix $I \in \mathbb{R}^{n_{\mathrm{p}} \times n_{\mathrm{p}}}$, Term 1 can be transformed into,

$$\sum_{k=0}^{n_{\mathrm{p}}-1} \frac{1}{2} \parallel u(k) - r_0 \parallel_{Q_0}^2 = \frac{1}{2} u' H_0 u + g_0' u + \rho_0, \tag{8.15}$$

with

$$b_0 = r_0 \Lambda, H_0 = I Q_0, g_0 = -I Q_0 b_0,$$

where $H_0 \in \mathbb{R}^{n_{\mathrm{p}} \times n_{\mathrm{p}}}$, $g_0 \in \mathbb{R}^{n_{\mathrm{p}} \times 1}$ are the Hessian and coefficient vector for Term 1, respectively. The constant part ρ_0 is ignored, as it does not affect the solution of the cost function. It is the same case for Terms 2 and 3.

Term 2 Similar to the calculation procedure for Term 1, Term 2 can be transformed into,

$$\sum_{k=0}^{n_{\mathrm{p}}-1} \frac{1}{2} \parallel y_1(k) - r_1 \parallel_{Q_1}^2 = \frac{1}{2} u' H_1 u + g_1' u + \rho_1, \tag{8.16}$$

with

$$b_1 = r_1 \Lambda - \Phi_1 x(0) - \Gamma_{1d} d \Lambda - \Gamma_{1f}, H_1 = \Gamma_{1u}' I Q_1 \Gamma_{1u}, g_1 = -\Gamma_{1u}' I Q_1 b_1,$$

where $H_1 \in \mathbb{R}^{n_{\mathrm{p}} \times n_{\mathrm{p}}}$, $g_1 \in \mathbb{R}^{n_{\mathrm{p}} \times 1}$ are the Hessian and coefficient vector for Term 2, respectively.

Term 3 With $\Delta y_2(k) = y_2(k) - y_2(k-1)$, the value of y_2 at the previous time step is marked as $y_2(-1)$. Then Term 3 can be transformed into,

$$\sum_{k=1}^{n_{\mathrm{p}}-1} \frac{1}{2} \parallel \Delta y_2(k) \parallel_{Q_2}^2 = \frac{1}{2} u' H_2 u + g_2' u + \rho_2, \tag{8.17}$$

with

$$M_1 = \begin{bmatrix} 1 & 0 & 0 & \cdots & 0 \\ -1 & 1 & 0 & \cdots & 0 \\ 0 & -1 & 1 & \cdots & 0 \\ \vdots & \vdots & \vdots & \ddots & \vdots \\ 0 & 0 & 0 & \cdots & 1 \end{bmatrix}, M_1 \in \mathbb{R}^{n_{\mathrm{p}} \times n_{\mathrm{p}}},\ M_2 = \begin{bmatrix} 1 \\ 0 \\ 0 \\ \vdots \\ 0 \end{bmatrix}, M_2 \in \mathbb{R}^{n_{\mathrm{p}} \times 1},$$

$$b_2 = -\Phi_2 x(0) - \Gamma_{2d} d \Lambda - \Gamma_{2f}, H_2 = \Gamma_{2u}' (M_1' I Q_3 M_1) \Gamma_{2u},$$
$$g_2 = -\Gamma_{2u}' (M_1' I Q_2 M_1) b_2 + \Gamma_{2u}' (M_2 Q_2 y_2(-1)),$$

where $H_2 \in \mathbb{R}^{n_{\mathrm{p}} \times n_{\mathrm{p}}}$, $g(2) \in \mathbb{R}^{n_{\mathrm{p}} \times 1}$ are the Hessian and coefficient vector for Term 3, respectively.

Based on (8.15)–(8.17), the equivalent standard QP format of the MPC problem for a single wind turbine can be written as:

$$\min_{u} \ \Psi = \left(\frac{1}{2}u'Hu + g'u\right), \tag{8.18}$$

with $H = H_0 + H_1 + H_2$, $g = g_0 + g_1 + g_2$.

Multiple Wind Turbines

For the multiple wind turbines case, the standard QP format of the MPC problem in (8.5) is derived from:

$$\min_{u} \Psi = \sum_{i=1}^{n_t} \Psi_i = \sum_{i=1}^{n_t} \left(\frac{1}{2}u_i'H_iu_i + g'u_i\right), \tag{8.19}$$

subject to

$$Gu = b, \tag{8.20}$$

$$u \in \mathcal{U}, \tag{8.21}$$

where $u = [u_1', ..., u_{n_t}']'$, $\mathcal{U}$ indicates the feasible region of u, and the equality constraint (8.20) is equivalent to (8.10). Accordingly, G and b can be derived as

$$G = [G_1, \cdots, G_{n_t}], G_i = [1, 0, \cdots, 0], G_i \in \mathbb{R}^{1 \times n_p},$$

$$b = P_{\text{ref}}^{\text{wfc}}.$$

8.2.5 Parallel Generalized Fast Dual Gradient Method

Properties of Dual Problem

In this section, the key properties required to apply the fast dual gradient method are described. It is obvious that the cost functions Ψ and Ψ_i are strongly convex with matrices H and H_i ($H = \text{diag}(H_1, ..., H_{n_t})$). By introducing the dual variable λ, the primal minimization problem (8.19) is transformed into the following Lagrange dual problem:

$$\sup_\lambda \inf_u \{\Psi(u) + \lambda(Gu - b)\} = \sup_\lambda \sum_{i=1}^{n_t} \left[\inf_{u_i} \{\Psi_i(u_i) + (\lambda G_i)u_i\} - \lambda\frac{b}{n_t}\right]. \tag{8.22}$$

With the following definitions of conjugate functions for Ψ and Ψ_i,

$$\Psi^\star(-G'\lambda) = \sup_u(-\lambda Gu - \Psi(u)), \Psi_i^\star = \sup_u(-\lambda Gu - \Psi_i(u_i)),$$

the dual problem in (8.22) can be rewritten as,

$$\sup_\lambda \{-\Psi^\star(-G'\lambda) - \lambda b\} = \sup_\lambda \sum_{i=1}^{n_t} \left\{-\Psi_i^\star(-G_i'\lambda) - \lambda\frac{b}{n_t}\right\}. \tag{8.23}$$

For simplicity, the following dual problem equations are defined,

$$d_i(\lambda) = -\Psi_i^\star(-G_i'\lambda) - \lambda\frac{b}{n_t}, \tag{8.24}$$

$$d(\lambda) = -\Psi^\star(-G'\lambda) - \lambda' b, d = \sum_{i=1}^{n_t} d_i. \tag{8.25}$$

Theorem 8.1 Assume that Ψ is strongly convex with H. The dual function d defined in (8.25) is concave, differentiable, and satisfies,

$$d(\lambda_1) \geq d(\lambda_2) + \nabla d(\lambda_2)(\lambda_1 - \lambda_2) - \frac{1}{2} \parallel \lambda_1 - \lambda_2 \parallel_{\mathbf{L}}^2 \tag{8.26}$$

for every λ_1, λ_2 and $\mathbf{L}$ with $\mathbf{L} \succeq GH^{-1}G'$. The proof has been presented by Giselsson [11].

Corollary 8.1 According to Theorem 1, assume that Ψ_i is strongly convex with matrix H_i, and the local dual function d_i defined in (8.24) is concave, differentiable, and satisfies,

$$d_i(\lambda_1) \geq d_i(\lambda_2) + \nabla d_i(\lambda_2)(\lambda_1 - \lambda_2) - \frac{1}{2} \parallel \lambda_1 - \lambda_2 \parallel_{\mathbf{L}_i}^2 \tag{8.27}$$

for every λ_1, λ_2 and $\mathbf{L}_i$ with $\mathbf{L}_i \succeq G_i H_i^{-1} G_i'$. The proof has been presented by Giselsson [11].

Theorem 1 and Corollary 1 imply a tighter quadratic lower bound to the dual function. It can be further proved that the obtained bound is the best obtained bound. Therefore, the approximation of the dual function is more accurate, which can improve the converge rate. In the following section, a generalized parallel optimization algorithm for the D-MPC is described.

Distributed Optimization Algorithm

The fast dual gradient method is implemented in parallel for wind farm control. Conventionally, the iteration terminates once the predefined tolerance is met. In this study, a fixed number of iterations $l_{\max}$ is selected as the stopping criterion in order to limit the online computation time. Dual variables λ, η, and φ are introduced.

Parallel generalized fast dual gradient method

Require: Initial guesses $\lambda^{[1]} = \eta^{[0]}, \varphi^{[1]} = 1$.

for: $l = 1, ..., l_{\max}$, **do**

1) Send $\lambda^{[l]}$ to wind turbines through communication from Central Unit to D-MPC.
2) Update and solve the local optimization with augmented cost function in individual D-MPC: $u_i^{[l]} = \arg\min_{u_i}\{\Psi_i + u_i' G' \lambda^{[l]}\}$.
3) Check the current operating region. If it changes, update $\mathbf{L}_i^{-1}$ in individual D-MPC.
4) Receive $u_i^{[l]}$ from each turbine and form $u^{[l]} = (u_1^{[l]}, ..., u_{n_t}^{[l]})$ in central unit.
5) Receive the updated $\mathbf{L}_i^{-1}$ in central unit.
6) Update $\mathbf{L}^{-1}$ according to $\mathbf{L}_i^{-1}$ and the dual variables in central unit:

$$\eta^{[l]} = \lambda^{[l]} + \mathbf{L}^{-1}(Gu^{[l]} - b)$$

$$\varphi^{[l+1]} = \frac{1 + \sqrt{1 + 4(\varphi^{[l]})^2}}{2}$$

$$\lambda^{[l+1]} = \eta^{[l]} + \left(\frac{\varphi^{[l]} - 1}{\varphi^{[l+1]}}\right)(\eta^{[l]} - \eta^{[l-1]})$$

end for

Proposition If the assumption in Theorem 1 holds and $\mathbf{L} \succeq GH^{-1}G'$, the algorithm converges at the rate,

$$d(\lambda^\star) - d(\lambda^l) \leq \frac{2 \parallel \lambda^\star - \lambda_2 \parallel_{\mathbf{L}}^2}{(l+1)^2}, \forall l \geq 1 \tag{8.28}$$

where l is the iteration number index. More details of the proof have been presented by Giselsson [11]. Compared with the standard gradient method, the convergence rate is improved, from $\mathcal{O}(1/k)$ to $\mathcal{O}(1/k^2)$, with a negligible increase in iteration complexity. $\mathbf{L} = GH^{-1}G'$ has the tightest upper bounds and is used in this study.

Due to the correlation of all the turbines, $\mathbf{L}^{-1}$ can be calculated as:

$$\mathbf{L}^{-1} = \sum_{i=1}^{n_t} \mathbf{L}_i^{-1} = \sum_{i=1}^{n_t} (G_i H_i^{-1} G_i)^{-1}. \tag{8.29}$$

It should be noticed that the model parameters of the individual turbine varies with changes of the operating region. Accordingly, the Hessian matrix H_i is also time-variant, which leads to the variation of $\mathbf{L}_i^{-1}$. In order to reduce the online computational complexity, the variables involved in the computation, including H_i and $\mathbf{L}_i^{-1}$, can be precomputed offline and stored according to the operating regions.

The computational burden of the central unit only includes the calculation of $\mathbf{L}^{-1}$ in (8.29) and the updates of the dual variable during iterations. Most computational tasks are distributed to the local D-MPCs. In addition, the communication burden between the D-MPCs and the central unit has been greatly reduced, due to the reduced number of iterations. Therefore, the proposed control structure is independent from the wind farm scale and appropriate for modern wind farm control applications.

8.3 Wind Farm Equipped with Energy Storage

Due to their flexible charging and discharging characteristics, energy storage system (ESSs) are considered effective tools to enhance the flexibility and controllability of wind farms. The ESS-type selection for a wind farm is dependent on the control purposes. Fast and short-term ESS can be used to improve the power quality and provide frequency control support, while slow and long-term ESS is more suitable for economic dispatch [1].

In this study, fast and short-term ESS units, such as batteries or supercapacitors, are used in a wind farm for real-time operational control. Besides the power quality improvement, ESSs have the potential to contribute to the alleviation of mechanical loads: in the coordinated control strategy described in Section 8.2, the power references distributed to individual wind turbines ($P^{\text{ref}}_{\text{wt\_}i}, i \in [1, n_t]$) are limited by the constraint in (8.10). With an additional ESS, the constraint can be relaxed to some extent. Accordingly, regulation is more flexible, which leads to better load reduction.

8.3.1 Wind Farm Control Structure

The control structure of wind farm with ESS is illustrated in Figure 8.2. Compared with the case without ESS (Figure 8.1), the additional ESS part is equipped with local D-MPC integrated into the wind farm controller. The power reference $P^{\text{ref}}_{\text{ess}}$ consists of two parts: $P^{\text{com}}_{\text{ess}}$ and $P^{\text{mpc}}_{\text{ess}}$, which correspond to different controllers:

- $P^{\text{mpc}}_{\text{ess}}$ is derived by the D-MPC, which aims to minimize the mechanical loads of wind turbines
- $P^{\text{com}}_{\text{ess}}$ is derived by calculating the mismatch between $P^{\text{ref}}_{\text{wf}}$ and $P^{\text{meas}}_{\text{wf}}$; that is, $P^{\text{com}}_{\text{ess}} = P^{\text{ref}}_{\text{wf}} - P^{\text{meas}}_{\text{wf}}$.

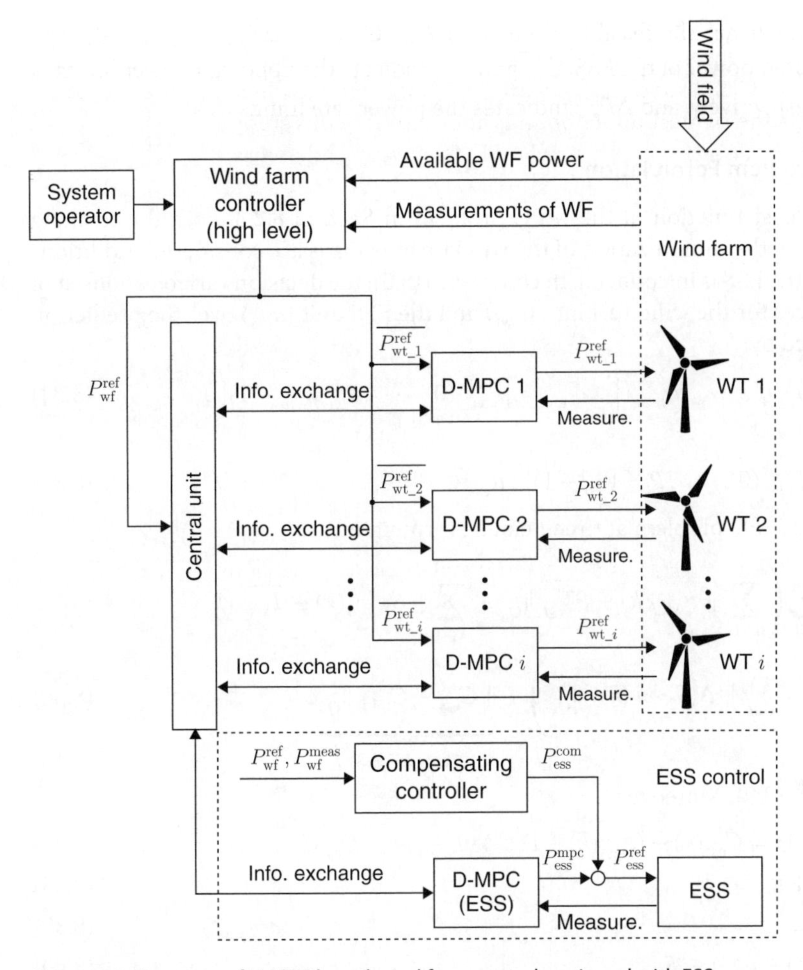

Figure 8.2 Structure of D-MPC based wind farm control equipped with ESS.

8.3.2 Modelling of ESS Unit

The ESS unit is modelled by an integrator, so as to describe its state of charge (SOC) in discrete time. This is expressed by:

$$C_{ess}(k+1) = C_{ess}(k) - P_{ess}^{ref} \cdot t_s - \eta_{loss} \cdot C_{ess}(k), \tag{8.30}$$

where t_s is the sampling time, C_{ess} denotes the SOC, and P_{ess}^{ref} denotes the discharging power. A loss term, modelled as $\eta_{loss} \times C_{ess}(k)$, is considered. η_{loss} is the loss coefficient, and is considered invariant in this study [15]. The constraints are as follows:

$$P_{ess}^{ref} \in \mathcal{P} = \left[-P_{ess}^{lim}, P_{ess}^{lim}\right], \tag{8.31}$$

$$C_{ess} \in \mathcal{C} = [\underline{C_{ess}}, \overline{C_{ess}}], \tag{8.32}$$

$$\Delta P_{ess}^{ref} \in \mathcal{R} = [-\Delta P_{ess}^{lim}, \Delta P_{ess}^{lim}], \tag{8.33}$$

where $\mathcal{P}$, $\mathcal{C}$, and $\mathcal{R}$ are the feasible regions for $P_{\text{ess}}^{\text{ref}}$, C_{ess}, and $\Delta P_{\text{ess}}^{\text{ref}}$, respectively, $P_{\text{ess}}^{\text{lim}}$ indicates the rated power of the ESS, $\overline{C_{\text{ess}}}$ and $\underline{C_{\text{ess}}}$ indicate the upper and lower limits of the capacity, respectively, and $\Delta P_{\text{ess}}^{\text{lim}}$ indicates the power rate limit.

8.3.3 MPC Problem Formulation

Similar to the cost function of the MPC problem in Section 8.2.3, both the tracking performance and the minimization of the wind turbine loads are considered. Additionally, a term for the ESS is introduced. In contrast to (8.5), the decision variables consist of power references for the wind turbines (u_{wt}) and the ESS unit (u_{ess}) over the prediction horizon, defined by:

$$u_{\text{wt}} = \left[u_{\text{wt\_1}}{}', \cdots, u_{\text{wt\_}n_{\text{t}}}{}'\right]', u_{\text{wt\_}i} = \left[P_{\text{wt\_}i}^{\text{ref}}(0), \cdots, P_{\text{wt\_}i}^{\text{ref}}(n_{\text{p}} - 1)\right]', \tag{8.34}$$
$$u_{\text{wt}} \in \mathbb{R}^{(n_{\text{p}} \cdot n_{\text{t}}) \times 1},$$

$$u_{\text{ess}} = \left[P_{\text{ess}}^{\text{mpc}}(0), \cdots, P_{\text{ess}}^{\text{mpc}}(n_{\text{p}} - 1)\right]', u_{\text{ess}} \in \mathbb{R}^{n_{\text{p}} \times 1}. \tag{8.35}$$

The modified MPC problem at time t can be formulated as:

$$\min_{u_{\text{wt}}, u_{\text{ess}}} \sum_{i=1}^{n_{\text{t}}} \left(\sum_{k=0}^{n_{\text{p}}-1} \| u_{\text{wt\_}i}(k) - \overline{P_{\text{wt\_}i}^{\text{ref}}} \|_{Q_P}^2 + \sum_{k=0}^{n_{\text{p}}-1} \| S_1 \cdot y_i(k) - \overline{T_{\text{s\_}i}} \|_{Q_T}^2 + \sum_{k=0}^{n_{\text{p}}-1} \| \Delta(S_2 \cdot y_i(k)) \|_{Q_F}^2 \right) + \underbrace{\| C_{\text{ess}} - C_{\text{mid}} \|_{Q_C}^2}_{\text{Term for ESS}}, \tag{8.36}$$

subject to (8.6)–(8.9). Moreover:

$$C_{\text{ess}}(k+1) = C_{\text{ess}}(k) - (u_{\text{ess}}(k) + P_{\text{ess}}^{\text{com}}) \cdot t_{\text{s}} - \eta_{\text{loss}} \cdot C_{\text{ess}}(k), \tag{8.37}$$
$$P_{\text{ess}}^{\text{com}}(t) + u_{\text{ess}} \in \mathcal{P}, \tag{8.38}$$
$$C_{\text{ess}} \in \mathcal{C}, \tag{8.39}$$
$$\Delta(P_{\text{ess}}^{\text{com}}(t) + u_{\text{ess}}) \in \mathcal{R}, \tag{8.40}$$
$$P_{\text{ess}}^{\text{com}} + u_{\text{ess}}(0) + \sum_{i=1}^{n_{\text{t}}} u_{\text{wt\_}i}(0) = P_{\text{wf}}^{\text{ref}}. \tag{8.41}$$

In order to facilitate the long-term and stable operation of the ESS unit, the SOC should be limited to within a range close to the medium SOC level C_{mid} as much as possible, where $C_{\text{mid}} = \frac{1}{2}(\underline{C_{\text{ess}}} + \overline{C_{\text{ess}}})$. The additional ESS term in (8.36) is used to prevent the charging level from being too low or too high. Q_C is the weighting factor.

As the decision variables (u_{wt}, u_{ess}), the first values are taken as the control inputs for each turbine and ESS unit: $P_{\text{wt}}^{\text{ref}}(t) = u_{\text{wt}}(0)$ and $P_{\text{ess}}^{\text{ref}}(t) = u_{\text{ess}}(0)$. The control inputs are coupled, and their sum equals $P_{\text{wf}}^{\text{ref}}$ (see (8.41)).

By substituting (8.41) into (8.38),

$$P_{\text{wf}}^{\text{ref}} - P_{\text{ess}}^{\text{lim}} \leq \sum_{i=1}^{n_{\text{t}}} u_{\text{wt\_}i}(0) \leq P_{\text{wf}}^{\text{ref}} + P_{\text{ess}}^{\text{lim}}. \tag{8.42}$$

Compared to 8.10, the control inputs of the wind turbines are relaxed and have more regulation flexibility. Accordingly, a better load minimization performance can be expected with use of an ESS unit. The extent to which load minimisation can be improved depends on the technical features of the ESS unit, including its response time, and its power and energy ratings.

8.4 Case Study

8.4.1 Wind Farm Control based on D-MPC without ESS

A wind farm comprising ten 5-MW wind turbines is used as the test system. The wind field modeling for the wind farm is generated from SimWindFarm [16], a MATLAB toolbox for dynamic wind farm modeling, simulation and control. The mean wind speed for each wind turbine is assumed to be known. All the wind turbines are in derated operation. The sampling time of the wind farm control is set at $t_s = 1$s. The prediction horizon for MPC is set at $n_p = 10$. The wind speed is considered to be a measurable disturbance and the value for the prediction horizon is predicted using a persistence assumption.

Different scenarios are designed to test the efficacy of the D-MPC. Firstly, the convergence of the fast dual gradient method is shown. A suitable maximum iteration number l_{max} is decided for the following simulation cases. Secondly, the operation of the wind farm under both high- and low-wind conditions is analyzed.

Convergence with the Fast Dual Gradient Method

Figure 8.3 illustrates the convergence of the fast dual gradient with different **L**. The y-axis denotes the deviation from the constraints (see (8.10)).

Apparently, the convergence rate with $\mathbf{L} \geq GH^{-1}G'$ is higher. When $\mathbf{L} = GH^{-1}G'$, which has the tightest upper bounds, the convergence is fastest. Only five iterations guarantees good performance of the D-MPC approach. Therefore, the maximum iteration number is selected to be $l_{max} = 5$ for the following simulations.

Operation Under High- and Low-wind Conditions

The operation of the wind farm is simulated under both high- and low-wind conditions. Accordingly, the power references of the wind farm P_{wf}^{ref} are defined as 40 MW

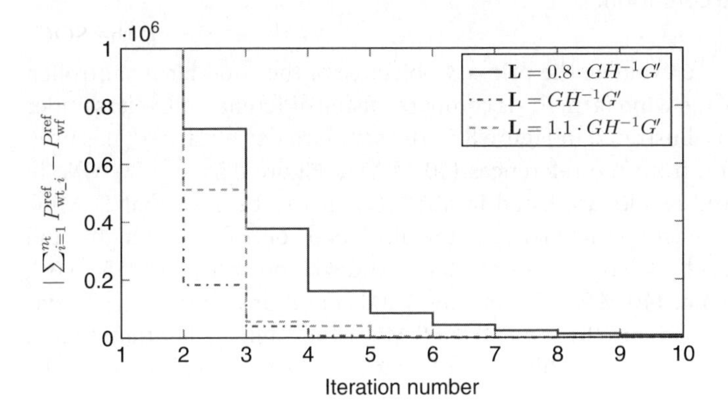

Figure 8.3 Convergence comparison with different **L**.

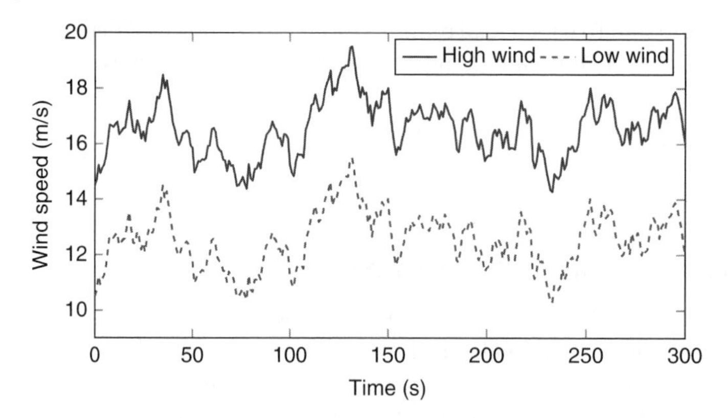

Figure 8.4 Wind speed variation of Wind Turbine 03.

and 30 MW, respectively. $P_{\mathrm{wf}}^{\mathrm{ref}}$ is kept constant during the simulation. The wind profiles of the high-speed case are derived by shifting low-speed wind profile upwards by 4 m/s. The simulation time is 300 s.

Wind Turbine 03 is taken as an example to illustrate the simulation results. Its wind profiles for both high- and low-wind conditions are shown in Figure 8.4. This covers the range between 11 and 20 m/s. The average wind speed values v_{avr} of all the wind turbines for both conditions are listed in Tables 8.3 and 8.5, respectively.

The weighting factors in (8.5) are set as $Q_P = 1$, $Q_T = 20$, and $Q_F = 5$. To avoid violent control and shaft load increases, Q_F is kept small in this study.

The control performances of three controllers are illustrated and compared:

- Centralized control
- C-MPC
- D-MPC.

For the centralized wind farm controller, the proportional distribution algorithm proposed of Sørensen et al. is applied [2]. The power reference for each wind turbine is the same: $P_{\mathrm{wt\_}i}^{\mathrm{ref}} = \frac{P_{\mathrm{wf}}^{\mathrm{ref}}}{n_t}$. Accordingly, $P_{\mathrm{wt\_}i}^{\mathrm{ref}} = 4$ MW for the high-wind condition and $P_{\mathrm{wt\_}i}^{\mathrm{ref}} =$ 3 MW for the low-wind condition.

Power reference tracking To evaluate the primary objective of the wind farm controller, the power generation of the wind farm P_{wf} is compared using different controllers under both wind conditions, as illustrated in Figure 8.5. The standard deviation $\sigma(P_{\mathrm{wf}})$ is used to quantify the deviation from the references (40 MW in Figure 8.5a and 30 MW in Figure 8.5b). The derived results are listed in Table 8.1. It can be seen that D-MPC shows almost identical control performance to the other two controllers. In high-wind conditions, all the controllers have the same standard deviation value (0.0085 MW), only 0.021% of the reference (40 MW). For the low-wind conditions, the standard deviations of C-MPC and D-MPC are the same (0.0058 MW) and a bit smaller than that of centralized controller (0.0059 MW). This tiny difference can be ignored, relative to the reference value (30 MW).

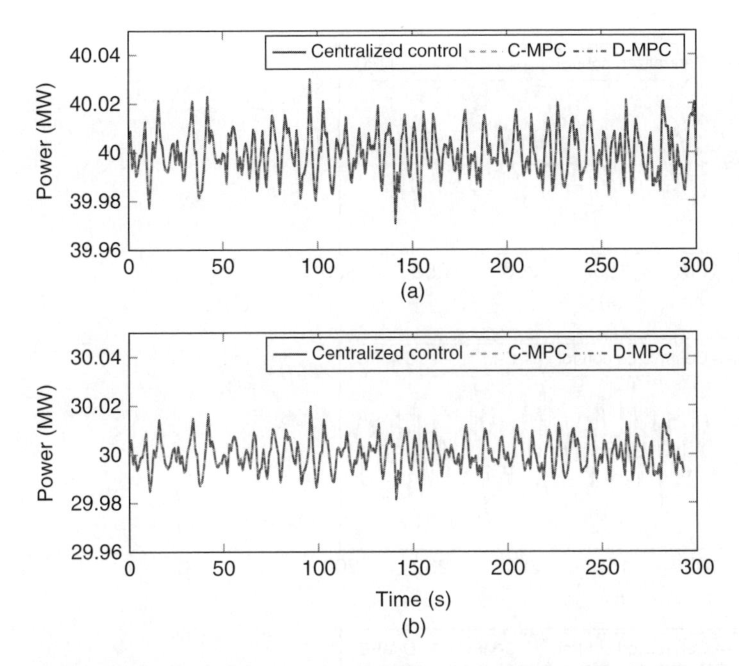

Figure 8.5 Power generation of the wind farm: (a) in high winds; (b) in low winds.

Table 8.1 Simulation statistics $\sigma(P_{wf})$

Wind condition	Centralized control (MW)	C-MPC (MW)	D-MPC (MW)
High wind	0.0085	0.0085 (0.00%)	0.0085 (0.00%)
Low wind	0.0059	0.0058 (−1.69%)	0.0058 (−1.69%)

Load alleviation of wind turbines It is unnecessary to illustrate the simulation results of all the wind turbines. As an example, T_s and F_t of a single wind turbine (WT 03) with different controllers are illustrated in Figure 8.6 (for high winds) and Figure 8.7 (for low winds). The power reference by MPC (C-MPC and D-MPC) varies following the wind speed (Figure 8.6a and Figure 8.7a, respectively). Accordingly, the deviation of T_s is significantly reduced (Figure 8.6b and Figure 8.7b). The alleviation of F_t is not obvious (Figure 8.6c and Figure 8.7c). Moreover, it can be observed that the control inputs of both C-MPC and D-MPC are almost identical, which implies that D-MPC has the same control performance as the C-MPC.

The standard deviations ($\sigma(T_s)$ and $\sigma(\Delta F_t)$) are used to quantify the shaft and thrust-induced loads, respectively. The results of all wind turbines are listed in Tables 8.2–8.5.

According to the results in Tables 8.2 and 8.4, compared with centralized control, the shaft torque deviation for each wind turbine is significantly reduced with D-MPC. By taking $\sigma(T_s)$ in centralized control as the reference, for the high-wind case, the reduction

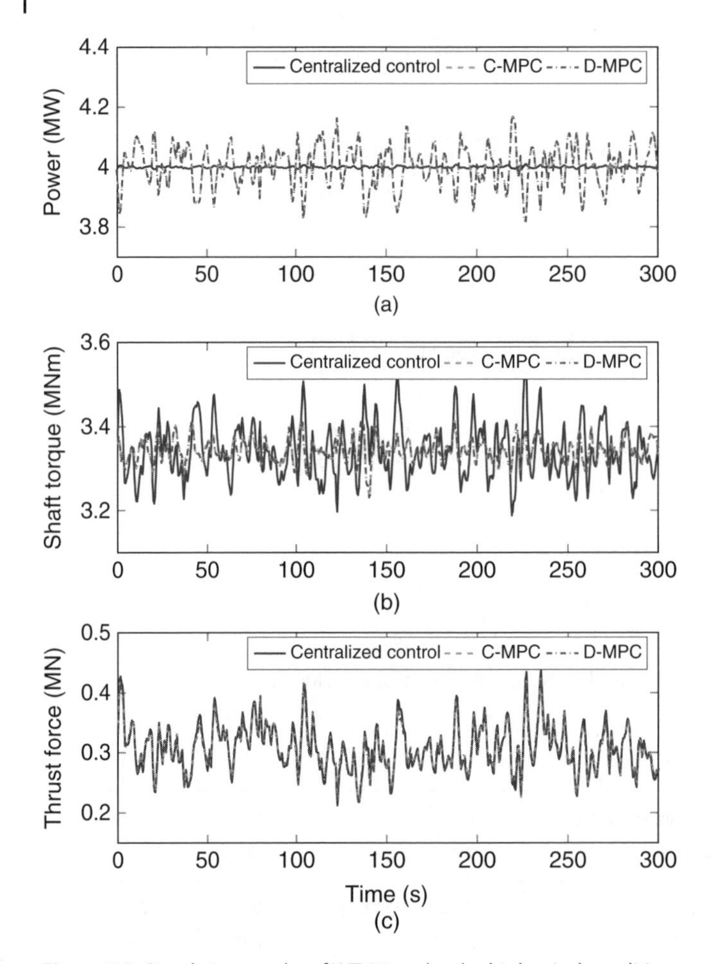

Figure 8.6 Simulation results of WT 03 under the high-wind condition.

percentages of $\sigma(\Delta T_s)$ range from 46.77% to 61.36% (Table 8.2). For the low-wind case, the reduction percentages are between 42.08% and 63.35% (Table 8.4).

The thrust force change is also damped to some extent in each wind turbine with the D-MPC, according to Tables 8.3 and 8.5. For the high-wind case, the reduction percentages of $\sigma(\Delta F_t)$ are between 0.00% and 3.43% (Table 8.3). For the low-wind case, the values are between 0.38% and 4.82% (Table 8.5).

8.4.2 Wind Farm Control based on D-MPC with ESS

In this study, a 800-kW/3-kWh ESS (say, as supercapacitor) is installed in the wind farm described in Section 8.4.1. The power rate limit of the ESS is set as $\Delta P_{ess}^{ref} = 300$ kW/s.

The operation of the wind farm is simulated under both high- and low-wind conditions. The power references of the wind farm P_{wf}^{ref} are defined as 40 MW and 30 MW, and these are kept constant during the simulation. The wind profiles of the high-speed case are derived by shifting the low-speed wind profile upwards by 4 m/s. The simulation time is 300 s.

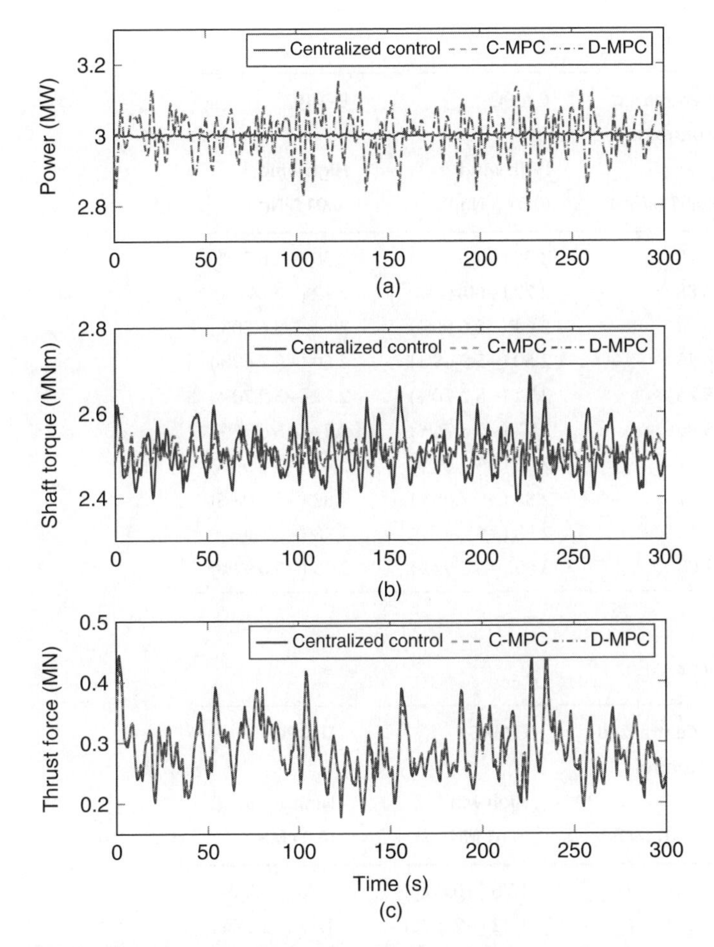

Figure 8.7 Simulation results of WT 03 under the low-wind condition.

The control performances of three controllers are illustrated and compared:

Centralized Control
No ESS is installed. The references $P^{ref}_{wt\_i}$ are calculated based on the proportional distribution algorithm: $P^{ref}_{wt\_i} = 4$ MW for the high-wind condition, and $P^{ref}_{wt\_i} = 3$ MW for the low-wind condition.

D-MPC
No ESS is installed. The references $P^{ref}_{wt\_i}$ are calculated based on the proposed D-MPC algorithm (Section 8.2.3).

D-MPC-ESS
ESS is installed and integrated into the D-MPC-based wind farm controller. The references $P^{ref}_{wt\_i}$ and P^{ref}_{ess} are calculated based on the proposed D-MPC algorithm (Section 8.3.3).

Table 8.2 Simulation statistics $\sigma(T_s)$.

Turbine	v_{avr}	Centralized control	C-MPC	D-MPC
			High wind	High wind
	(m/s)	(0.01 MNm)	(0.01 MNm)	(0.01 MNm)
WT 01	17.78	6.20	3.30 (−4 6.77%)	3.30 (−4 6.77%)
WT 02	17.11	6.88	2.72 (−60.47%)	2.72 (−60.47%)
WT 03	16.58	6.00	2.64 (−5 6.00%)	2.64 (−5 6.00%)
WT 04	16.68	6.45	2.60 (−5 9.69%)	2.60 (−5 9.69%)
WT 05	17.62	5.75	2.72 (−5 2.70%)	2.72 (−5 2.70%)
WT 06	17.08	5.99	2.77 (−5 3.76%)	2.77 (−5 3.76%)
WT 07	16.59	5.71	2.45 (−5 7.09%)	2.45 (−5 7.09%)
WT 08	17.41	6.71	2.88 (−5 7.08%)	2.88 (−5 7.08%)
WT 09	17.28	6.16	2.38 (−61.36%)	2.38 (−61.36%)
WT 10	16.33	7.08	3.05 (−5 6.92%)	3.05 (−5 6.92%)

Table 8.3 Simulation statistics $\sigma(\Delta F_t)$.

Turbine	v_{avr}	Centralized control	C-MPC	D-MPC
			High wind	High wind
	(m/s)	(0.01 MN)	(0.01 MN)	(0.01 MN)
WT 01	17.78	1.76	1.76 (−0.00%)	1.76 (−0.00%)
WT 02	17.11	1.76	1.72 (−2.27%)	1.72 (−2.27%)
WT 03	16.58	1.66	1.65 (−0.60%)	1.65 (−0.60%)
WT 04	16.68	1.75	1.69 (−3.43%)	1.69 (−3.43%)
WT 05	17.62	1.56	1.53 (−1.92%)	1.53 (−1.92%)
WT 06	17.08	1.61	1.59 (−1.24%)	1.59 (−1.24%)
WT 07	16.59	1.70	1.67 (−1.76%)	1.67 (−1.76%)
WT 08	17.41	1.86	1.83 (−1.61%)	1.83 (−1.61%)
WT 09	17.28	1.53	1.48 (−3.27%)	1.48 (−3.27%)
WT 10	16.33	2.04	2.00 (−1.96%)	2.00 (−1.96%)

Since the mismatch between P_{wf}^{ref} and P_{wf}^{meas} can be well compensated by the ESS unit through the additional power command P_{ess}^{com}, the power reference tracking performance is not illustrated.

Load Alleviation of Wind Turbines

$\sigma(T_s)$ and $\sigma(\Delta F_t)$ are used to quantify the shaft and thrust-induced loads, respectively. The results for all the wind turbines are listed in Tables 8.6–8.9.

Table 8.4 Simulation statistics $\sigma(T_s)$.

Turbine	v_{avr}	Centralized control	C-MPC	D-MPC
			High wind	High wind
	(m/s)	(0.01 MNm)	(0.01 MNm)	(0.01 MNm)
WT 01	13.78	5.16	2.12 (−5 8.91%)	2.12 (−5 8.91%)
WT 02	13.11	5.72	2.18 (−61.89%)	2.18 (−61.89%)
WT 03	12.58	4.82	2.33 (−5 1.66%)	2.33 (−5 1.66%)
WT 04	12.68	6.06	3.51 (−4 2.08%)	3.51 (−4 2.08%)
WT 05	13.62	4.80	1.79 (−62.71%)	1.79 (−62.71%)
WT 06	13.08	5.00	1.85 (−63.00%)	1.85 (−63.00%)
WT 07	12.59	4.70	2.37 (−4 9.57%)	2.37 (−4 9.57%)
WT 08	13.41	5.62	2.06 (−63.35%)	2.06 (−63.35%)
WT 09	13.28	4.98	2.63 (−4 7.19%)	2.63 (−4 7.19%)
WT 10	12.33	6.48	3.56 (−4 5.06%)	3.56 (−4 5.06%)

Table 8.5 Simulation statistics $\sigma(\Delta F_t)$.

Turbine	v_{avr}	Centralized control	C-MPC	D-MPC
			High wind	High wind
	(m/s)	(0.01 MN)	(0.01 MN)	(0.01 MN)
WT 01	13.78	1.83	1.82 (−0.55%)	1.82 (−0.55%)
WT 02	13.11	1.88	1.82 (−3.19%)	1.82 (−3.19%)
WT 03	12.58	1.80	1.74 (−3.33%)	1.74 (−3.33%)
WT 04	12.68	2.56	2.51 (−1.95%)	2.51 (−1.95%)
WT 05	13.62	1.63	1.60 (−1.84%)	1.60 (−1.84%)
WT 06	13.08	1.71	1.67 (−2.34%)	1.67 (−2.34%)
WT 07	12.59	1.79	1.76 (−1.68%)	1.76 (−1.68%)
WT 08	13.41	1.95	1.90 (−2.56%)	1.90 (−2.56%)
WT 09	13.28	1.66	1.58 (−4.82%)	1.58 (−4.82%)
WT 10	12.33	2.66	2.65 (−0.38%)	2.65 (−0.38%)

According to the results in Tables 8.6 and 8.8, compared with centralized control, $\sigma(T_s)$ is greatly reduced with the D-MPC. By taking $\sigma(T_s)$ in centralized control as the reference, the reduction percentages range from 46.77% to 61.36% under the high-wind condition (Table 8.6), and from 42.08% to 63.35% under the low-wind condition (Table 8.8). On the basis of this, the additional ESS further increases the load reduction of individual wind turbines. With D-MPC-ESS, the reduction percentages range from 53.55% to 74.35% under the high-wind condition (Table 8.6), and from 48.18% to

Table 8.6 Simulation statistics $\sigma(T_s)$.

Turbine	v_{avr}	Centralized control	D-MPC (without ESS)	D-MPC (with ESS)
			High wind	High wind
	(m/s)	(0.01 MNm)	(0.01 MNm)	(0.01 MNm)
WT 01	17.78	6.20	3.30 (−4 6.77%)	2.88 (−5 3.55%)
WT 02	17.11	6.88	2.72 (−60.47%)	2.15 (−68.75%)
WT 03	16.58	6.00	2.64 (−5 6.00%)	1.66 (−72.33%)
WT 04	16.68	6.45	2.60 (−5 9.69%)	1.83 (−71.63%)
WT 05	17.62	5.75	2.72 (−5 2.70%)	2.44 (−5 7.57%)
WT 06	17.08	5.99	2.77 (−5 3.76%)	2.42 (−5 9.60%)
WT 07	16.59	5.71	2.45 (−5 7.09%)	1.51 (−73.56%)
WT 08	17.41	6.71	2.88 (−5 7.08%)	2.50 (−62.74%)
WT 09	17.28	6.16	2.38 (−61.36%)	1.58 (−74.35%)
WT 10	16.33	7.08	3.05 (−5 6.92%)	2.55 (−63.98%)

Table 8.7 Simulation statistics $\sigma(\Delta F_t)$.

Turbine	v_{avr}	Centralized control	D-MPC (without ESS)	D-MPC (with ESS)
			High wind	High wind
	(m/s)	(0.01 MN)	(0.01 MN)	(0.01 MN)
WT 01	17.78	1.76	1.76 (−0.00%)	1.76 (−0.00%)
WT 02	17.11	1.76	1.72 (−2.27%)	1.70 (−3.41%)
WT 03	16.58	1.66	1.65 (−0.60%)	1.63 (−1.81%)
WT 04	16.68	1.75	1.69 (−3.43%)	1.68 (−4.00%)
WT 05	17.62	1.56	1.53 (−1.92%)	1.54 (−1.28%)
WT 06	17.08	1.61	1.59 (−1.24%)	1.58 (−1.86%)
WT 07	16.59	1.70	1.67 (−1.76%)	1.66 (−2.35%)
WT 08	17.41	1.86	1.83 (−1.61%)	1.82 (−2.15%)
WT 09	17.28	1.53	1.48 (−3.27%)	1.48 (−3.27%)
WT 10	16.33	2.04	2.00 (−1.96%)	2.00 (−1.96%)

73.49% under the low-wind condition (Table 8.8). These results verify the expectation created in Section 8.3.3.

According to the results in Tables 8.7 and 8.9, $\sigma(F_t)$ is also reduced to some extent with D-MPC, as compared to centralized control. By taking $\sigma(F_t)$ in centralized control as the reference, the reduction percentages range from 0.00% to 3.43% under the high-wind condition (Table 8.7), and from 0.38% to 4.82% under the low-wind condition (Table 8.9). On the basis of this, the additional ESS further increases the load reduction. With D-MPC-ESS, the reduction percentages range from 0.00% to 4.00% under the

Table 8.8 Simulation statistics $\sigma(T_s)$.

Turbine	v_{avr}	Centralized control	D-MPC (without ESS)	D-MPC (with ESS)
			High wind	High wind
	(m/s)	(0.01 MNm)	(0.01 MNm)	(0.01 MNm)
WT 01	13.78	5.16	2.12 (−58.91%)	1.49 (−71.12%)
WT 02	13.11	5.72	2.18 (−61.89%)	1.62 (−71.68%)
WT 03	12.58	4.82	2.33 (−51.66%)	2.10 (−56.43%)
WT 04	12.68	6.06	3.51 (−42.08%)	3.14 (−48.18%)
WT 05	13.62	4.80	1.79 (−62.71%)	1.17 (−75.63%)
WT 06	13.08	5.00	1.85 (−63.00%)	1.38 (−72.40%)
WT 07	12.59	4.70	2.37 (−49.57%)	1.81 (−61.49%)
WT 08	13.41	5.62	2.06 (−63.35%)	1.49 (−73.49%)
WT 09	13.28	4.98	2.63 (−47.19%)	2.13 (−57.23%)
WT 10	12.33	6.48	3.56 (−45.06%)	3.19 (−50.77%)

Table 8.9 Simulation statistics $\sigma(\Delta F_t)$.

Turbine	v_{avr}	Centralized control	D-MPC (without ESS)	D-MPC (with ESS)
			High wind	High wind
	(m/s)	(0.01 MN)	(0.01 MN)	(0.01 MN)
WT 01	13.78	1.83	1.82 (−0.55%)	1.81 (−1.09%)
WT 02	13.11	1.88	1.82 (−3.19%)	1.81 (−3.72%)
WT 03	12.58	1.80	1.74 (−3.33%)	1.72 (−4.44%)
WT 04	12.68	2.56	2.51 (−1.95%)	2.50 (−2.34%)
WT 05	13.62	1.63	1.60 (−1.84%)	1.60 (−1.84%)
WT 06	13.08	1.71	1.67 (−2.34%)	1.67 (−2.34%)
WT 07	12.59	1.79	1.76 (−1.68%)	1.76 (−1.68%)
WT 08	13.41	1.95	1.90 (−2.56%)	1.89 (−3.08%)
WT 09	13.28	1.66	1.58 (−4.82%)	1.57 (−5.42%)
WT 10	12.33	2.66	2.65 (−0.38%)	2.58 (−3.01%)

high-wind condition (Table 8.7), and from 1.09% to 5.42% under the low-wind condition (Table 8.9).

8.5 Conclusion

In this chapter, the D-MPC algorithm using the fast dual gradient method is proposed for the active power control of a wind farm.

Compared with conventional centralized wind farm control, the D-MPC strikes a balance between the power reference tracking and the minimization of the wind turbine loads. In contrast to C-MPC, in D-MPC, most of computation tasks are distributed to local D-MPCs at each actuator (wind turbine or ESS). The computational burden of the central unit is therefore significantly reduced. This control structure is independent of the scale of the wind farm. Moreover, with a properly calculated Lipschitz constant **L**, the fast dual gradient method can significantly improve the convergence rate, from $\mathcal{O}(1/k)$ to $\mathcal{O}(1/k^2)$, which reduces the communication burden between the local D-MPCs and central unit.

Through different case studies, the power tracking control performance of the developed D-MPC is shown to be identical to C-MPC. The mechanical loads experienced by individual wind turbines are largely alleviated without affecting the tracking of the power reference of the wind farm. D-MPC is suitable for real-time control of modern wind farms.

The proposed D-MPC is implemented for two wind farm configurations: with and without ESS. It is expected that load reduction will be more significant with additional ESS, and this is verified by the simulation results. In this chapter, only a single ESS unit was considered. However, the developed algorithm can also be extended to the case of multiple ESS units.

References

1 Lubosny, Z. and Bialek, J.W. (2007) Supervisory control of a wind farm. *IEEE Transactions on Power Systems*, **22** (3), 985–994.

2 Sørensen, P.E., Hansen, A.D., Iov, F., Blaabjerg, F., and Donovan, M.H. (2005) *Wind farm models and control strategies, Tech. rep.*, Risø National Laboratory, Denmark.

3 Munteanu, I., Bratcu, A.I., Cutululis, N.A., and Ceanga, E. (2008) *Optimal Control of Wind Energy Systems: Towards a global approach*, Springer Science & Business Media.

4 Biegel, B., Madjidian, D., Spudić, V., Rantzer, A., and Stoustrup, J. (2013) Distributed low-complexity controller for wind power plant in derated operation, in *Control Applications (CCA), 2013 IEEE International Conference on*, IEEE, pp. 146–151.

5 Madjidian, D., Mårtensson, K., and Rantzer, A. (2011) A distributed power coordination scheme for fatigue load reduction in wind farms, in *Proceedings of the 2011 American Control Conference*, IEEE, pp. 5219–5224.

6 Spudic, V., Jelavic, M., Baotic, M., and Peric, N. (2010), Hierarchical wind farm control for power/load optimization, *The Science of making Torque from Wind* (Torque2010).

7 Teleke, S., Baran, M.E., Bhattacharya, S., and Huang, A.Q. (2010) Rule-based control of battery energy storage for dispatching intermittent renewable sources. *IEEE Transactions on Sustainable Energy*, **1** (3), 117–124.

8 Everett III, H. (1963) Generalized Lagrange multiplier method for solving problems of optimum allocation of resources. *Operations Research*, **11** (3), 399–417.

9 Doan, M.D., Keviczky, T., and de Schutter, B. (2011) An iterative scheme for distributed model predictive control using Fenchel's duality. *Journal of Process Control*, **21** (5), 746–755.

10 Giselsson, P., Doan, M.D., Keviczky, T., De Schutter, B., and Rantzer, A. (2013) Accelerated gradient methods and dual decomposition in distributed model predictive control. *Automatica*, **49** (3), 829–833.

11 Giselsson, P. (2013) Improving fast dual ascent for MPC – Part I: The distributed case. *arXiv preprint arXiv:1312.3012.*

12 Giselsson, P. (2013) Improving fast dual ascent for MPC – Part II: The embedded case. *arXiv preprint arXiv:1312.3013.*

13 Spudić, V., Jelavić, M., and Baotić, M. (2011) Wind turbine power references in coordinated control of wind farms. *Automatika – Journal for Control, Measurement, Electronics, Computing and Communications*, **52** (2), 82–94.

14 Barlas, T.K. and Van Kuik, G. (2010) Review of state of the art in smart rotor control research for wind turbines. *Progress in Aerospace Sciences*, **46** (1), 1–27.

15 Hovgaard, T.G., Larsen, L.F., Jørgensen, J.B., and Boyd, S. (2013) MPC for wind power gradients – utilizing forecasts, rotor inertia, and central energy storage, in *Control Conference (ECC), 2013 European*, IEEE, pp. 4071–4076.

16 Grunnet, J.D., Soltani, M., Knudsen, T., Kragelund, M.N., and Bak, T. (2010) Aeolus toolbox for dynamics wind farm model, simulation and control, in *The European Wind Energy Conference & Exhibition, EWEC 2010.*

9

Model Predictive Voltage Control of Wind Power Plants

Haoran Zhao and Qiuwei Wu

Technical University of Denmark

In this chapter, an autonomous wind farm voltage controller based on model predictive control (MPC) is proposed. The commonly used devices for reactive power compensation and voltage regulation of the wind farm are coordinated to keep the voltages of all the buses within the available range. These devices include static Var compensators (SVCs), static Var generators (SVGs), wind turbine generators (WTGs) and on-load tap changing (OLTC) transformers. By optimization of the reactive power distribution throughout the wind farm, the dynamic reactive power reserve can be maximized. Two control modes are designed: for voltage-violated and normal operating conditions. A wind farm with 20 wind turbines was used as the test case to verify the proposed control scheme.

9.1 Introduction

Nowadays, the increasing penetration of wind power and growing size of the wind farms have introduced technical challenges to the maintenance of voltage stability [1]. Due to the long distances between large wind farms and load centers, the short-circuit ratios (SCR) are small and the grid at the connection points is weak [2], which can cause large voltage fluctuations. Moreover, grid disturbances may cause cascading trips of wind turbine generators (WTGs). Therefore, the technical requirements of modern wind farms for voltage support, as specified by system operators, have become more stringent. Specifically, the requirements specify a reactive power capability and a voltage operating range at the point of connection (POC) [3].

In order to fulfill these requirements, the reactive power (Var) or voltage (Volt) regulation devices installed at wind farms must be coordinated. The devices include static var compensators (SVCs), static var generators (SVGs), on-load tap changing (OLTC) transformers. In addition, modern WTGs equipped with power electronic converters (Type 3 and Type 4), which can control the reactive power, are involved in the voltage control [4].

So far, grid codes for wind power integration have often specified different modes for wind farms so that their reactive power can be controlled. These modes include power factor control, reactive power control, and voltage control [5]. The voltage control mode

Modeling and Modern Control of Wind Power, First Edition. Edited by Qiuwei Wu and Yuanzhang Sun.

Companion website: www.wiley.com/go/wu/modeling

exhibits superior performance for the transmission system; the wind farm controls the voltage at the POC specified by the system operator. This chapter focuses on wind farm control under this mode.

Two issues for the wind farm voltage control will be addressed. Firstly, the wind farm collector system connects a large number of WTGs through several medium voltage (MV) feeders. Because of the long length and low X/R ratio ($X/R \leq 1$) of these feeders, the voltage change along the feeder is significant and should not be neglected. The voltages at the end of the feeders may be close to their limits and the WTGs therefore have a high risk of being tripped. Secondly, the various voltage regulation devices should be coordinated. Without proper coordination among these devices, conflicts may occur between their different control performances and objectives. The dynamic responses of these devices are different. Specifically, the response of SVCs and SVGs is quite fast, with time constants of the order of milliseconds (50–200 ms for SVCs and 20–100 ms for SVGs) [6]. The response time of WTGs is around 1–10 s [7]. For OLTCs, the time required to move from one tap position to another largely depends on the tap changer design, and may vary from a few seconds to several minutes.

Conventionally, the total required reactive power reference is calculated according to the voltage at the POC and then dispatched to all WTGs proportionally. The proportional dispatch algorithm is based on either the maximum or available reactive power [8–10]. Although these methods are easy to implement, the voltages of WTG buses are not taken into account. Guo et al. chose to optimally distribute the reactive power [11]. A detailed wind farm collector system model is used to calculate the sensitivity coefficients. However, the proposed optimal control is only based on the current status. The coordination of fast and slow devices over a longer period is not considered.

Model predictive control (MPC) is suitable for coordinated control of Var devices in a wind farm. By using the receding horizon principle, a finite-horizon optimal control problem is solved over a fixed interval of time. In this study, an MPC-based wind farm voltage controller (WFVC) is designed. This aims to maintain all the bus voltages within their feasible ranges and maximize the fast dynamic Var reserve. To improve the computation efficiency and overcome possible convergence problems, the calculation of the sensitivity coefficients is based on an analytical method. The OLTC is incorporated into the MPC without changing the control structure.

9.2 MPC-based WFVC

Figure 9.1 shows the configuration of a wind farm. The buses within the wind farm include a bus at the POC, which corresponds to the high-voltage (HV) side of the main substation transformer, a bus at the medium voltage (MV) side of the main substation transformer, and buses of WTGs.

Figure 9.2 shows the structure of the proposed MPC-based WFVC. It has two control modes, used in different operating conditions:

- *corrective control mode* is an emergency control mode, which aims to correct any bus voltage of the wind farm that violates relevant limits
- *preventive control mode* aims to maximize the fast Var reserve to handle potential disturbances in the future, and further minimize the voltage deviation at the POC V_{POC} from the reference value $V_{\mathrm{POC}}^{\mathrm{ref}}$ issued by the system operator.

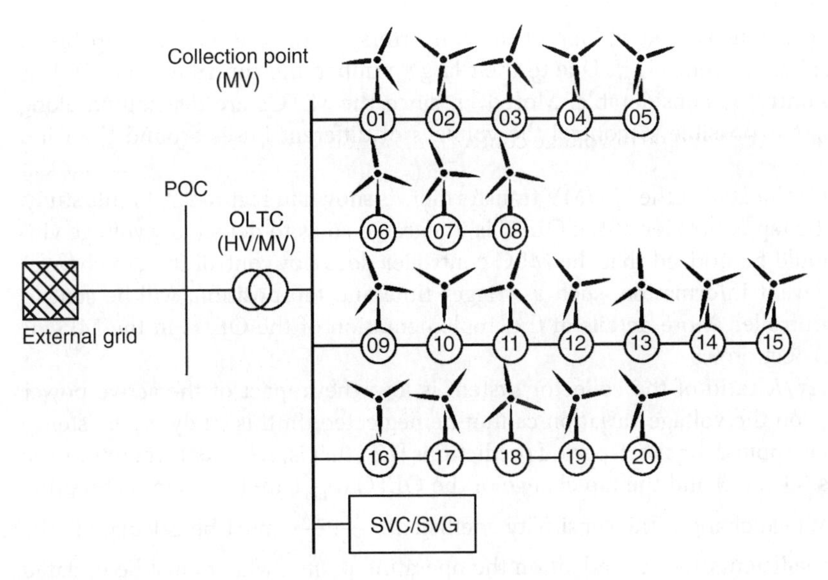

Figure 9.1 Wind farm configuration.

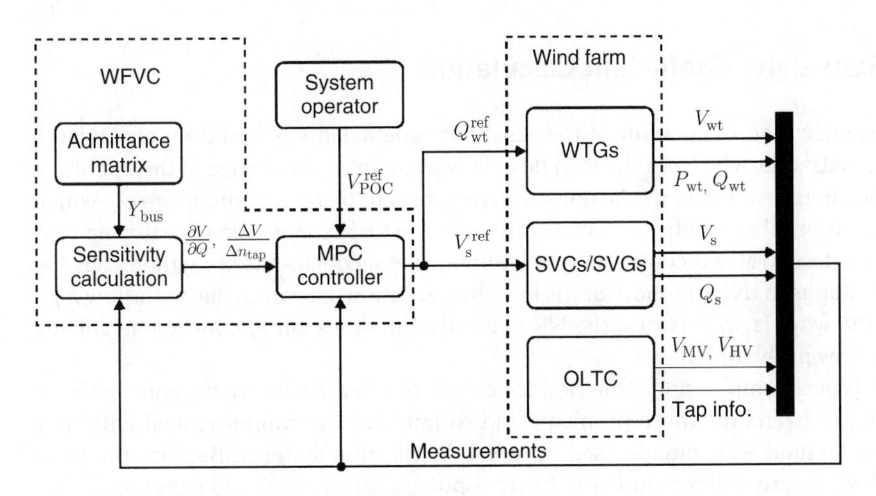

Figure 9.2 Control structure of a wind farm.

The regulation commands for all WTGs (Q_{wt}^{ref}) and SVCs/SVGs (V_s^{ref}) are determined by the MPC controller.

The SVCs/SVGs can operate in either constant-V mode or constant-Q mode. In constant-Q mode, it is easier to coordinate with the WTGs. However, the SVCs/SVGs cannot provide dynamic Var support to regulate the voltage of the controlled bus (the POC in this study) in time, when the voltage at the controlled bus violates the set limits. In this study, constant-V mode is adopted. Based on the prediction model and V_s^{ref}, the equivalent Var reference of the SVCs/SVGs, Q_s^{ref}, can be calculated and coordinated with the WTGs.

With the constant-Q control loop of the converters, modern WTGs are capable of tracking the Var reference $Q_{\mathrm{wt}}^{\mathrm{ref}}$. Due to their large number, the contribution of WTGs to voltage control is considerable. Moreover, since the WTGs are distributed along the feeders, it is possible to control the voltages of different buses around the wind farm.

The OLTC is located at the HV/MV transformer, as shown in Figure 9.2. In this study, the automatic tap controller of the OLTC is integrated so as to detect any voltage violation. It should be noticed that the MPC controller does not control the tap changer directly. Relevant information, such as trigger time and tap position, will be sent to the MPC controller. More details of the implementation of the OLTC in the MPC is described in Section 9.5.

Since the X/R ratio of the collector system is low, the impact of the active power change ΔP_{wt} on the voltage variation cannot be neglected. In this study, a persistence assumption is applied for short-period predictions [12]; that is, $\Delta P_{\mathrm{wt}} \approx 0$. Therefore, the Var outputs (Q_{wt}, Q_{s}) and the tap change of the OLTC (n_{tap}) are the main factors that affect the voltage change. The sensitivity coefficients $\frac{\partial V}{\partial Q}, \frac{\partial V}{\partial n_{\mathrm{tap}}}$ must be calculated. The sensitivity coefficients are dependent on the operating points, which must be updated for each control step.

9.3 Sensitivity Coefficient Calculation

Conventionally, the calculation of the sensitivity coefficients is based on an updated Jacobian matrix derived from the load flow. However, for every change of the operating conditions in the network, the Jacobian matrix needs to be rebuilt and inversed, which creates non-trivial computational constraints for the implementation of real-time centralized or decentralized controllers. In addition, the Jacobian-based method uses the Newton–Raphson (NR) method for the load-flow solution. However, due to the low X/R ratio of the wind farm network, the NR method sometimes fail to converge in solving the load-flow problem [13, 14].

An analytical computational method for calculating sensitivity coefficients has been developed to overcome these problems and to improve the computational efficiency [15]. This method was initially used in radial distribution systems. Because the wind farm collector system has a similar network topology, it has been used here too.

9.3.1 Voltage Sensitivity to Reactive Power

Define $\mathcal{N}$ as the set of all buses of a wind farm $\mathcal{N} = \{1, 2, \cdots N_{\mathrm{b}}\}$, where N_{b} is the bus number. The PQ injections at each bus are assumed to be constant and their dependencies on the voltage are ignored [15]. The relation between the power injection S and voltage V (both in complex form) is,

$$\overline{S_i} = \overline{V_i} \sum_{j \in \mathcal{N}} (Y_{\mathrm{bus}}(i,j) V_j), \tag{9.1}$$

where i and j are the bus indexes, Y_{bus} is the admittance matrix, and $\overline{S}$ and $\overline{V}$ are the conjugates of S and V, respectively.

The partial derivatives of the voltage at Bus $i \in \mathcal{N}$ with respect to reactive power Q_l at Bus $l \in \mathcal{N}$ satisfy the following equations:

$$\frac{\partial \overline{S_i}}{\partial Q_l} = \frac{\partial \{P_i - \mathrm{j}Q_i\}}{\partial Q_l} = \frac{\partial \overline{V_i}}{\partial Q_l} \sum_{j \in \mathcal{N}} Y_{\text{bus}}(i,j) V_j + \overline{V_i} \sum_{j \in \mathcal{N}} Y_{\text{bus}}(i,j) \frac{\partial V_j}{\partial Q_l}$$
$$= \begin{cases} -\mathrm{j}1, & \text{if } i = l. \\ 0, & \text{otherwise.} \end{cases} \tag{9.2}$$

Obviously, (9.2) is a system of equations that is linear to the unknown variables $\frac{\partial V_i}{\partial Q_l}, \frac{\partial \overline{V_i}}{\partial Q_l}$. It has a unique solution for radial electrical networks [15]. Based on the obtained $\frac{\partial V_i}{\partial Q_l}$ and $\frac{\partial \overline{V_i}}{\partial Q_l}$, the partial derivatives of the voltage magnitude $\frac{\partial |V_i|}{Q_l}$ can be calculated as:

$$\frac{\partial |V_i|}{\partial Q_l} = \frac{1}{|V_i|} \mathrm{Re}\left(\overline{V_i} \frac{\partial V_i}{\partial Q_l} \right). \tag{9.3}$$

9.3.2 Voltage Sensitivity to Tap Position

In this subsection, the analytical expressions for the voltage sensitivity to tap positions of a transformer is introduced. The power injections at the buses are assumed to be constant and their dependencies on the voltage are ignored.

The voltage at Bus l is defined as $V_l = |V_l| e^{\mathrm{j}\theta_l}$, where θ is the phase angle. The tap changer is located at Bus k. For a bus $i \in \mathcal{N}$, the partial derivatives with respect to the voltage magnitude $|V_k|$ of Bus k are

$$-\overline{V_i} Y_{\text{bus}}(i,k) e^{\mathrm{j}\theta_k} = \overline{W_{ik}} \sum_{j \in \mathcal{N}} (Y_{\text{bus}}(i,j) V_j) + \overline{V_i} \sum_{j \in \mathcal{N}} Y_{\text{bus}}(i,j) W_{jk}, \tag{9.4}$$

where

$$W_{ik} = \frac{\partial V_i}{\partial |V_k|} = \left(\frac{1}{|V_i|} \frac{\partial |V_i|}{\partial |V_k|} + \mathrm{j} \frac{\partial \theta_i}{\partial |V_k|} \right) V_i, i \in \mathcal{N}.$$

Obviously, (9.4) is a system of equations that is linear with respect to $\overline{W_{ik}}$ and W_{ik}. Similarly, (9.4) has a unique solution for a radial electrical network. Based on the obtained $\overline{W_{ik}}$ and W_{ik}, the voltage sensitivities to the tap position of the transformer at Bus k can be calculated as:

$$\frac{\partial |V_i|}{\partial |V_k|} = |V_i| \mathrm{Re}\left(\frac{W_{ik}}{V_i} \right). \tag{9.5}$$

Since the tap position of the transformers n_{tap} is an integer, the voltage sensitivities to each tap change $\frac{\Delta |V_i|}{\Delta n_{\text{tap}}}$ can be calculated as:

$$\frac{\Delta |V_i|}{\Delta n_{\text{tap}}} = \frac{\partial |V_i|}{\partial |V_k|} \Delta V_{\text{tap}}, \tag{9.6}$$

where ΔV_{tap} is the voltage change per tap.

9.4 Modeling of WTGs and SVCs/SVGs

In this section, the discrete models for the WTGs and SVCs/SVGs are described. These are used as the prediction model for the MPC.

9.4.1 WTG Modeling

Suppose the current time is t_0, and Q_{wt}^{ref} is defined as the Var reference for the WTG, calculated by,

$$Q_{wt}^{ref} = Q_{wt}(t_0) + \Delta Q_{wt}^{ref}, \tag{9.7}$$

where $Q_{wt}(t_0)$ is the Var measurement at t_0.

The dynamic behaviour of the constant-Q control loop of the WTG can be described by a first-order function,

$$\Delta Q_{wt} = \frac{1}{1 + s\tau_{wt}} \Delta Q_{wt}^{ref}, \tag{9.8}$$

where τ_{wt} is the time constant and s is the complex variable. The corresponding state-space model is,

$$\Delta \dot{Q}_{wt} = -\frac{1}{\tau_{wt}} \Delta Q_{wt} + \frac{1}{\tau_{wt}} \Delta Q_{wt}^{ref}. \tag{9.9}$$

In practical operation, the Var output of the WTG (Type 3 and Type 4) Q_{wt} is constrained by the operating limits of the converters [16]. Comparably, the Var capability range for full-converter WTGs (Type 4) is larger due to the increased rating of the converter. The Var capability is dependent on the terminal voltage and active power P_{wt}. Figure 9.3 illustrates a typical PQ curve of a full-converter WTG.

During the short prediction horizon, P_{wt} is assumed to be constant. The constraint of ΔQ_{wt} can be determined according to $Q_{wt}(t_0)$ and its PQ curve,

$$Q_{wt}^{min} \leq \Delta Q_{wt} + Q_{wt}(t_0) \leq Q_{wt}^{max}, \tag{9.10}$$

where Q_{wt}^{min} and Q_{wt}^{max} indicate the minimum and maximum Var capacities of the WTG, respectively.

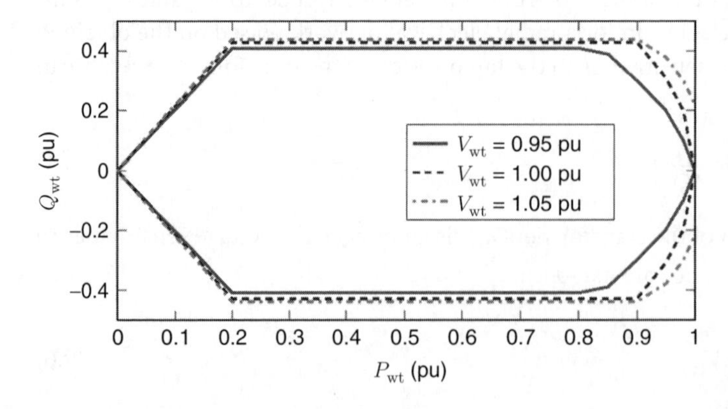

Figure 9.3 *PQ* curve of a full-converter WTG.

9.4.2 SVC/SVG Modeling

The voltage reference for the SVC/SVG is V_s^{ref}. This reference is sent to the local PI controller. Accordingly, the equivalent Var reference Q_s^{ref} can be calculated as:

$$Q_s^{ref} = Q_s(t_0) + \Delta Q_s^{ref}, \tag{9.11}$$

$$\Delta Q_s^{ref} = K_P(V_s^{ref} - V_s) + K_I\frac{1}{s}(V_s^{ref} - V_s), \tag{9.12}$$

where K_P and K_I are the proportional and integral gains of the PI controller, respectively. Here, the voltage at the controlled bus (the POC) V_s is largely dependent on Q_s and Q_{wt}. As the sensitivity value is assumed to be constant during the prediction horizon, the following equation can be derived:

$$V_s = V_s(t_0) + \frac{\partial|V_s|}{\partial Q_s}\Delta Q_s + \frac{\partial|V_s|}{\partial Q_{wt}}\Delta Q_{wt}, \tag{9.13}$$

where $\frac{\partial|V_s|}{\partial Q_{wt}}$ and ΔQ_{wt} are the vectors including all WTGs.

The dynamic of the constant-Q control loop of the SVC/SVG can be described by a first-order function,

$$\Delta Q_s = \frac{1}{1 + s\tau_s}\Delta Q_s^{ref}, \tag{9.14}$$

where τ_s is the time constant and s is the complex variable.

By defining,

$$\Delta V_s^{ref} \triangleq V_s^{ref} - V_s(t_0), \tag{9.15}$$

$$\Delta V_{int} \triangleq \frac{V_s^{ref} - V_s}{s}, \tag{9.16}$$

where ΔV_s^{ref} is the reference of voltage change and ΔV_{int} is the integral of the deviation between V_s^{ref} and V_s, (9.11)–(9.16) can be transformed into a state-space form. In the following, the derivation of the state-space model of the SVC/SVG is described. This is performed in three steps.

Step 1: Calculation of ΔQ_s^{ref} Based on (9.13) and (9.16), (9.12) can be rewritten as:

$$\begin{aligned}\Delta Q_s^{ref} &= K_P(V_s^{ref} - V_s) + K_I\frac{1}{s}(V_s^{ref} - V_s)\\ &= K_P\left(V_s^{ref} - V_s(t_0) - \frac{\partial|V_s|}{\partial Q_s}\Delta Q_s - \frac{\partial|V_s|}{\partial Q_{wt}}\Delta Q_{wt}\right) + K_I\Delta V_{int}.\end{aligned} \tag{9.17}$$

Substituting (9.15) into (9.17),

$$\Delta Q_s^{ref} = K_P\left(\Delta V_s^{ref} - \frac{\partial|V_s|}{\partial Q_s}\Delta Q_s - \frac{\partial|V_s|}{\partial Q_{wt}}\Delta Q_{wt}\right) + K_I\Delta V_{int}. \tag{9.18}$$

Step 2: Derivation of the differential equation of ΔQ_s The transfer function (9.14) can be expressed in the following differential equation:

$$\Delta\dot{Q}_s = -\frac{1}{\tau_s}\Delta Q_s + \frac{1}{\tau_s}\Delta Q_s^{ref}, \tag{9.19}$$

Substituting (9.18) into (9.19),

$$\Delta\dot{Q}_{\mathrm{s}} = -\frac{1}{\tau_{\mathrm{s}}}\left(1+\frac{K_{\mathrm{P}}}{\tau_{\mathrm{s}}}\frac{\partial|V_{\mathrm{s}}|}{\partial Q_{\mathrm{s}}}\right)\Delta Q_{\mathrm{s}} + \frac{K_{\mathrm{I}}}{\tau_{\mathrm{s}}}\Delta V_{\mathrm{int}} - \frac{K_{\mathrm{P}}}{\tau_{\mathrm{s}}}\frac{\partial|V_{\mathrm{s}}|}{\partial Q_{\mathrm{wt}}}\Delta Q_{\mathrm{wt}} + \frac{K_{\mathrm{P}}}{\tau_{\mathrm{s}}}\Delta V_{\mathrm{s}}^{\mathrm{ref}}. \quad (9.20)$$

Step 3: Derivation of the differential equation of ΔV_{int} The transfer function (9.16) can be transformed into the following differential equation,

$$\Delta\dot{V}_{\mathrm{int}} = (V_{\mathrm{s}}^{\mathrm{ref}} - V_{\mathrm{s}}). \quad (9.21)$$

Substituting (9.13) and (9.15) into (9.21),

$$\begin{aligned}\Delta\dot{V}_{\mathrm{int}} &= (V_{\mathrm{s}}^{\mathrm{ref}} - V_{\mathrm{s}}) \\ &= V_{\mathrm{s}}^{\mathrm{ref}} - V_{\mathrm{s}}(t_0) - \frac{\partial|V_{\mathrm{s}}|}{\partial Q_{\mathrm{s}}}\Delta Q_{\mathrm{s}} - \frac{\partial|V_{\mathrm{s}}|}{\partial Q_{\mathrm{wt}}}\Delta Q_{\mathrm{wt}} \\ &= \Delta V_{\mathrm{s}}^{\mathrm{ref}} - \frac{\partial|V_{\mathrm{s}}|}{\partial Q_{\mathrm{s}}}\Delta Q_{\mathrm{s}} - \frac{\partial|V_{\mathrm{s}}|}{\partial Q_{\mathrm{wt}}}\Delta Q_{\mathrm{wt}}.\end{aligned} \quad (9.22)$$

According to (9.21) and (9.22), the state-space model of the SVC/SVG is,

$$\begin{bmatrix}\Delta\dot{Q}_{\mathrm{s}} \\ \Delta\dot{V}_{\mathrm{int}}\end{bmatrix} = \boldsymbol{A}_{\mathbf{S}}\begin{bmatrix}\Delta Q_{\mathrm{s}} \\ \Delta V_{\mathrm{int}}\end{bmatrix} + \boldsymbol{E}_{\mathbf{s}}\Delta Q_{\mathrm{wt}} + \boldsymbol{B}_{\mathbf{s}}\Delta V_{\mathrm{s}}^{\mathrm{ref}}, \quad (9.23)$$

with

$$\boldsymbol{A}_{\mathbf{s}} = \begin{bmatrix} -\frac{1}{\tau_{\mathrm{s}}}\left(1+K_{\mathrm{P}}\frac{\partial|V_{\mathrm{s}}|}{\partial Q_{\mathrm{s}}}\right) & \frac{K_{\mathrm{I}}}{\tau_{\mathrm{s}}} \\ -\frac{\partial|V_{\mathrm{s}}|}{\partial Q_{\mathrm{s}}} & 0 \end{bmatrix}, \quad \boldsymbol{E}_{\mathbf{s}} = \begin{bmatrix} -\frac{K_{\mathrm{P}}}{\tau_{\mathrm{s}}}\frac{\partial|V_{\mathrm{s}}|}{\partial Q_{\mathrm{wt}}} \\ -\frac{\partial|V_{\mathrm{s}}|}{\partial Q_{\mathrm{wt}}} \end{bmatrix}, \quad \boldsymbol{B}_{\mathbf{S}} = \begin{bmatrix} \frac{K_{\mathrm{P}}}{\tau_{\mathrm{s}}} \\ 1 \end{bmatrix}.$$

The following constraints should be respected,

$$Q_{\mathrm{s}}^{\min} \le \Delta Q_{\mathrm{s}} + Q_{\mathrm{s}}(t_0) \le Q_{\mathrm{s}}^{\max}, \quad (9.24)$$

$$V_{\mathrm{s}}^{\min} \le V_{\mathrm{s}}^{\mathrm{ref}} \le V_{\mathrm{s}}^{\max}, \quad (9.25)$$

where $Q_{\mathrm{s}}^{\min}$ and $Q_{\mathrm{s}}^{\max}$ indicate the minimum and maximum Var capacities, respectively, and $V_{\mathrm{s}}^{\min}$ and $V_{\mathrm{s}}^{\max}$ are the minimum and maximum feasible voltages.

9.4.3 General Composite Model

According to the derived state-space models of the WTG (9.9) and SVC/SVG (9.23), the general composite state-space model, including N_{s} SVCs/SVGs and N_{wt} WTGs, is formulated as,

$$\dot{x} = \boldsymbol{A}x + \boldsymbol{B}u \quad (9.26)$$

with

$$x = [\underbrace{\Delta Q_{\mathrm{s\_1}}, \Delta V_{\mathrm{int\_1}}, \cdots, \Delta Q_{\mathrm{s\_}N_{\mathrm{s}}}, \Delta V_{\mathrm{int\_}N_{\mathrm{s}}}}_{\mathrm{SVC/SVG}}, \underbrace{\Delta Q_{\mathrm{wt\_1}}, \cdots, \Delta Q_{\mathrm{wt\_}N_{\mathrm{wt}}}}_{\mathrm{WTG}}]',$$

$$u = [\underbrace{\Delta V_{\mathrm{s\_1}}^{\mathrm{ref}}, \cdots, \Delta V_{\mathrm{s\_}N_{\mathrm{s}}}^{\mathrm{ref}}}_{\mathrm{SVC/SVG}}, \underbrace{\Delta Q_{\mathrm{wt\_1}}^{\mathrm{ref}}, \cdots, \Delta Q_{\mathrm{wt\_}N_{\mathrm{wt}}}^{\mathrm{ref}}}_{\mathrm{WTG}}]',$$

$$A = \left[\begin{array}{ccc|ccc} \boldsymbol{A}_{s\_1} & \cdots & 0 & \boldsymbol{E}_{s\_1} & \cdots & 0 \\ \vdots & \ddots & \vdots & \vdots & \ddots & \vdots \\ 0 & \cdots & \boldsymbol{A}_{s\_N_s} & 0 & \cdots & \boldsymbol{E}_{s\_N_{wt}} \\ \hline 0 & \cdots & 0 & -\dfrac{1}{\tau_{wt\_1}} & \cdots & 0 \\ \vdots & \ddots & \vdots & \vdots & \ddots & \vdots \\ 0 & \cdots & 0 & 0 & \cdots & -\dfrac{1}{\tau_{wt\_N_{wt}}} \end{array}\right],$$

$$\boldsymbol{B} = \left[\begin{array}{ccc|ccc} \boldsymbol{B}_{s\_1} & \cdots & 0 & 0 & \cdots & 0 \\ \vdots & \ddots & \vdots & \vdots & \ddots & \vdots \\ 0 & \cdots & \boldsymbol{B}_{s\_N_s} & 0 & \cdots & 0 \\ \hline 0 & \cdots & 0 & \dfrac{1}{\tau_{wt\_1}} & \cdots & 0 \\ \vdots & \ddots & \vdots & \vdots & \ddots & \vdots \\ 0 & \cdots & 0 & 0 & \cdots & \dfrac{1}{\tau_{wt\_N_{wt}}} \end{array}\right].$$

Based on the sampling time Δt_p, (9.26) can be transformed into a discrete model using the discretization method introduced by Tóth [17]:

$$x(k+1) = \boldsymbol{A}_d x(k) + \boldsymbol{B}_d u(k) \tag{9.27}$$

with the following constraints,

$$\begin{aligned} &Q_{wt\_i}^{min} \leq \Delta Q_{wt\_i}(k) + Q_{wt\_i}(t_0) \leq Q_{wt_i}^{max},\ i \in [1, \cdots, N_{wt}], \\ &Q_{s\_i}^{min} \leq \Delta Q_{s\_i}(k) + Q_{s\_i}(t_0) \leq Q_{s\_i}^{max},\ i \in [1, \cdots, N_s], \\ &V_{s\_i}^{min} \leq V_{s\_i}^{ref}(k) \leq V_{s\_i}^{max},\ i \in [1, \cdots, N_s], \end{aligned}$$

where $\boldsymbol{A}_d$, $\boldsymbol{B}_d$ are the discrete forms of $\boldsymbol{A}$, $\boldsymbol{B}$ in (9.26), respectively.

9.5 Coordination with OLTC

The working principle of the local automatic tap changer controller is illustrated in Figure 9.4. A deadband V_{db} is introduced in order to avoid unnecessary switching around the reference voltage V_{ref}. V_{db} is symmetrical around V_{ref}. Under normal operating conditions, the bus voltage V operates within the deadband and no actions are taken by the controller. If V violates the deadband, a timer is triggered at $t = t_{tri}$. If the voltage violation persists longer than a preset time delay T_{delay}, the tap n_{tap} will increase ($n_{tap} + 1$) or decrease ($n_{tap} - 1$) according to the voltage condition [18]. T_{delay} is largely dependent on the tap changer design.

The trigger time t_{tri} and tap position n_{tap} will be sent to the MPC controller. The MPC controller will check if there exists a potential tap action t_{act} within the prediction period T_p, indicated by Sign_{tap}. Suppose the current time is $t = t_0$,

$$\text{Sign}_{tap} = \begin{cases} 1, & \text{if } t_0 \leq t_{act} \leq t_0 + T_p. \\ 0, & \text{otherwise.} \end{cases} \tag{9.28}$$

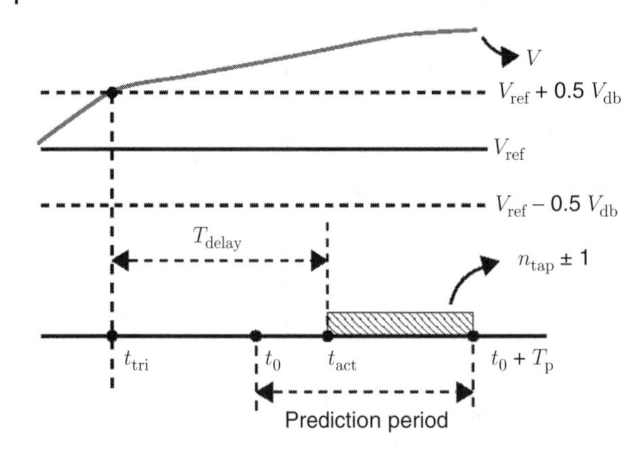

Figure 9.4 Working principle of OLTC in MPC.

If $t_0 \leq t_{\text{act}} \leq t_0 + T_{\text{p}}$, the tap will change in the remaining of the prediction period $t_{\text{act}} \sim t_0 + T_{\text{p}}$. As the tap changer is located at the main transformer, which is the root bus of the collector system, the tap change will affect the bus voltages along the feeders. Based on the derived $\frac{\Delta|V|}{\Delta n_{\text{tap}}}$ in (9.6), the degree of the effect can be calculated, and this will be included in the calculation of the predicted voltages. More details are described in Section 9.6.

9.6 Formulation of MPC Problem for WFVC

As mentioned in Section 9.2, two control modes are designed, appropriate for different operating conditions. In this section, the cost function as well as the constraints of the MPC-based WFVC are formulated for both control modes.

It should be noted that the control sampling time of the WFVC, ΔT_{c}, is normally measured in seconds. It is larger than the time constants of the fast Var devices (SVCs/SVGs). Thus the fast dynamics should also be captured to coordinate the fast and slow devices. Accordingly, the sampling period of the prediction, ΔT_{p}, should be smaller. In this study, ΔT_{c} is further divided into N_{s} steps. Accordingly, for a prediction horizon T_{p}, the total number of prediction steps can be calculated as $N_{\text{p}} = \frac{T_{\text{p}}}{\Delta T_{\text{c}}} \times N_{\text{s}}$.

The T_{p} selection is important for the control performance. If T_{p} is too small, the various devices cannot be well coordinated. If T_{p} is too large, the accuracy of the persistence assumption of P_{wt} and sensitivity coefficients will decrease.

9.6.1 Corrective Voltage Control Mode

In this study, V_{poc} and V_{mv} are the measured voltages at the POC and the MV side of the main substation transformer, respectively. V_{wt} is the vector of the measured voltages at the WTG buses, defined as $V_{\text{wt}} \triangleq [V_{\text{wt\_1}}, \cdots, V_{\text{wt\_}N_{\text{wt}}}]'$, $V_{\text{wt}} \in \mathbb{R}^{N_{\text{wt}} \times 1}$. $V_{\text{poc}}^{\text{ref}}$ is the voltage reference issued by system operator (typically 1.0 pu), and $V_{\text{wt}}^{\text{ref}}$ is the nominal voltage of each WTG (typically 1.0 pu). $V_{\text{poc}}^{\text{th}}$ and $V_{\text{mv}}^{\text{th}}$ are the thresholds of V_{poc} and V_{mv}. In this study, $V_{\text{poc}}^{\text{th}} = 0.01$ pu and $V_{\text{mv}}^{\text{th}} = 0.03$ pu. $V_{\text{wt}}^{\text{th}}$is the threshold of V_{wt}. Since the protection configuration is usually set [0.9, 1.1], to ensure sufficient operation margins, $V_{\text{wt}}^{\text{th}} = 0.08$ pu.

If any bus voltage of the wind farm violates its threshold; that is, $\| V_{\text{poc}} - V_{\text{poc}}^{\text{ref}} \| \geq V_{\text{poc}}^{\text{th}}$ or $\| V_{\text{mv}} - V_{\text{mv}}^{\text{ref}} \| \geq V_{\text{mv}}^{\text{th}}$ or $\| V_{\text{wt}} - V_{\text{wt}}^{\text{ref}} \| \geq V_{\text{wt}}^{\text{th}}$, the WFVC will be switched to this mode.

Cost Function

This control mode aims to ensure all the terminal voltages – throughout the whole farm – remain within the thresholds. The cost function is expressed by,

$$\min_{\Delta Q_{\text{wt}}, V_s^{\text{ref}}} \sum_{k=1}^{N_p} (\| \Delta V_{\text{poc}}^{\text{pre}}(k) \|_{W_{\text{poc}}}^2 + \| \Delta V_{\text{mv}}^{\text{pre}}(k) \|_{W_{\text{mv}}}^2 + \| \Delta V_{\text{wt}}^{\text{pre}}(k) \|_{W_{\text{wt}}}^2),$$

where $\Delta V_{\text{poc}}^{\text{pre}}(k)$, $\Delta V_{\text{mv}}^{\text{pre}}(k)$, and $\Delta V_{\text{wt}}^{\text{pre}}(k)$ are the deviations of the predicted V_{poc}, V_{mv}, and V_{wt} from their references at the kth step, respectively; W_{poc}, W_{mv}, and W_{wt} are the weighting factors. It should be noted that only the buses with violated voltage are considered in the cost function. Once the voltage deviation is within the threshold, the corresponding weighting factor is set to 0 in the newly formulated MPC problem.

$\Delta V_{\text{mv}}^{\text{pre}}(k)$ and $\Delta V_{\text{wt}}^{\text{pre}}(k)$ are decided by the Var injection of SVCs/SVGs, WTGs, and tap change of OLTC,

$$\begin{aligned} \Delta V_{\text{mv}}^{\text{pre}}(k) = {} & V_{\text{mv}} + \frac{\partial |V_{\text{mv}}|}{\partial Q_{\text{wt}}} \Delta Q_{\text{wt}}(k) + \frac{\partial |V_{\text{mv}}|}{\partial Q_s} \Delta Q_s(k) \\ & + \text{Sign}_{\text{tap}} \left(\frac{\Delta |V_{\text{mv}}|}{\Delta n_{\text{tap}}} \Delta n_{\text{tap}} \right) - V_{\text{mv}}^{\text{ref}}, \end{aligned} \tag{9.29}$$

$$\begin{aligned} \Delta V_{\text{wt}}^{\text{pre}}(k) = {} & V_{\text{wt}} + \frac{\partial |V_{\text{wt}}|}{\partial Q_{\text{wt}}} \Delta Q_{\text{wt}}(k) + \frac{\partial |V_{\text{wt}}|}{\partial Q_s} \Delta Q_s(k) \\ & + \text{Sign}_{\text{tap}} \left(\frac{\Delta |V_{\text{wt}}|}{\Delta n_{\text{tap}}} \Delta n_{\text{tap}} \right) - V_{\text{wt}}^{\text{ref}}. \end{aligned} \tag{9.30}$$

Due to electrical coupling with the external grid, the impact of a tap change at the V_{POC} is quite limited, and it is neglected in this study. $\Delta V_{\text{poc}}^{\text{pre}}(k)$ can be obtained by,

$$\Delta V_{\text{poc}}^{\text{pre}}(k) = V_{\text{poc}} + \frac{\partial |V_{\text{poc}}|}{\partial Q_{\text{wt}}} \Delta Q_{\text{wt}}(k) + \frac{\partial |V_{\text{poc}}|}{\partial Q_s} \Delta Q_s(k) - V_{\text{poc}}^{\text{ref}}.$$

Constraints

Besides (9.19), the bus voltages should respect the following conditional constraints,

$$-V_{\text{poc}}^{\text{th}} \leq \Delta V_{\text{poc}}^{\text{pre}}(k) \leq V_{\text{poc}}^{\text{th}}, \text{ if } \| V_{\text{poc}} - V_{\text{poc}}^{\text{ref}} \| \leq V_{\text{poc}}^{\text{th}}, \tag{9.31}$$

$$-V_{\text{mv}}^{\text{th}} \leq \Delta V_{\text{mv}}^{\text{pre}}(k) \leq V_{\text{mv}}^{\text{th}}, \text{ if } \| V_{\text{mv}} - V_{\text{mv}}^{\text{ref}} \| \leq V_{\text{mv}}^{\text{th}}, \tag{9.32}$$

$$-V_{\text{wt}}^{\text{th}} \leq \Delta V_{\text{wt}}^{\text{pre}}(k) \leq V_{\text{wt}}^{\text{th}}, \text{ if } \| V_{\text{wt}} - V_{\text{wt}}^{\text{ref}} \| \leq V_{\text{wt}}^{\text{th}}. \tag{9.33}$$

Once the voltage violates a constraint, this constraint needs to be relaxed to guarantee a feasible solution of the MPC.

The control inputs ΔQ_{wt}, V_s^{ref} could only be updated at the control points. Their values are maintained within the control period,

$$\Delta Q_{\text{wt}}(iN_s + k) = \Delta Q_{\text{wt}}(iN_s), \tag{9.34}$$

$$V_s^{\text{ref}}(iN_s + k) = V_s^{\text{ref}}(iN_s), \tag{9.35}$$

$$i \in [0, \cdots N_p - 1], \quad k \in [0, \cdots N_s - 1].$$

9.6.2 Preventive Voltage Control Mode

If all the bus voltages are within their thresholds, the WFVC will be switched to the this mode.

Cost Function

Two control objectives are considered in this mode. Firstly, to better fulfill the requirements of the system operator, the deviation between the measured voltage at the POC V_{POC} and its reference value must be further minimized. Secondly, fast dynamic Var support capabilities, including both up- and down-regulation, must be maximized to handle potential disturbances. This can be implemented by minimizing the deviation between Q_s and the middle level of its operating range $0.5(Q_s^{max} + Q_s^{min})$. Var insufficiency can be compensated through other, slower Var sources for maintaining the voltage of buses throughout the wind farm. Accordingly, the cost function is expressed by,

$$\min_{\Delta Q_{wt}, V_s^{ref}} \sum_{k=1}^{N_p} (\| \Delta V_{poc}^{pre}(k) \|_{W_{poc}}^2 + \| \Delta Q_s^{pre}(k) \|_{W_s}^2), \tag{9.36}$$

where W_s is the weighting factor, and $\Delta Q_s^{pre}(i)$ is calculated by,

$$\Delta Q_s^{pre}(k) = Q_s + \Delta Q_s(k) - \frac{1}{2}(Q_s^{max} + Q_s^{min}). \tag{9.37}$$

Constraints

The constraints of this mode include (9.19) and (9.31)–(9.35), which are similar to those of the corrective control mode.

The MPC problem formulated can be transformed to a standard quadratic programming (QP) problem, which can be efficiently solved by commercial QP solvers in milliseconds [19]. It should be noted that chattering may occur when the WFVC switches between these two modes. A hysteresis loop can be used to avoid the chattering.

9.7 Case Study

In this section, a wind farm, comprising twenty 5-MW full-converter WTGs, a 7 MVar SVG and an OLTC with a $\pm 8 \times 1.25\%$ tap changer, was used as the test system. Its configuration is shown in Figure 9.1. The Thevenin equivalent model is used to model the external grid. The wind field modeling was generated from SimWindFarm [20], a toolbox for dynamic wind farm modeling, simulation and control. Turbulence and wake effects for the wind farm are considered.

In order to test the efficacy of the proposed WFVC, two case scenarios were designed:

- Scenario 1: the wind farm is under normal operating conditions and the internal wind power fluctuation is considered.
- Scenario 2: besides internal power fluctuation, the impact of the external grid on the wind farm is considered.

In both scenarios, the results of the optimal controller (OPT) based on the current measurement was compared with those of proposed MPC controller. ΔT_c and T_p are set as 1 s and 5 s, respectively.

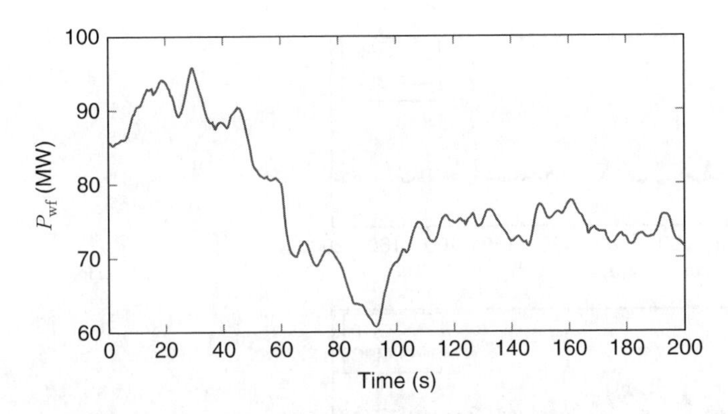

Figure 9.5 Power output of the wind farm.

9.7.1 Scenario 1: Normal Operation

All WTGs were regulated to capture the maximum power. The simulation time is 200 s. The total power output of the wind farm P_{wf} is shown in Figure 9.5.

The fluctuation of active power has an impact on the prediction accuracy of P_{wt} and may further affect the control performance. In order to adequately test the proposed WFVC, different wind power conditions should be included. In this study, the whole operational period is divided into two parts. From 0–100 s the wind power fluctuates from 60 to 95 MW, which is close to the full capacity of the wind farm. From 100 to 200 s, the wind power output becomes smoother, fluctuating around 73 MW.

As shown in Figure 9.1, WT15, the furthest bus along the feeder, is selected as the presentative WTG bus. The simulation results of voltages at three key buses – V_{poc}, V_{mv}, and V_{wt_15} – are shown in Figure 9.6. All the voltages are within their thresholds and therefore the WFVC operates in preventive control mode.

As the primary control objective of this mode, it can be observed that V_{poc} is regulated around its reference value $V_{poc}^{ref} = 1$ pu (see Figure 9.6a). Only small deviations are detected when P_{wf} is close to the wind farm capacity or when the power fluctuates strongly. In the former case, since P_{wt} is almost full, the Var support of the WTGs is limited by the *PQ* curve in Figure 9.3. For the latter case, the fast variation of P_{wt} has a considerable impact on the voltage deviation. However, the standard deviations $\sigma(V_{poc})$ of both controllers are quite small: 0.038% for OPT and 0.018% for MPC. Both controllers show good control performance. The performance of MPC is better.

As the other control objective of this mode, the simulation results for the Q_s of both controllers are shown in Figure 9.7. For both controllers, only small Q_s are detected; that is, the fast Var reserve has been maximized. Both controllers show good control performance. The mean value $\overline{Q}_s$ and standard deviation $\sigma(Q_s)$ are 1.32% and 1.47% for OPT, while for the MPC the values are 0.30% and 0.03%, respectively. The performance of MPC is better.

9.7.2 Scenario 2: Operation with Disturbances

In this case scenario, the disturbances of the external grid are considered. The simulation results are illustrated in Figures 9.8–9.10.

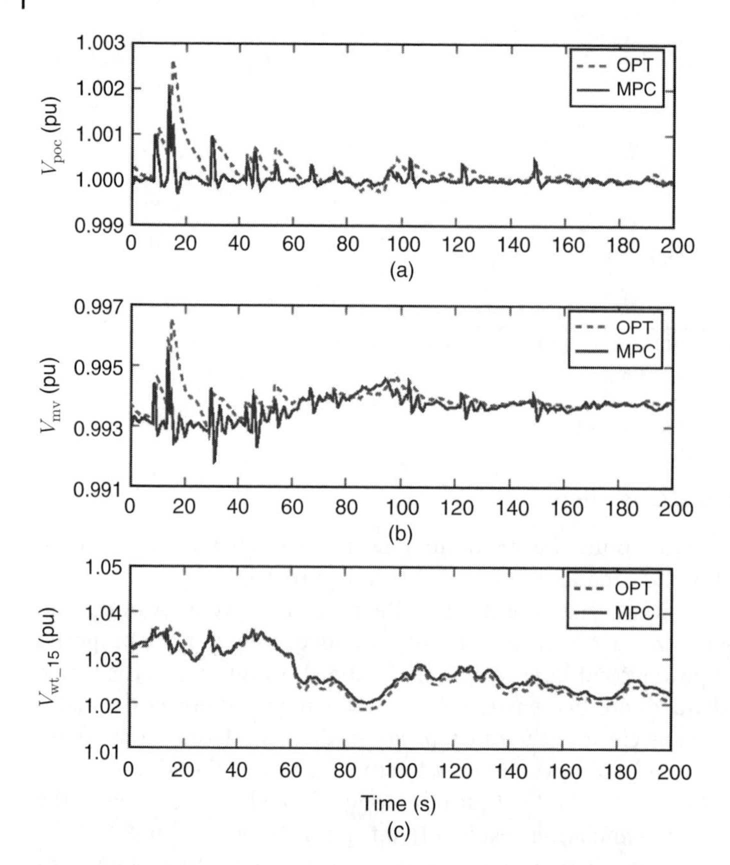

Figure 9.6 Voltages of different buses within the wind farm (Scenario 1): (a) V_{poc}, (b) V_{mv}, (c) V_{wt_15}.

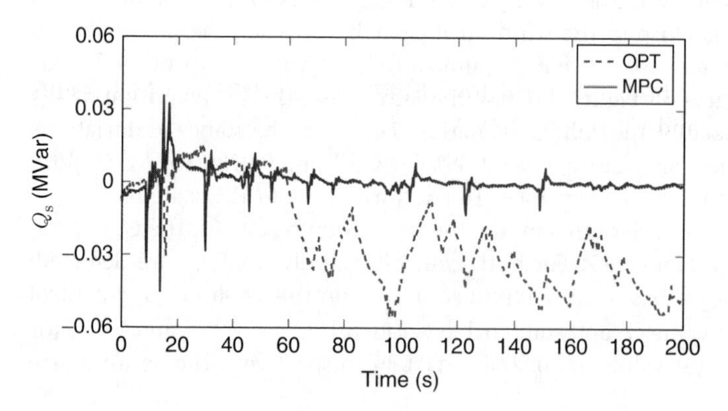

Figure 9.7 Var output of SVG (Scenario 1).

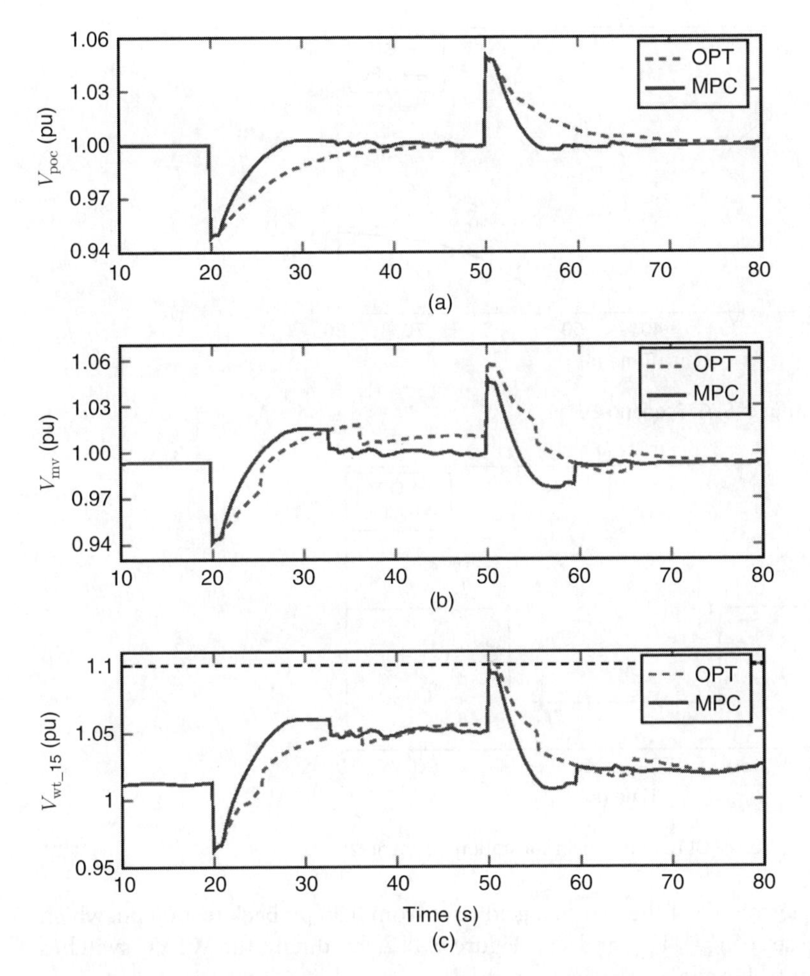

Figure 9.8 Voltages of different buses within the wind farm (Scenario 2): (a) V_{poc}, (b) V_{mv}, (c) V_{wt}.

At $t = 20$ s, the voltage of the external grid drops from 1.00 pu to 0.95 pu, which results in decreases of V_{poc}, V_{mv}, and V_{wt} (Figure 9.8). Accordingly, the WFVC switches to corrective control mode.

For the MPC, the recovery is fast. At $t = 30$ s, ΔV_{poc} has returned to within its threshold (Figure 9.8a). With the coordination with the tap changer, ΔV_{mv} and ΔV_{wt_15} also return to their thresholds at about $t = 32$ s (Figure 9.8b,c). The WFVC switches back to preventive control mode and $|Q_s|$ decreases in order to maximize its reserve, as shown in Figure 9.9. At $t = 50$ s, $|Q_s|$ has reached around 0 MVar. The maximum Var injection is 4 MVar. There is only one tap action, at $t = t_5$ (Figure 9.10).

For the OPT, the recovery is much slower; ΔV_{poc} returns to within its threshold at $t = 40$ s (Figure 9.8a). ΔV_{mv} and ΔV_{wt_15} return to their thresholds at about $t = 35$ s (Figure 9.8b,c). The maximum Var injection is 6.5 MVar, which reaches almost the Var upper limit, as shown in Figure 9.9. The decrease of $|Q_s|$ is also slow. At $t = 50$ s, $|Q_s|$ is still quite high, at around 4 MVar. Two tap actions are detected, at $t = t_1$ and $t = t_2$ (Figure 9.10).

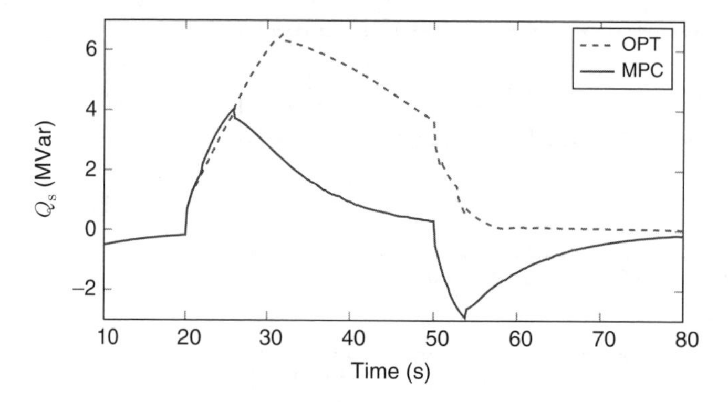

Figure 9.9 Var output of SVG (Scenario 2).

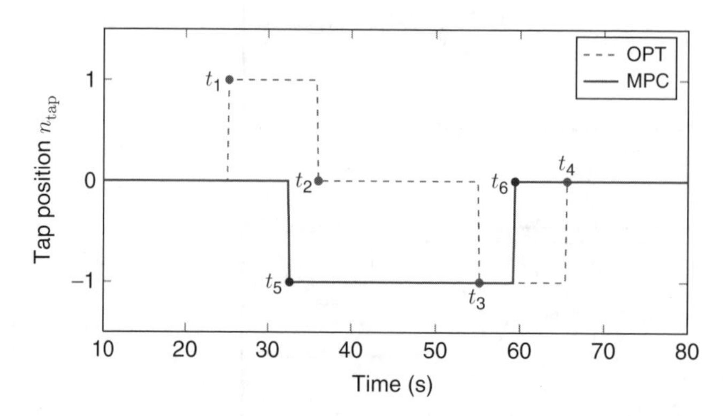

Figure 9.10 Tap position of OLTC at the main substation (Scenario 2).

At $t = 50$ s, the voltage of the external grid rises from 0.95 pu back to 1.00 pu, which results in increases of V_{poc}, V_{mv}, and V_{wt} (Figure 9.8). Accordingly, the WFVC switches to corrective control mode.

For the MPC, the recovery is quite fast. At $t = 55$ s, ΔV_{poc} has returned to within its threshold (Figure 9.8a). With coordination with the tap changer, ΔV_{mv} and ΔV_{wt_15} also return to their thresholds, at about $t = 60$ s (Figure 9.8b,c). The WFVC switches back to preventive control mode and $|Q_s|$ decreases, as shown in Figure 9.9. The maximum Var absorption is 2.5 MVar. Only one tap action occurs, at $t = t_6$ (Figure 9.10).

For the OPT, the recovery is much slower; ΔV_{poc} returns to its threshold at $t = 65$ s (Figure 9.8a). ΔV_{mv} and ΔV_{wt_15} return to their thresholds at around $t = 65$ s (Figure 9.8b,c). The maximum Var absorption is 4 MVar (Figure 9.9). Two tap actions are detected, at $t = t_3$ and $t = t_4$ (Figure 9.10).

9.8 Conclusion

In this study, Var/Volt regulation devices with different time constants are optimally coordinated by the an MPC-based WFVC. According to the different operating conditions, two control modes are designed:

- In preventive voltage control mode, the dynamic Var support capabilities are maximized to prevent potential disturbances and the voltage of the POC is improved to fulfill the requirements of the system operator.
- In corrective voltage control mode, all bus voltages throughout the whole wind farm are regulated to be within the set limits.

An analytical method is used to calculate the voltage sensitivities to improve the computational efficiency. The control performance of the proposed MPC was verified by the case studies.

References

1 Hossain, M.J., Pota, H.R., Mahmud, M.A., and Ramos, R.A. (2012) Investigation of the impacts of large-scale wind power penetration on the angle and voltage stability of power systems. *IEEE Systems Journal*, **6** (1), 76–84.

2 Neumann, T., Feltes, C., and Erlich, I. (2011) Response of DFG-based wind farms operating on weak grids to voltage sags, in *2011 IEEE Power and Energy Society General Meeting*, IEEE, pp. 1–6.

3 Wu, Q., Xu, Z., and Østergaard, J. (2010) Grid integration issues for large scale wind power plants (WPPs), in *IEEE PES General Meeting*, IEEE, pp. 1–6.

4 Chen, Z., Guerrero, J.M., and Blaabjerg, F. (2009) A review of the state of the art of power electronics for wind turbines. *IEEE Transactions on Power Electronics*, **24** (8), 1859–1875.

5 Fortmann, J., Wilch, M., Koch, F.W., and Erlich, I. (2008) A novel centralised wind farm controller utilising voltage control capability of wind turbines. *16th PSCC, Glasgow, Scotland.*

6 Jovcic, D. and Ahmed, K. (2015) *High Voltage Direct Current Transmission: Converters, Systems and DC Grids*, John Wiley & Sons.

7 Larsen, E.V. and Achilles, A.S. (2016), System and method for voltage control of wind generators. US Patent 9,318,988.

8 Sørensen, P.E., Hansen, A.D., Iov, F., Blaabjerg, F., and Donovan, M.H. (2005) Wind farm models and control strategies, *Tech. rep.*, Risø National Laboratory, Denmark.

9 Hansen, A.D., Sørensen, P., Iov, F., and Blaabjerg, F. (2006) Centralised power control of wind farm with doubly fed induction generators. *Renewable Energy*, **31** (7), 935–951.

10 Karthikeya, B. and Schutt, R.J. (2014) Overview of wind park control strategies. *IEEE Transactions on Sustainable Energy*, **2** (5), 416–422.

11 Guo, Q., Sun, H., Wang, B., Zhang, B., Wu, W., and Tang, L. (2015) Hierarchical automatic voltage control for integration of large-scale wind power: Design and implementation. *Electric Power Systems Research*, **120**, 234–241.

12 Zhao, H., Wu, Q., Guo, Q., Sun, H., and Xue, Y. (2015) Distributed model predictive control of a wind farm for optimal active power control Part II: Implementation with clustering-based piece-wise affine wind turbine model. *IEEE Transactions on Sustainable Energy*, **6** (3), 840–849.

13 Khatod, D.K., Pant, V., and Sharma, J. (2006) A novel approach for sensitivity calculations in the radial distribution system. *IEEE Transactions on Power Delivery*, **21** (4), 2048–2057.

14 Haque, M. (2000) A general load flow method for distribution systems. *Electric Power Systems Research*, **54** (1), 47–54.

15 Christakou, K., LeBoudec, J.Y., Paolone, M., and Tomozei, D.C. (2013) Efficient computation of sensitivity coefficients of node voltages and line currents in unbalanced radial electrical distribution networks. *IEEE Transactions on Smart Grid*, **4** (2), 741–750.

16 Amaris, H., Alonso, M., and Ortega, C.A. (2012) *Reactive Power Management of Power Networks with Wind Generation*, Springer Science & Business Media.

17 Tóth, R. (2010) *Modeling and Identification of Linear Parameter-varying Systems*, Springer.

18 Tchokonte, Y.N.N. (2009) *Real-time Identification and Monitoring of the Voltage Stability Margin In Electric Power Transmission Systems Using Synchronized Phasor Measurements*, Kassel University Press.

19 Ferreau, H.J. (2011) *Model predictive control algorithms for applications with millisecond timescales*, PhD thesis, Arenberg Doctoral School of Science, University of Leuven.

20 Grunnet, J.D., Soltani, M., Knudsen, T., Kragelund, M.N., and Bak, T. (2010) Aeolus toolbox for dynamics wind farm model, simulation and control, in *The European Wind Energy Conference & Exhibition, EWEC 2010.*

10

Control of Wind Farm Clusters

Yan Li[1]*, Ningbo Wang*[2]*, Linjun Wei*[1]*, and Qiang Zhou*[2]

[1] *China Electric Power Research Institute, Beijing, China*
[2] *Gansu Electric Power Corporation Wind Power Technology Center, Gansu, China*

10.1 Introduction

Due to the high intermittence of active power production of large wind farms, the integration of large scale wind power may cause a series of problems, such as the fluctuating power grid voltage, transmitted power exceeding set limits, increases of system short circuit capacity, and changes in system transient stability.

Normally, wind turbines (WTs) in wind farms try to maximize active power production. To handle the intermittency of wind power and balance the power system, sufficient spinning reserves have to be put in place. With the increase of installed wind power capacity, the reserve capacity of the power system also needs to increase. This not only leads to cost increases, but also to a decrease in system power generation efficiency. Therefore, control strategies have been developed to smooth the active power output by using the rotational inertia of the WTs and pitch control. The intention is to establish a centralized power control mode based on the correlation of wind farms.

Another problem brought about by the grid integration of wind farms is the decrease of the system inertia and the deterioration of the frequency characteristic of the system. With conventional power plants replaced by wind farms, the rotational inertia of the power system is reduced, and the regulating characteristic of the system frequency becomes worse [1, 2].

With increases in wind power capacity, large wind farms must become controllable. The active power prediction and schedules of wind farms must be considered for the active power control of wind farms. An active power output value should be effectively allocated to each WT to make sure that the wind farm's power output is close to the schedule, and that wind farm can contribute to the frequency regulation of power system, just as conventional power plants do, and so reduce the impact on power systems [3–5].

Large-scale and centralized wind farms are mostly connected to power grids with radial long collecting lines; there is little or no load in the local power grid, which is

Modeling and Modern Control of Wind Power, First Edition. Edited by Qiuwei Wu and Yuanzhang Sun.

Companion website: www.wiley.com/go/wu/modeling

a weak sending power grid with strong voltage sensitivity. The weak sending power grid is weak in voltage support and has significant reactive power problems. The variation of wind power output and disturbances of the power grid will influence the voltage stability of the power grid. In particular when the wind power output is high, it will likely lead to a decrease of voltage stability of the power system. Moreover, since WTs absorb reactive power during fault periods, leading to high voltage drop of wind farms, many WTs will be tripped by low voltage protection measures. A large number of capacitive reactive power compensation devices will still be operating if there is no fast automatic voltage regulation. This will make the system voltage rise sharply and over-voltage protection will cause the tripping of many WTs. The large and frequent tripping of WTs seriously impacts the secure and reliable operation of the power grid. Therefore, on the one hand, the reactive power and voltage control of wind farms needs to ensure that the key bus voltage meets requirements; on the other hand, voltage stability of wind farms should be improved to minimize the wind power waste caused by voltage stability issues.

This chapter introduces the active power control requirements of wind power, and the active power control and automatic generation control (AGC) of wind farm clusters. It also describes the impact of wind power on system voltage stability and the three-level voltage control of wind farm clusters.

10.2 Active Power and Frequency Control of Wind Farm Clusters

The active power control of wind farm clusters is a key technology for the controllable operation of wind farms. The development of active power control technology will ensure more efficient utilization of wind energy and play an important role in the secure and reliable operation of power systems.

Seven modes of active power control are introduced in Section 10.2.1. In Sections 10.2.2 and 10.2.3, the active power control strategies for large wind power bases and the relevant system design are described in order to meet the requirements of power system operation.

10.2.1 Active Power Control Mode of Wind Farms

Active power control of wind farms is specified in the grid codes of many countries to enable wind power integration. There are two basic requirements. Firstly, to control the maximum change rate of wind power and secondly to control the output power of wind farms according to the set-points issued by the system operator. In some grid codes, it is specified that wind farms should be capable of reducing the active power and participating in the primary frequency regulation of the power system [6]. The range of power reduction, the response time, and technical parameters of control system participating in primary frequency modulation (dead zone, the differential coefficient, response time, and so on) are also specified. The ways to control the active output of wind farms include cutoff of WTs, cutoff of the entire wind farm, and adjusting WT output levels (for variable-speed WTs). The Danish transmission system operator defines seven ways to control active power in its technical regulations for wind farm integration. The requirements are set out below [7]:

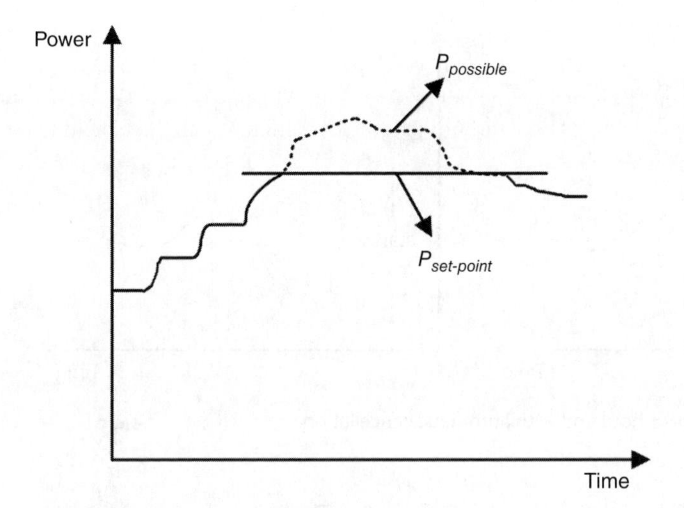

Figure 10.1 Absolute production constraint [7].

Absolute Production Constraint

The adjustable absolute production constraint is shown in Figure 10.1.

It must be possible to limit the production of a wind farm to a set-point in the 20–100% range of the rated power. The deviation between the set-point and the measured 5-min mean value at the connection point must not exceed ±5% of the rated power of the wind farm. It must be possible to set the regulation speed for upward and downward regulation of 10–100% of the rated power per minute.

Delta Production Constraint

The delta production constraint is shown in Figure 10.2. It must be possible to limit active power output to a required constant value in proportion to the possible active power.

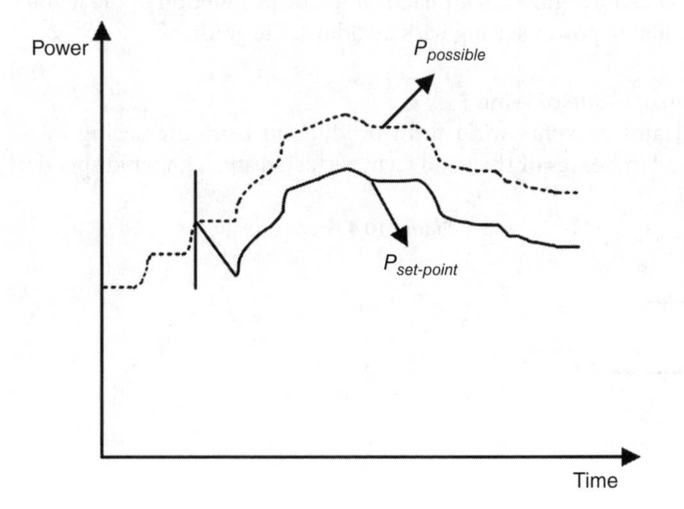

Figure 10.2 Delta production constraint [7].

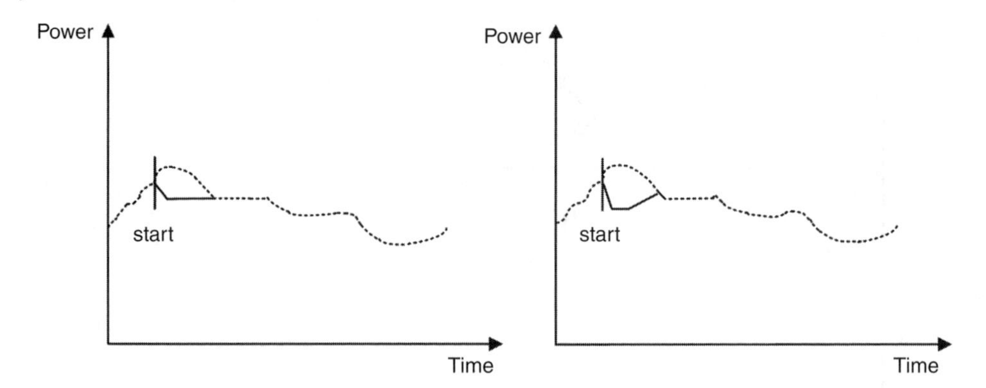

Figure 10.3 Balance regulation without and with automatic cancellation.

Balance Regulation

The balance regulation requirement is shown in Figure 10.3. The balance regulation must be implemented as a rapid power regulation ensuring upward/downward regulation of the wind farm production when required. A power change must be set according to balance regulation orders, as a desired change (±MW change) to the current production and a desired power gradient (MW/min).

It must be possible to reset balance power automatically after a predefined time, and the active power of the wind farm should return to the available power setting with an adjustable gradient. In connection with the balance regulation, it may be permitted that the production exceeds the limit a little. How large the over-production that is allowed must be set carefully. It must be possible to enable and disable the balance regulation.

Stop Regulation

Stop regulation maintains the current production of the wind farm as much as possible. The requirement is shown in Figure 10.4. If the wind speed decreases, stop regulation will not be required. After the stop regulation is deactivated, the production of the wind farm must return to the available power setting with an adjustable gradient.

Power Gradient Constraint

The power gradient constraint prevents wind farm production from increasing too quickly when the wind speed increases or the wind farm is started at a high wind speed.

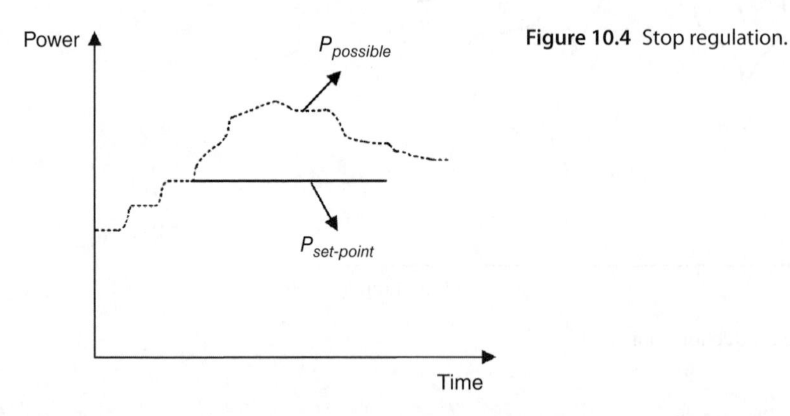

Figure 10.4 Stop regulation.

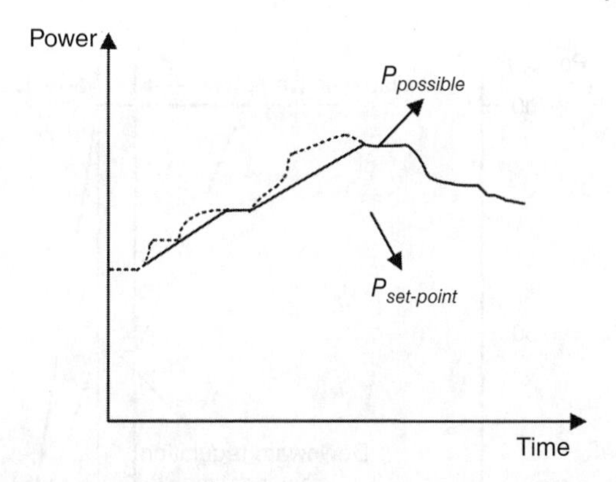

Figure 10.5 Power gradient constraint.

The constraint is shown in Figure 10.5. If the wind speed decreases, the constraint will not be required unless the delta production constraint is activated. It must be possible to set the maximum permissible speed at production increase and decrease individually. It must also be possible to enable and disable the function. The system operator will determine the maximum permissible values for gradient constraints.

System Protection

Via an external control signal to the wind farm controller, it must be possible to make a rapid downregulation of the wind farm for system protection regulation. The requirement is shown in Figure 10.6. Downregulation must take place at a preset speed. It must be possible to set the maximum amount of power that the system protection regulation can downregulate. As long as the external system protection signal is active and the maximum power change has not been reached, the downregulation can continue. When the external signal is deactivated, the system protection regulation will stop and the current power production must be maintained.

It must be possible to reset the system protection function manually. When this takes place, the regulation will return to maintain the current power. If the external system protection signal is still active when the function is reset, a new power limit is calculated on the basis of the current production, and the wind farm may be further downregulated.

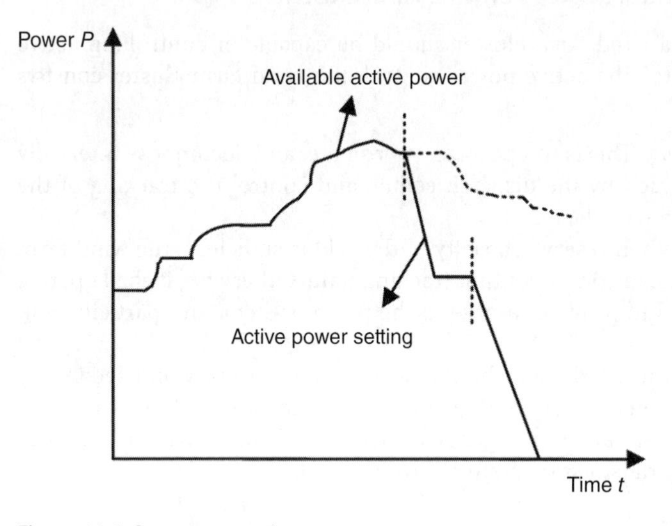

Figure 10.6 System protection.

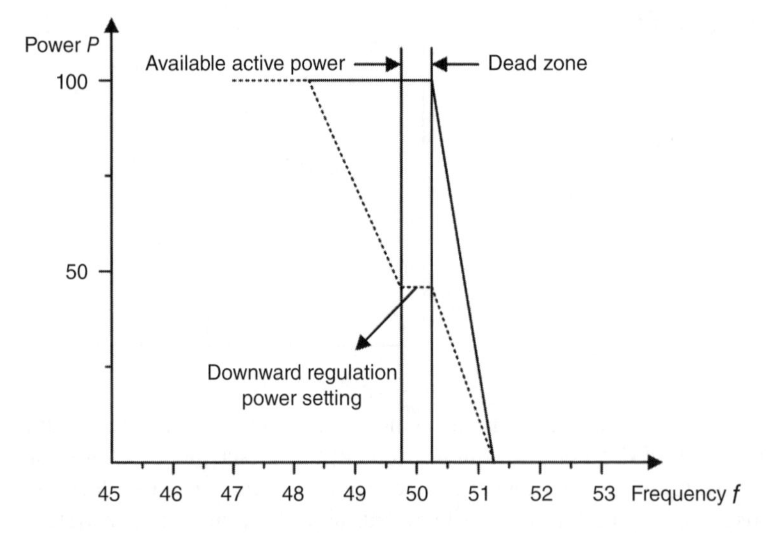

Figure 10.7 Frequency regulation with and without previous downward regulation.

In the case of system protection regulation, it shall be possible to down regulate the production from the full load to stop in a maximum of 30 seconds. It shall be possible to activate / deactivate the system protection regulation function separately.

Frequency-controlled Regulation

For automatic frequency regulation, the WTs must change the production depending on the system frequency. With the wind farm controller, it must be possible to set the total regulation of the wind farm. Figure 10.7 shows two examples of frequency-controlled regulation. When the wind farm generates the available power, it will only provide downregulation. When the wind farm operates in derated mode – that is, the power production is less than the available power – it can provide both upregulation and downregulation.

10.2.2 Active Power Control Strategy of Wind Farm Cluster

Active power control of a wind farm cluster should be capable of controlling active power and the change rate. The active power control of a wind farm cluster consists of four modes:

- *Output schedule tracking:* This is to optimize, coordinate, and decompose internally the output schedule issued by the dispatch center, and control the tracking of the output schedule by each wind farm.
- *Peak load regulating*: When reserve capacity of the grid is sufficient, the wind farm cluster will operate at the maximum output tracking state. Otherwise, it shall operate at a derated power level to provide reserve to the power system, and participate in peak load regulation.
- *Frequency regulation:* The wind farm should participate in primary and secondary frequency regulation when the system frequency exceeds the preset dead-band.
- *Emergency active power support:* Emergency control should be activated to reduce active output when the transmission system is overloaded.

Table 10.1 Wind farm classification for real-time dispatch.

Wind farm state	Predicted data	Operating data	Controllable or not	Remarks
0	—	—	—	Shutdown
1	—	—	×	Incapable of real-time control
2	×	×	✓	Communication failure for data uploading
3	✓	✓	✓	Observable, controllable
4	×	✓	✓	Failing to acquire ultra-short-time power prediction data, according to day-ahead/ rolling plan
5	✓	×	✓	Failing to acquire operating data, without regard to operating hours

Identification of the Control Priority of Wind Farms

For a wind farm cluster, wind farms should be classified to make control easier. The classification is made according to the observability, controllability, and the operating mode of the wind farms. The details of the classification are listed in Table 10.1.

The control priority is based on the following principles:

Wind farms at State 0 are stopped. A wind farm shut down earlier should be restarted before a wind farm shut down later, and these farms should not start until the downtime exceeds $T_{stop\_\min}$.

Wind farms at State 1 don't participate in the real-time regulation, and should operate according to the day-ahead/rolling schedule. Assess the low-voltage ride through (LVRT), dynamic reactive power compensation, technical countermeasures against accidents, operating voltage/flow/overload, operating economy, and cumulative damage of direct shutdown of WTGs of the wind farms at State 1 (referred to hereafter as wind farms of Category I), calculate the weights of all wind farms and rank them from small to large, and form a "shutdown priority sequence".

Wind farms at States 2–5 (hereafter referred to as wind farms of Category II) can be controlled in real time. Based on the limit of wind power, wind farms of Category II have two active real-time control modes:

- In free-generation mode, a wind farm does not have a limit on its output as long as the ramping rate meets requirements.
- In output-tracking mode, a wind farm realizes power tracking control by pitch control or by starting or shutting down WTs according to the active power setpoint from the master station of the cluster.

Wind farms of Category II WF_{2_j} will be in the free-generation mode when their annual total full load hours (FLH) T_{WF2_j} is less than the preset FLH $T_{\min}$; otherwise, they will be in output-tracking mode. Wind farms of Category II follow the priority generation principle and form a "priority sequence of reducing output", in which wind farms in free-generation mode are put at the rear end.

Real-time Active Power Control Strategy

Let $P(t)$ be the active power set-point of the wind farm cluster at t from the dispatch center and $P(t-1)$ the actual active power output of the wind farm cluster in the previous control time step. The overall output of wind farms of Category II participating in the real-time control is that $P(t-1)$ minus the output of M wind farms of Category I.

$$P_{WFCII}(t-1) = P(t-1) - \sum_{j=1}^{M} P_j(t-1) = P(t-1) - \sum_{j=1}^{M} P_j^{real}(t-1) \tag{10.1}$$

$$P_{WFCII}(t) = P(t) - \sum_{j=1}^{M} P(t) = P_{WFC}(t) - \sum_{j=1}^{M} P_j^{plan}(t) \tag{10.2}$$

The flowchart of the real-time active power control of the wind farm cluster is shown in Figure 10.8. In the figure, active power regulation to be realized in time step t is,

$$\Delta P(t) = P_{WFCII}(t) - P_{WFCII}(t-1) \tag{10.3}$$

10.2.3 AGC of Wind Farm Cluster

Automatic generation control (AGC) is adjustment of the output of multiple generators at different power plants in response to load changes. An AGC scheme for a wind farm cluster is proposed here. It uses the characteristics of the wind farm system and is shown in Figure 10.9.

The AGC system consists of a dispatch center, wind farm remote terminal unit (RTU), AGC communication management units, and turbine control systems:

Wind farm RTU The wind farm RTU is a connection between the dispatch center, the WTs and the substations of wind farms. It distributes the set-point from the dispatch center. It may communicate with the dispatch center through protocol 101 and 104. The connection between the RTU and the communication management units can be realized through an optical fiber. Monitoring, display, and recording of the status of WTs and transformer stations can be performed at the RTU.

AGC communication management units The communication management unit can coordinate the communication between a wind farm controller and its turbines. The dispatch centre sends the command through a standard protocol to the communication management units of the wind farm. It is suggested that IEC 61400-25 and IEC61850 protocols should be used for communication with the WTs and substations, respectively.

10.3 Reactive Power and Voltage Control of Wind Farms

10.3.1 Impact of Wind Farm on Reactive Power Margin of the System

A wind farm is usually connected to the bulk power system through step-up substations. Most wind farms are located far from load centers and their connection lines are usually tens of kilometers and sometimes even a hundred kilometers long. Consequently, the grid at the point of connection (POC) is weak. Changes of either the active or reactive power of wind farms will make the voltage at the POC fluctuate. Thus, the voltage stability margin at the POC is small.

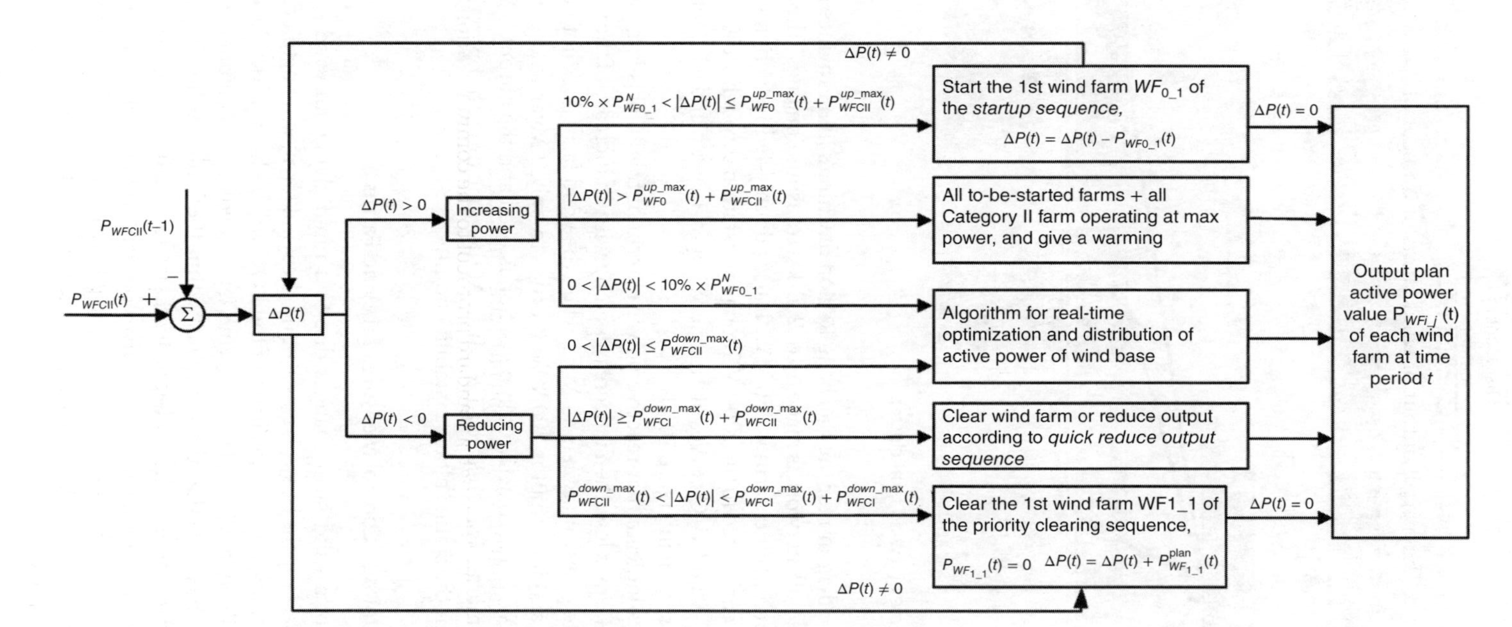

Figure 10.8 Flow of real-time active power control strategy of the wind farm cluster.

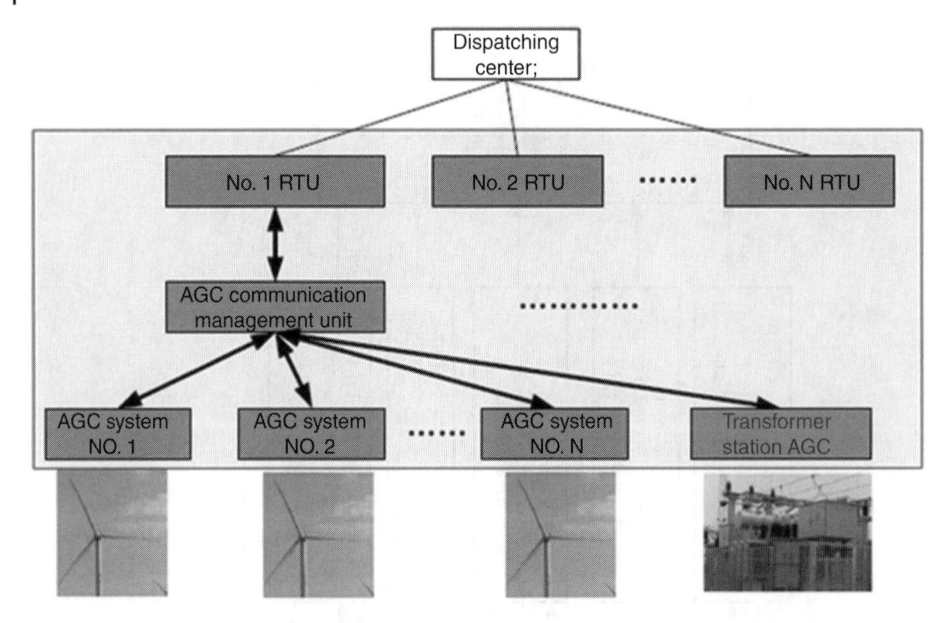

Figure 10.9 AGC scheme of a wind farm cluster.

The grid structure diagram for the Baicheng Region of China's Jilin Province is shown in Figure 10.10. The bulk network is made up of 220-kV transmission lines. The installed capacity of the Tuanjie Wind Farm is 250 MW. It uses doubly-fed induction generator (DFIG) WTs and is connected to the 220-kV Taonan substation through a 100-km transmission line. The P–V and V–Q curves of Tuanjie Wind Farm, with and without reactive compensation, are shown in Figure 10.11.

Simulation results indicate that reactive power margin of the POC is large when the output of the wind farm is low, then it decreases along with an increase of the wind farm output. The reactive power margin is 44 Mvar when the wind farm does not have reactive compensation and its output is 200 MW. It drops to only 9 Mvar when the output increases to 250 MW. When reactive compensation is used and the output is 250 MW, the reactive power margin is increased to 26 Mvar. At the same time, the operating point is further away from the static voltage instability point.

10.3.2 Reactive Voltage Control Measures for Wind Farms

When the output of a wind farm changes due to a rapid change in wind speed, the system voltage, especially the bus voltages around the wind farm, will fluctuate. Usually, when the wind farm output is low, voltage of the POC will be high; when the output is high, the voltage will be low. Therefore, the wind farm needs to be equipped with reactive compensation to ensure that the voltage is within the limits. When high levels of wind power are integrated into the grid, it is necessary to install reactive compensation on the grid side and the wind farm side. Reactive compensation on the grid side consists of series compensation and high-voltage controllable reactors; reactive compensation on the wind-farm side includes reactive power from WTs and dynamic reactive compensation equipment.

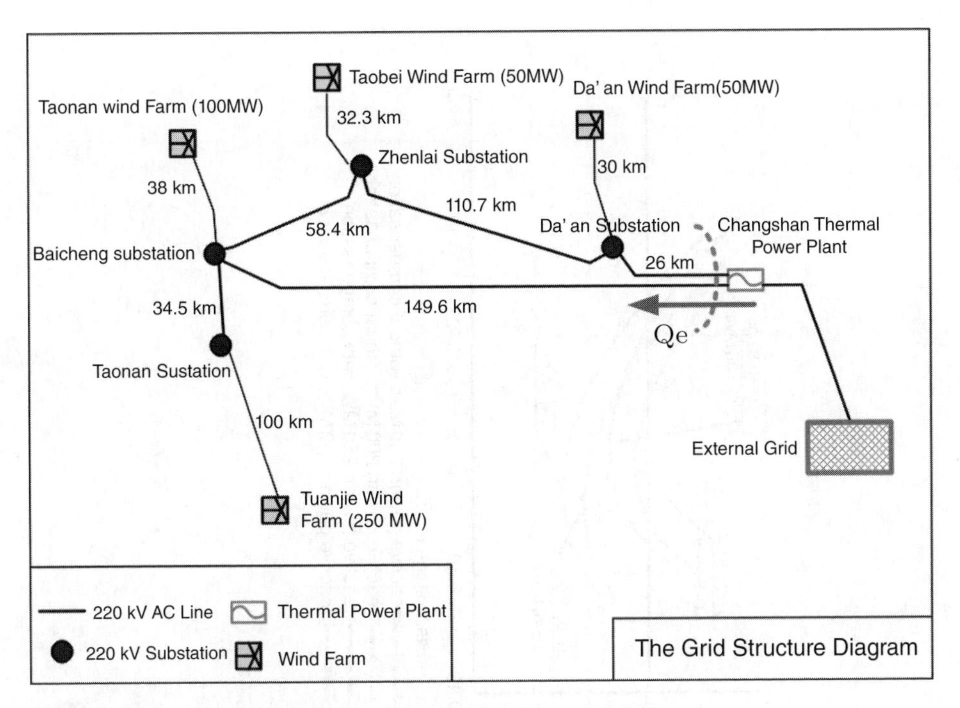

Figure 10.10 Grid structure diagram of Baicheng region.

Series Compensation for the Transmission Line

The inductive reactance of transmission lines is a major determinant of the stability limit. Series capacitor compensation for remote transmission lines reduces their reactance effectively, reinforces the grid structure, and improves voltage and transient stability [8].

Series capacitors are conventionally used to compensate for long overhead lines. Recently, however, series capacitors have been used to effectively compensate short but heavily loaded lines. In transient stability improvement applications, when a fault or power swing is detected, a set of series capacitors will be switched on, and then switched off after 0.5 s or so. Such switchable capacitor banks can be situated in a substation such that they can be used by more than one line.

Slow grid construction is a bottleneck for the rapid development of wind power in some area of North China. As a result, generation of green electricity has been curtailed. For this reason, the grid company built the Guyuan 500 kV series compensation device, which has a total capacity of 2 × 663.39 Mvar. The series compensation increases the transmission capacity, so that it can handle the 1-GW wind power output of the entire region. It has also significantly improved the transmission of electricity from adjacent Inner Mongolia, thereby improving energy security in the North China power grid.

High-voltage Controllable Reactor

The volatility of wind power has a great influence on the voltage of the grid. Traditional 500-kV, or even 750-kV, high-voltage reactors cannot be adjusted online, and they only compensate the charging power on high-voltage transmission lines. However, in an area with a large installed wind power capacity, the voltage of neighboring 500-kV or 750 kV

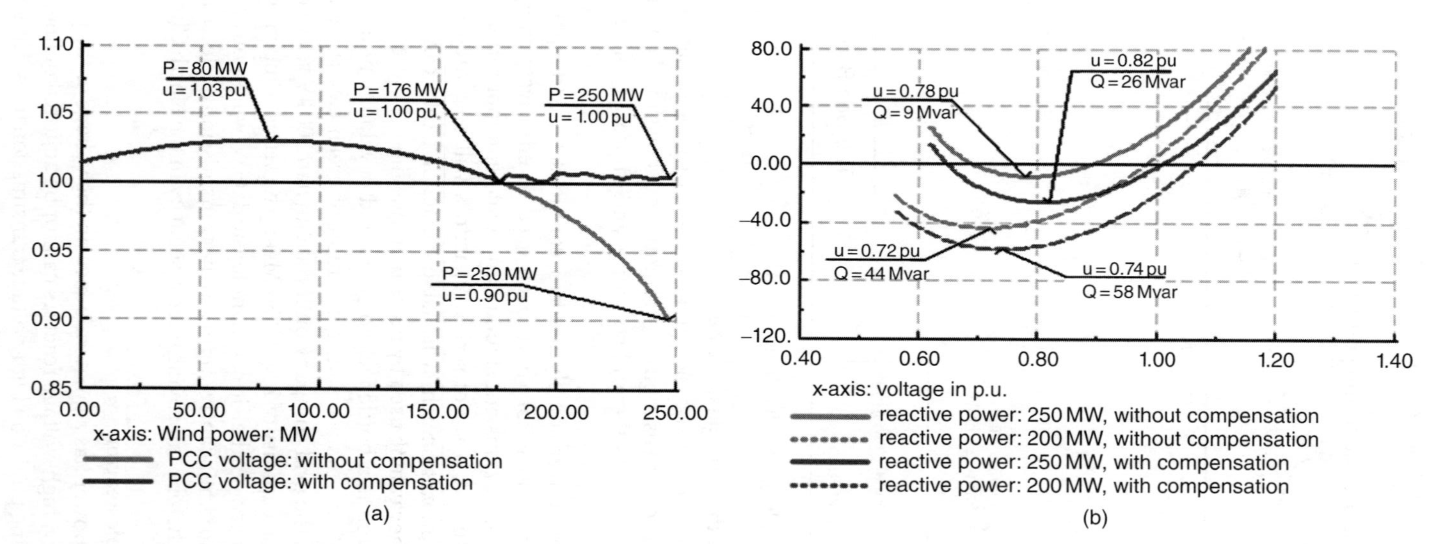

Figure 10.11 Tuanjie Wind Farm: (a) P–V curve and (b) V–Q curve.

busbars will drop a lot due to weak control over the voltage of wind farms. At present, reactors on the low-voltage side of transformers are not able to control the busbar voltage on the high-voltage side. If reactors on the high-voltage side were adjustable online, the grid voltage will be more stable and controllable.

A 10-GW wind power base has been developed in Jiuquan, China. To solve problems of violating voltage limits at the Gansu-Xinjiang interconnection project and to enable the large-scale transport of wind power, the State Grid Corporation of China established the first 750-kV controllable reactor project in 2012, and installed it on the 750-kV busbar side of the Dunhuang Substation. It has a capacity of 300 Mvar, controllable at 100%, 75%, 50%, and 25%. With this project in operation, the system has effectively solved the contradiction between reactive power regulation and overvoltage, relieved the voltage stability problem caused by intermittency of wind power and long-distance power transmission of the northwestern grid, reduced transmission losses, improved transmission capacity and grid safety and stability, and thereby provided favorable conditions for wind power transport from the Hexi region.

SVG in Wind Farms

Static Var generators (SVGs) are usually used for large-scale power transmission networks to strengthen dynamic reactive power compensation and improve voltage stability at the end of the grid or heavy load centers. SVGs can respond to voltage changes in a few cycles. The quick response can reduce the voltage drops of wind farms when the remote AC system fails, and improves the fault ride through capability of wind farms. Similarly, SVGs can reduce the overvoltage magnitude when clearing a system fault, and reduce the generator tripping caused by overvoltages. With a strong overload capacity, SVGs may provide double or triple rated capacity and reactive power support for 2 s during a fault, thus improving the quality of the bus voltage [9].

When the wind farm is located at the end of the grid, with reactive compensation devices of capacitor banks, the voltage of the relevant buses will fluctuate greatly when the wind power output is at a high level. For instance, the P–V curve of the Shangyi Wind Power Collection Station, part of the Zhangbei 10-GW wind power base, is shown in Figure 10.12. The left-hand figure represents the P–V curve with the capacitor banks and the right-hand figure shows the P–V curve with the SVG with the same capacity installed at the collection station.

It can be seen that, because the grid structure is weak, the grid operation is close to the static instability point, and a small power change may even lead to a big voltage change; wind power fluctuation results in the frequent switching of capacitor banks. The installation of SVG effectively reduces the voltage fluctuations caused by power fluctuations, and improves voltage stability.

Control of WT Converter

Variable-speed WTs are capable of controlling reactive power and voltage. The P–V curves of a wind farm connected through a 50-km line are shown in Figure 10.13. The WTs of the wind farm have different power factors. The left-hand figure shows the POC voltage and the right-hand figure is the voltage curves at 40 km. This shows that when the WT power factor is 0.98 (lagging), the voltages of both the POC and WTs first rise and then fall and the POC voltage remains around 1.05 pu; when the WT power factor is 1.0 or 0.98 leading, the voltage of each point falls along with an increase of wind power;

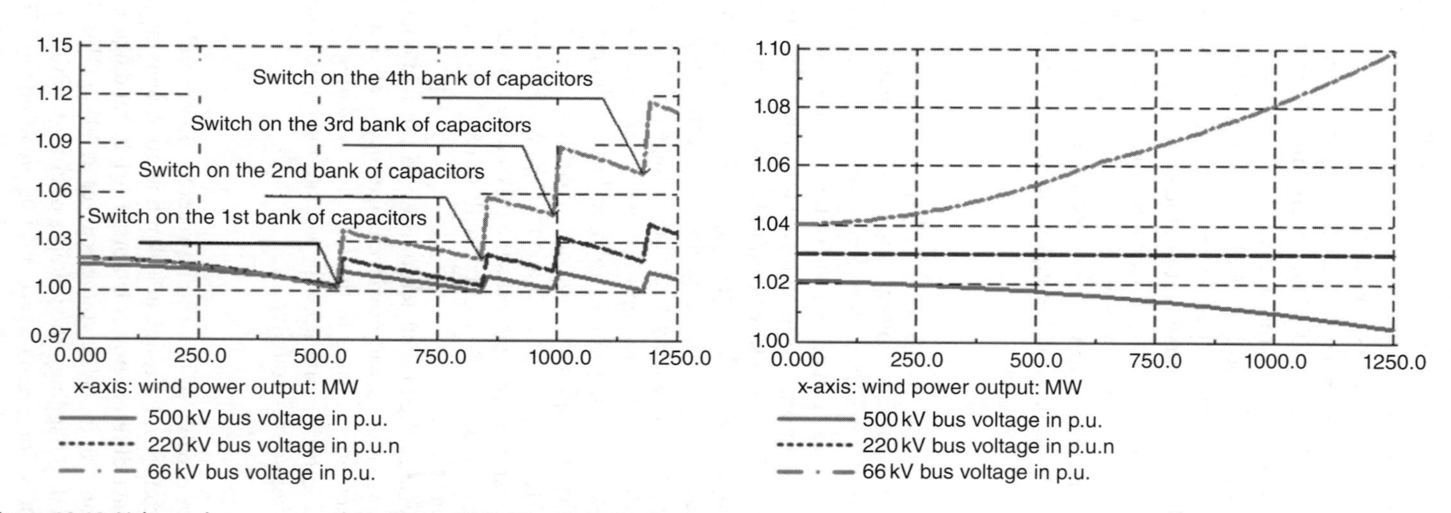

Figure 10.12 Voltage change curve of the Shangyi Wind Power Collection Station.

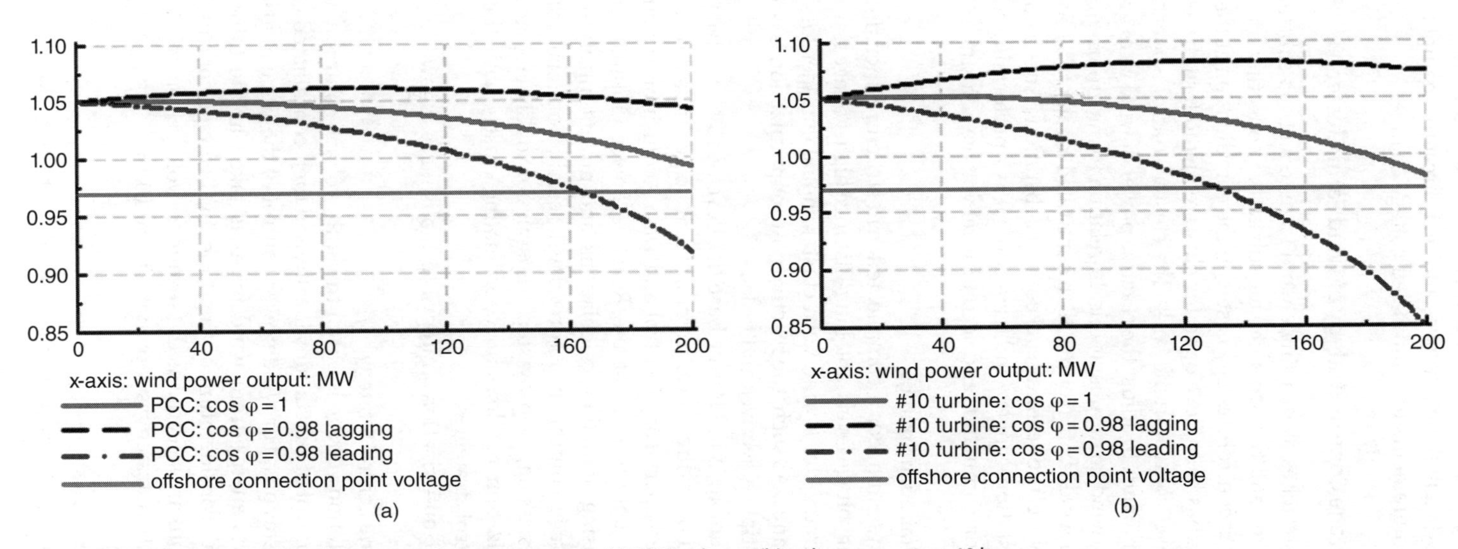

Figure 10.13 Influence of wind turbine power factor on voltage: (a) POC voltage, (b) voltage curves at 40 km.

the POC voltage change is between 0.05 and 0.15 pu. Therefore, the WT power factor should be set as 0.989(lagging), which saves additional reactive compensation capacity and also meets the requirements for minimized voltage change.

10.3.3 Reactive Voltage Control Strategy of Wind Farm Cluster

A wind farm cluster stretches from a single wind farm to a group substation collecting energy from wind farms and then to a cluster substation collecting energy from wind farm groups. A reactive control strategy based on the coordination at the levels of wind farms/groups/clusters has been developed for clustered wind farm bases. The strategy is based on a three-level control framework. The cluster control is located at the third level, which sends commands to adjust the reactive power of wind farm group collection substations and single wind farm substations. In relation to the group control, if there is no voltage violation, a wind farm group will be controlled locally as a second-level control substation; otherwise, the second-level control substation issues a command to the first-level control substation. Single wind farm control as a performing station executes third-level optimization commands and second-level voltage adjustment commands.

Wind Farm Cluster Optimization Strategy

A wind farm cluster system needs a complicated grid structure, including wind farms, substations at different voltage levels, and long-distance high-voltage transmission lines. The voltage control strategy for a wind farm cluster should consider loss minimization and dynamic performance. As such, the reactive power optimization is complicated and must incorporate multiple objectives and tasks.

The China Gansu Jiuquan Wind Power Base has developed a reactive power optimal control strategy based on a three-level framework. It acquires data online throughout the network using the Supervisory Control and Data Acquisition system, then brings about whole network optimization at the Reactive Power Optimal Control Master Station at Jiuquan, using load flow calculations and an optimization algorithm. It achieves optimal voltage control in a three-level reactive optimal control strategy, performing fuzzy processing, step-by-step transmission of signals, and hierarchical control of the optimization results, and coordinating closely with the second- and first-level voltage control strategies.

In this way, a voltage optimization strategy is established, as shown in Figure 10.14.

Wind Farm Group Voltage Control Strategy

When the group substation voltage is out of the limit, voltage control of the wind farm group first involves control of substations locally, and then coordinated control among wind farm collection stations. For wind power control of the Jiuquan grid, the 330-kV substation that collects energy from wind farms is taken as the third level control substation, with overall consideration of the three-level voltage control. Three or four other 330 kV substations are selected as the secondary voltage control substations. The secondary voltage control strategy is shown in Figure 10.15.

Wind Farm Voltage Control Strategy

The voltage control strategy for wind farms is shown in Figure 10.16.

Figure 10.14 Reactive optimal control strategy diagram.

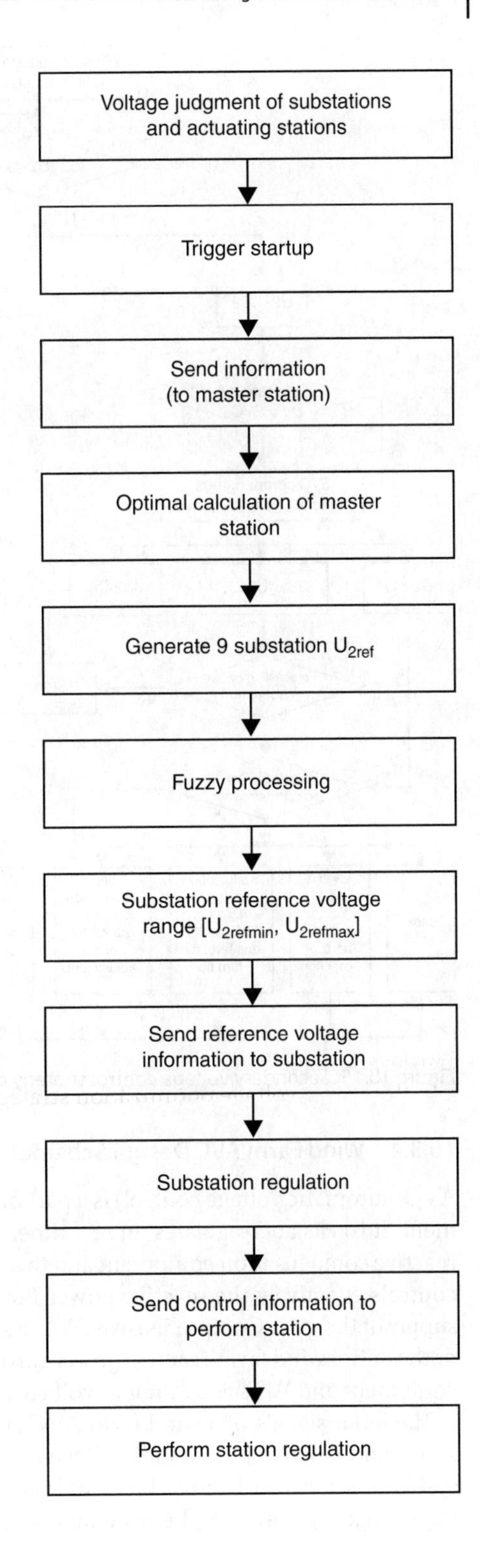

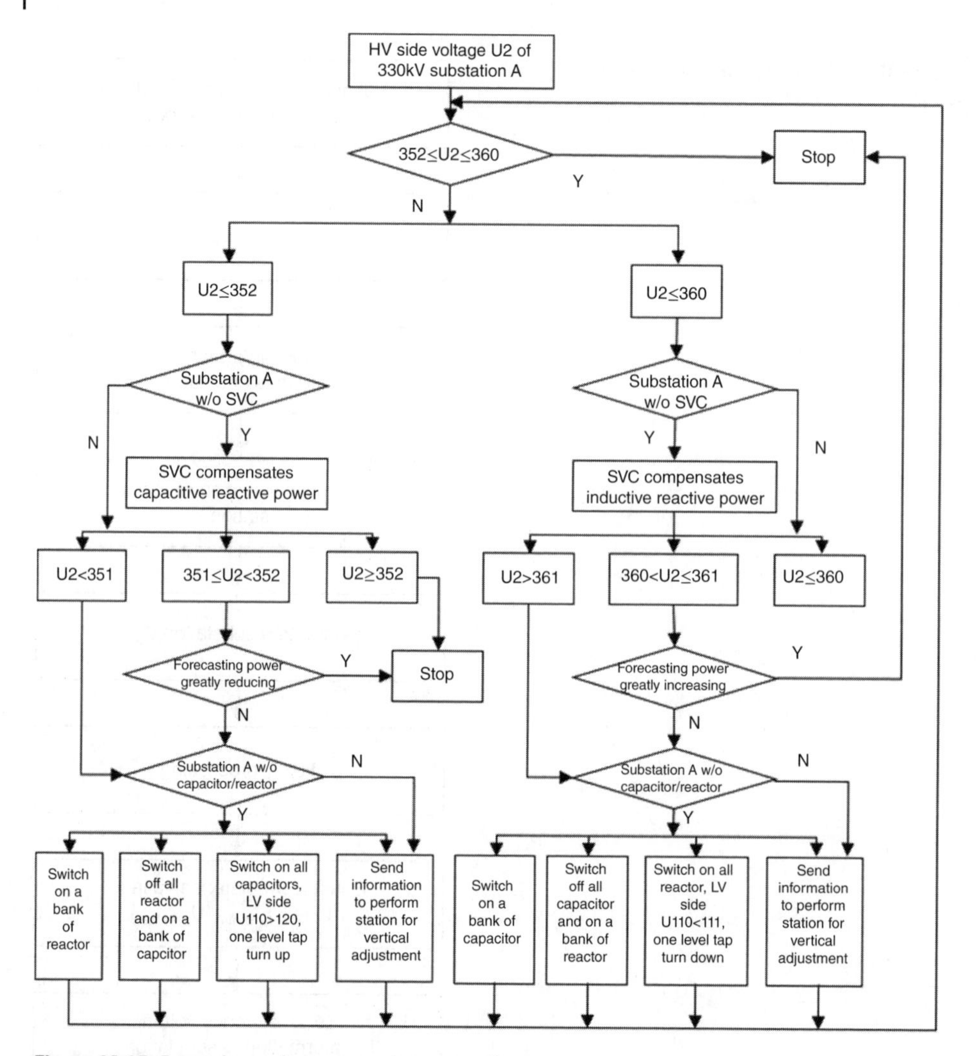

Figure 10.15 Secondary voltage control strategy diagram.

10.3.4 Wind Farm AVC Design Scheme

AVC (automatic voltage control) is a part of the automation of electric network management. It tracks and regulates, in real time, the reactive power of generators, adjusts the reactive compensation equipment and the on-load tap changer (OLTC) of transformers, controls effectively the reactive power flow of regional grids, and improves electricity supply of the grid. Comprehensive AVC system for wind farms achieves optimal control of the entire wind farm reactive power through regulation of the reactive compensation equipment and WTs according to voltage set-points at the POC.

The input signals of a wind farm AVC system include voltage set-point, wind speed, active and reactive power, and voltage at the POC. The structure of a wind farm AVC system is shown in Figure 10.17. WTGs capable of voltage control can participate in the voltage regulation of the system; otherwise, the reactive power can be adjusted only

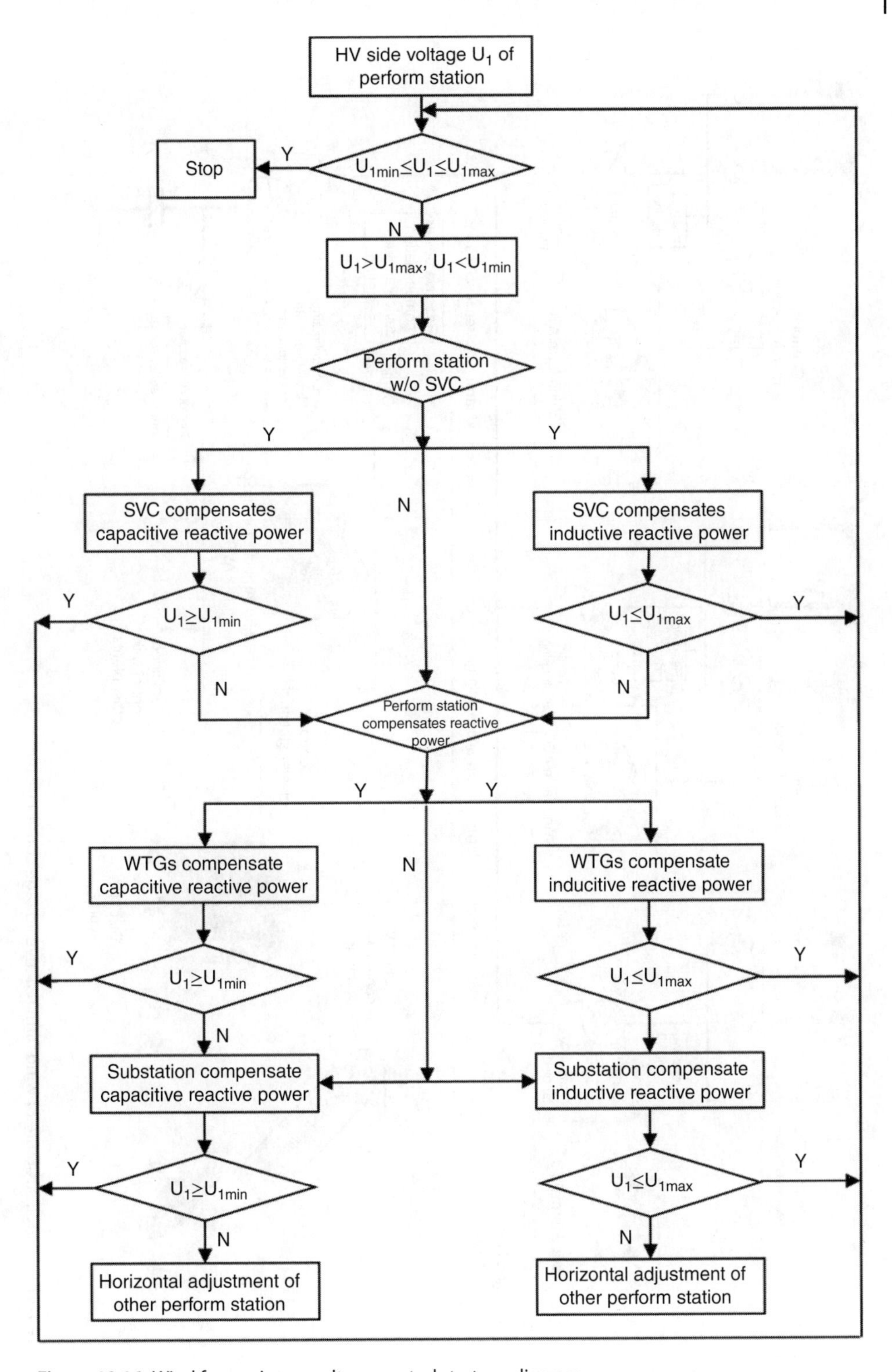

Figure 10.16 Wind farm primary voltage control strategy diagram.

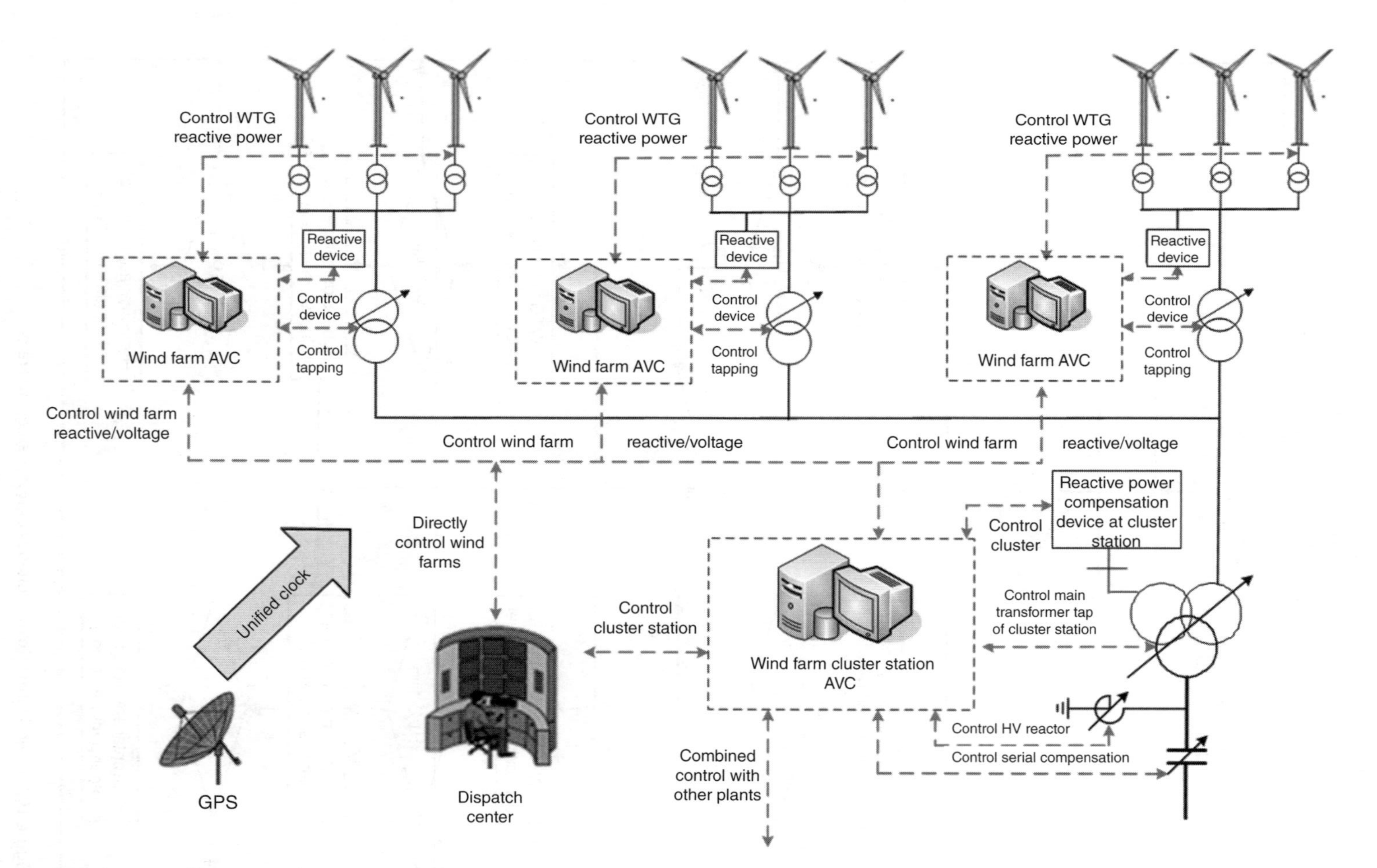

Figure 10.17 Overall structure of wind farm reactive voltage control system.

through regulation of reactive compensation devices and the OLTCs of transformers [10].

The wind farm AVC performance requirements of the Gansu Jiuquan Renewable Energy Base include:

- control cycle of the voltage control command should not exceed 5 min
- control cycle of reactive control command should not exceed 1 min
- voltage control response should not exceed 120 s and the deviation should be smaller than 0.5 kV
- reactive control response time should not exceed 30 s and the deviation should be smaller than 5 Mvar.

10.4 Conclusion

With the increasing penetration level of wind power, it wind farms must be able to participate in the frequency and voltage control of power systems. By coordination of the wind farm cluster, wind farm groups, and wind farms, a wind farm cluster is able to provide AGC and AVC services. This will help secure reliable operation of the whole power system.

References

1 Chi, Y. (2006) Studies on the stability issues about large scale wind farm grid integration, PhD thesis, China Electric Power Research Institute, Beijing, China.
2 Wei, L. (2009) Studies on the low voltage ride through of variable speed wind power generator, MSc thesis, Shandong University, Jinan, China.
3 Niu, D., Li, J., Wei, L., and Chi, Y. (2016) Study on technical factors analysis and overall evaluation method regarding wind power integration in trans-provincial power grid, *Power System Technology*, **40** (4), 1087–1093.
4 Janssens, N.A. (2007) Active power control strategies of DFIG wind turbines, *IEEE Power Technology*, **1** (5), 516–521.
5 Lubosny, Z. (2007) Supervisory control of a wind farm, *IEEE Power Systems*, **22** (3), 985–994.
6 Dai, H. and Chi, Y. (2012) Comparison study on grid codes for connecting wind farm into power system, *Electric Power*, **45** (10), 1–11.
7 Elkraft System and Eltra (2004) Wind turbines connected to grids with voltages above 100 kV: Technical Regulation for the Properties and the Regulation of Wind Turbines, *Regulation TF* **3**.2.5.
8 Tang, Y. (2011) *Power System Voltage Stability*, Science Press, Beijing, China.
9 Zhu, Y., Chi, Y., and Li, Y. (2012) *Reactive Power Compensation and Voltage Control for Wind Farms*, Publishing House of Electronics Industry, Beijing, China.
10 Wang, W.,Fan, G.,and Zhao, H. (2007) Comparison of technical regulations for connecting wind farm to power grid and preliminary research on its integrated control system, *Power System Technology*, **31** (18), 73–77.

three-phase [illegible] compensation [illegible]

[illegible]

[illegible]

[illegible]

[illegible]

[illegible]

References

1 [illegible] China.
2 Wei, [illegible] (2009) Study on [illegible] low voltage [illegible] variable speed wind power generator [illegible]
3 [illegible] (2008) [illegible] reactive power [illegible] analysis [illegible] wind farm [illegible]
4 [illegible]
5 [illegible] (2007) [illegible] 22 [illegible]
6 Dai [illegible] (2012) [illegible] grid codes for [illegible] wind [illegible] power [illegible]
7 [illegible] Technical Regulation for [illegible]
8 Tang [illegible] (2011) [illegible] Science Press, Beijing.
9 [illegible] China [illegible] GB/T [illegible]
10 Wang, W., [illegible] and Zhang [illegible] (2007) Comparison of [illegible] power [illegible] and [illegible] control [illegible] system. Power System [illegible] 31(18), 75–79.

11

Fault Ride Through Enhancement of VSC-HVDC Connected Offshore Wind Power Plants

Ranjan Sharma[1], Qiuwei Wu[2], Kim Høj Jensen[1], Tony Wederberg Rasmussen[2], and Jacob Østergaard[2]

[1] *Siemens Wind Power, Denmark*
[2] *Centre for Electric Power and Energy, Technical University of Denmark*

11.1 Introduction

Voltage source converter–high voltage direct current (VSC-HVDC) connections have become a new trend for long-distance offshore wind power transmission. It has been confirmed by a lot of research that the maximum distance of a high-voltage AC (HVAC) submarine cable transmission system is limited, due in particular to surplus charging currents of the cables [1, 2]. The VSC-HVDC transmission system has the ability to overcome these limitations and offers other advantages over the HVAC transmission system [3–6]. It has been demonstrated that it is generally more efficient to transport power over an HVDC line for high power and distances larger than 100 km [1]. A commonly discussed technique for long-distance power transmission is to use a mixed AC/DC system with a medium-voltage AC (MVAC) collector network at the wind power plant (WPP) and an HVDC transmission line.

One of the most challenging issues in connecting an offshore WPP via a VSC-HVDC transmission line to a host power system is to fulfill the grid code requirements during power system faults. It is essential to ensure that offshore WPPs can meet the fault ride through (FRT) requirements specified by various grid codes. This topic has been widely investigated [7–16]. Feltes et al. used a controlled demagnetization method to realize fast voltage drop at the sending end converter (SEC) and fast power reduction from the WPP during the three phase fault in host system [8]. Ramtharang et al presented a control structure for full-converter wind turbine based WPPs, including de-loading the fully rated converter wind turbines and emulating short circuits in the WPP collector system by reducing the voltage at the offshore HVDC converter terminal [9]. Vrionis et al. developed a control strategy to define the optimal converter valve blocking time according to the severity of the fault and to alleviate the oscillations at the electrical system after de-blocking the converter valves [12]. Xu et al. proposed control strategies for FRT of HVDC-connected offshore WPPs through offshore converter control by sending output power orders to individual wind turbines or by frequency modulation in the offshore AC collector system [13, 14]. DC voltage control and power dispatch schemes for a multi-terminal HVDC system integrating large offshore wind farms were proposed

Modeling and Modern Control of Wind Power, First Edition. Edited by Qiuwei Wu and Yuanzhang Sun.

Companion website: www.wiley.com/go/wu/modeling

by Xu and Yao [15] for both normal operation and onshore grid FRT. A new hybrid HVDC connection for large wind farms with doubly fed induction generators (DFIGs) was proposed by Zhou et al., and the system's performance under both normal and fault conditions was demonstrated for such a hybrid HVDC-connected wind farm [16].

This chapter presents a feedforward DC voltage control based FRT scheme for VSC-HVDC-connected offshore WPPs. The proposed DC voltage control scheme considers both DC voltage error between the reference value and the measured value, and the AC current measurement from the WPP collector system. It achieves fast and robust FRT of VSC-HVDC-connected offshore WPPs.

The usual FRT methods for VSC-HVDC-connected WPPs can be divided into five categories [8]:

- Method A: Dissipation of the waste energy by a fully rated DC chopper
- Method B: Active current reduction through WPP-side converter power control
- Method C: Active current reduction through power set-point adjustment
- Method D: Active current reduction through frequency control
- Method E: Voltage reduction in wind farm through WPP-side converter control.

The proposed FRT technique using DC voltage control is based on Method E. Compared to Method A, the proposed method is more cost-effective, and is also better than Methods B, C, and D since no communication is involved between the WPP-side converter and the individual wind turbines. It can also achieve fast voltage control in the WPP collector system. The method proposed by Feltes et al. [8] is also based on Method E and uses controlled demagnetization to achieve fast voltage reduction without producing the typical generator short-circuit currents and the related electrical and mechanical stress to wind turbines and converter. It has been verified that it works very well for DFIG-based WPPs. However, it might not be necessary for fully rated wind turbine generator (WTG) based WPPs because the fully rated WTGs are decoupled from the WPP AC collector system. Compared to the method of Feltes et al. [8], the proposed feedforward DC voltage control based FRT technique considers the measured current from the WPP collector system and can realize very fast voltage reduction within the WPP collector system. Here, the control is further extended to unbalanced faults, where the DC voltage oscillations are directly used to reflect the grid condition at the WPP collector network, especially faults at the collector network.

The chapter is arranged as follows. The model of a VSC-HVDC-connected offshore WPP is described in detail in Section 11.1. The control of the grid-side and WPP-side converters of the VSC-HVDC connection is also explained in Section 11.1. The feedforward DC voltage control based FRT scheme is described in Section 11.2. Time-domain simulation results are presented and discussed in Section 11.3. In Section 11.4, conclusions are drawn.

11.2 Modeling and Control of VSC-HVDC-connected Offshore WPPs

In order to facilitate the derivation of the feedforward DC voltage control based FRT technique, the model of a VSC-HVDC-connected offshore WPP with an external grid is described in Section 11.2.1. The control of the grid-side converter of the HVDC connection under steady-state conditions and during a voltage dip in the host power system is explained in Section 11.2.2. The control of the WPP-side converter under steady-state conditions and during a fault in the WPP collector system is described in Section 11.2.3.

11.2.1 Modeling of VSC-HVDC-connected WPP with External Grid

The model of a VSC-HVDC-connected offshore WPP with an external grid is shown in Figure 11.1.

The host power system (grid) is represented by a Thévenin voltage source with equivalent impedance. The grid-side VSC of the HVDC transmission line connects the power system through a grid-side transformer. The HVDC transmission cables are represented by equivalent proportional integral (PI) models. For a transmission distance of 100 km, eight PI equivalent sections were used to represent each cable. Increasing the number of sections would add little extra detail at the expense of a considerable increase in computation time. The WPP-side VSC is connected to the WPP collector grid via a park transformer.

The WPP is modeled as an up-scaled single wind turbine, representing an aggregated model. Individual wind turbines are based on a fully rated converter system and are Type-4-WTGs according to the IEEE definition [17, 18]. The fully rated converter WTG model has the following parts:

- an aerodynamic model, including power coefficient, tip-speed ratio, and pitch
- a shaft model, with inertia and damping of two-mass rotors, gearbox, and generator
- a turbine controller, including all necessary mechanical controls
- converters and control to regulate the active and reactive power as well as the DC-link voltage,
- a DC-link, including the DC-link capacitance,
- FRT-limiting active current injection to the network upon fault detection and a negative sequence control during unbalanced faults (negative sequence current references set to 0 so that only positive sequence currents are injected).

It has been shown that a WPP-based variable speed wind turbine can be represented by a single up-scaled wind turbine model for dynamic and transient studies with acceptable electrical behavior accuracy [19–22]. As the WPP is modeled as a single up-scaled wind turbine, an equivalent circuit is used to represent the collector network. The total cable length of the collector network is relatively small. Therefore, it is reasonable to model it using a single equivalent network. The VSCs at the two ends of the HVDC transmission system are represented by an average-value model together with a detailed control system.

11.2.2 Modeling of VSC-HVDC-connected WPP

The control of the grid-side VSC can be divided into two loops. The inner current loop controls the converter current by sending reference voltages to the converter modulation unit. The inner current loop operates in a *dq* rotating reference frame synchronized to the AC voltage by a phase-locked loop (PLL). The reference values for the inner active current and reactive current controller are derived from the relatively slower outer

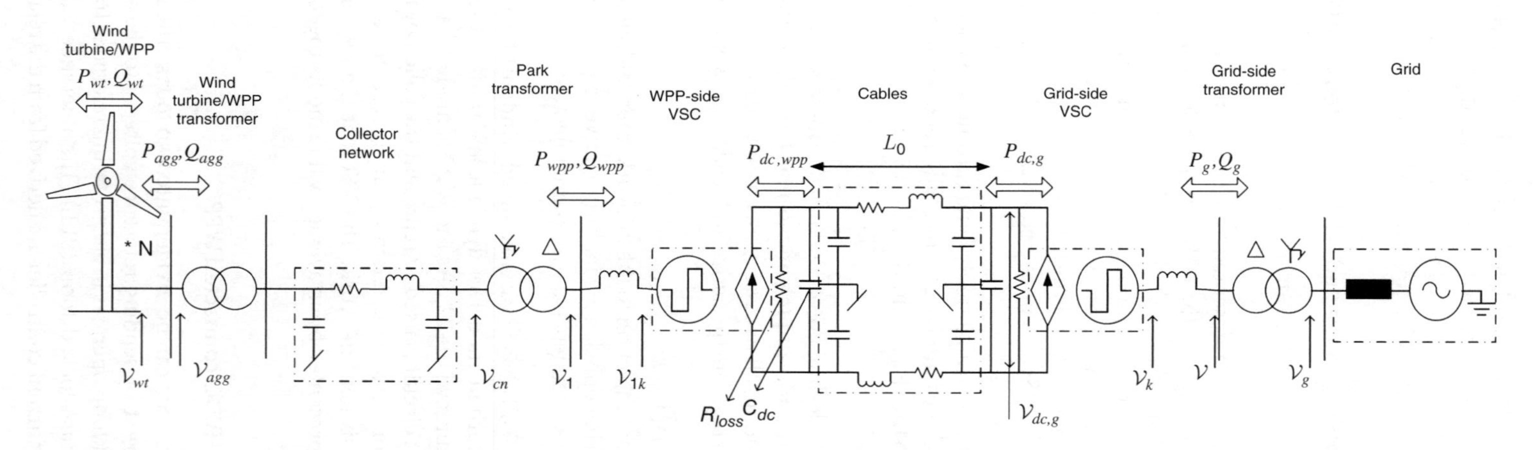

Figure 11.1 Flowchart of EV grid impact study.

loops. The outer loop consists of a DC voltage controller (i^r_{kd} as the active current reference) and an AC voltage controller (i^r_{kq} as the reactive current reference). The DC voltage controller is provided a reference of 1.0 pu, whereas the AC voltage controller varies depending upon the grid demand, but is typically set to 1.0 pu.

The inner current control loop can be further divided into two parts, with positive and negative sequence current control. Control of negative sequence current is required during AC grid voltage imbalances. The reference values of the negative sequence current can either be set to 0 or can be actively derived to minimize the DC voltage oscillations. These issues, of unbalanced voltage and DC voltage oscillation, are considered by Song and Nam [23]. In this paper, active negative sequence control is used to minimize DC voltage oscillations during small voltage imbalances, and the negative sequence control is set with the lowest priority during fault events, which implies that negative sequence current references are forced to 0 during severe fault events.

A VSC with an output inductance $L_\mathrm{k} = 0.15$ pu and resistance $R_\mathrm{k} = 0.015$ pu can be expressed as:

$$v_k(t) - v(t) = L_k \cdot \frac{di_k(t)}{dt} + R_k \cdot i_k(t) \tag{11.1}$$

where $v_\mathrm{k}(t)$ is the VSC terminal voltage, $v(t)$ is the grid voltage at the low voltage(LV) side of the grid-side transformer, and $i_\mathrm{k}(t)$ is the phase current.

The expressions in the stationary reference frame can be directly derived from (11.1) to control both the positive and negative sequence currents. A standard Park's transformation matrix is applied, for a abc/dq transformation. A delayed signal cancellation method is used for the sequence separation. Saccomando and Svensson conclude that the delayed signal cancellation method with a fixed delay of 1/4 of the fundamental period is the best of the available approaches.

$$\begin{aligned} v^r_{kdp}(n) = {} & \frac{R_k}{2} \cdot (i^r_{kdp}(n) + i_{kdp}(n)) + \frac{L_k}{T_s} \cdot (i^r_{kdp}(n) - i_{kdp}(n)) \\ & - \frac{\omega \cdot L_k}{2} \cdot (i^r_{kqp}(n) + i_{kqp}(n)) + v_{dp}(n) \end{aligned} \tag{11.2}$$

$$\begin{aligned} v^r_{kqp}(n) = {} & \frac{R_k}{2} \cdot (i^r_{kqp}(n) + i_{kqp}(n)) + \frac{L_k}{T_s} \cdot (i^r_{kqp}(n) - i_{kqp}(n)) \\ & - \frac{\omega \cdot L_k}{2} \cdot (i^r_{kdp}(n) + i_{kdp}(n)) + v_{qp}(n) \end{aligned} \tag{11.3}$$

$$\begin{aligned} v^r_{kdn}(n) = {} & \frac{R_k}{2} \cdot (i^r_{kdn}(n) + i_{kdn}(n)) + \frac{L_k}{T_s} \cdot (i^r_{kdn}(n) - i_{kdn}(n)) \\ & - \frac{\omega \cdot L_k}{2} \cdot (i^r_{kqn}(n) + i_{kqn}(n)) + v_{dn}(n) \end{aligned} \tag{11.4}$$

$$\begin{aligned} v^r_{kqn}(n) = {} & \frac{R_k}{2} \cdot (i^r_{kqn}(n) + i_{kqn}(n)) + \frac{L_k}{T_s} \cdot (i^r_{kqn}(n) - i_{kqn}(n)) \\ & - \frac{\omega \cdot L_k}{2} \cdot (i^r_{kdn}(n) + i_{kdn}(n)) + v_{qn}(n) \end{aligned} \tag{11.5}$$

where $v^r_{kdp}(n)$ is the reference of the positive sequence d-axis voltage at the grid-side converter, $v^r_{kqp}(n)$ is the reference of the positive sequence q-axis voltage at the grid-side converter, $v^r_{kdn}(n)$ is the reference of the negative sequence d-axis voltage at the grid-side converter, and $v^r_{kqn}(n)$ is the reference of the negative sequence q-axis voltage at the grid-side converter.

Equations 11.2–11.5 can be further simplified. For brevity, only the simplified (11.2) is presented below. A similar simplification process can be applied to the remaining three equations.

$$v^r_{kdp}(n) = R_k \cdot i_{kdp}(n) + G_c \cdot (i^r_{kdp}(n) - i_{kdp}(n)) - \frac{\omega \cdot L_k}{2} \cdot (i^r_{kqp}(n) + i_{kqp}(n)) + v_{dp}(n) \tag{11.6}$$

where G_c is a PI controller with $k_p = 50$ and $k_i = 400$.

The PI controller processes the error signal in order to stabilize the current control, and the remaining terms in the above equations are feedforward terms, which are mainly responsible for decoupling the dynamics of $i_{kdp}(n)$ from $i_{kqp}(n)$ (or vice-versa) and decoupling the dynamics of $i_{kdp}(n)$ and $i_{kqp}(n)$ from those of the host power system. Inner current control and outer voltage control are common in VSC applications and therefore are not discussed in further detail here. A current limit of $i_{\lim} = \sqrt{(i^r_{kd})^2 + (i^r_{kq})^2} = 1.1$ pu has been applied with the maximum allowable $i^r_{kq} = 1.0$ pu and $i^r_{kd} = 1.1$ pu.

When a three-phase fault occurs at the high voltage (HV) side of the grid transformer, the grid-side VSC will respond by injecting higher active current into the system to balance the HVDC voltage and hence the active power flows. However, if the voltage dip is large enough, the current limits of the grid-side VSC will restrict any further increase in active current. At the same time during a voltage dip, the reactive current control takes the highest priority, as per grid code requirements, and this further reduces the active power capacity [25]. To calculate the reference value of reactive current delivered to the grid, the measured voltage of the LV-side grid transformer is compared against a reference value and a voltage gain of $k_{v,gain} = 2$ is applied to the difference.

$$i^r_{kqp}(n) = k_{v,gain} \cdot i_{q,\lim} \cdot (v_{gd,ref} - v_{gd,eq}(n)) \tag{11.7}$$

where

$$v_{gd,eq}(n) = v_{dp}(n) - i_{dq}(n) \cdot X_{tx} \tag{11.8}$$

where $i_{q,\lim}$ is the maximum q-axis current limit, $v_{gd,ref}$ is the reference grid voltage, $v_{gd,eq}(n)$ is the equivalent grid voltage measured from the LV-side transformer, v_{dp} is the voltage magnitude of the LV-side grid transformer, and X_{tx} is the equivalent impedance of transformer.

The reactive current control of the grid-side converter during grid faults is shown in Figure 11.2.

As a result, the active power transfer will be reduced and limited, and the DC voltage will increase if the WPP is producing the same amount of active power as under pre-fault conditions. The rate of the DC voltage increase dv_{dc}/dt is mainly determined by the amount of active power generated by the WPP and the HVDC-side equivalent capacitors.

11.2.3 Control of WPP-side VSC

The WPP-side VSC is responsible for providing synchronizing grid voltage to the entire WPP. During normal operation, an open-loop control system is used to provide a constant AC voltage magnitude and frequency reference to the WPP collector network

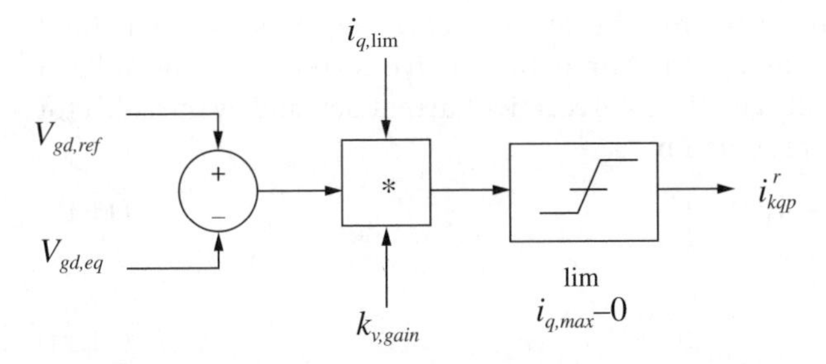

Figure 11.2 Reactive current control of grid-side converter during a voltage dip in host power system.

via the WPP-side VSC. To implement a constant voltage, the voltage references are defined in the synchronous reference frame. A dq/ abc transformation constant is used so that the voltage vector in the d- axis equals the peak value of the per-phase AC voltage. The three voltage references are $v^r_{1kd} = 1.0$ pu, $v^r_{1kq} = 0$ pu, and $v^r_{1k0} = 0$ pu. The modulation index m can be calculated from the measured DC voltage (the DC voltage is kept relatively constant by the grid-side VSC during normal steady-state operation) and the peak AC voltage reference. The three-phase AC voltage references can be further calculated as below:

$$m = \frac{2 \cdot v^r_{1kd}}{v_{\text{DC,WPP}}} \tag{11.9}$$

$$v^r_{1ka} = m \sin(\omega t) \tag{11.10}$$

$$v^r_{1kb} = m \sin\left(\omega t + \frac{2\pi}{3}\right) \tag{11.11}$$

$$v^r_{1kc} = m \sin\left(\omega t + \frac{4\pi}{3}\right) \tag{11.12}$$

where m is the modulation index, $v_{\text{dc,wpp}}$ is the WPP-side DC voltage, and v^r_{1ka}, v^r_{1kb}, and v^r_{1kc} are the AC voltage references of A, B, and C phases from the WPP-side VSC, respectively.

A constant frequency ($\omega = 2\pi f_{\text{m}}$) can be generated with the help of a virtual PLL or the frequency measurement signal communicated from the grid end. This ensures that the AC voltage amplitude and the frequency at the collector network are all constants regardless of the DC voltage fluctuation. The fluctuations in the WPP-side DC voltage can be due to the variations in active power transfer over the HVDC cables as wind speed varies or AC voltage imbalances in the host power system. Therefore, the WPP-side VSC can be compared to an equivalent AC slack bus (infinite bus), which absorbs the power produced by the WPP without altering the voltage amplitude and angle.

This provides a control mechanism to the WPP-side VSC and as such there is no direct current control implemented. In a normal steady-state situation, the total current in the WPP-side electrical system is controlled by wind turbine converters (provided that the wind turbines in the WPP are based on full-converter technology). Therefore, the WPP-side VSC must have a current capacity equal to or slightly higher than the WPP ratings (to provide some tolerance).

During disturbances in the collector network, indirect current control (a current limit) is implemented by using the voltage drop technique. As the current through the converter exceeds an upper threshold limit, the terminal voltage of the VSC is reduced sufficiently in a controlled manner to limit the fault current. The upper threshold limit of the current is mainly determined by the sum of total active current from the WPP at the maximum rated wind velocity and the total reactive current demand by the collector network at rated active power production.

$$v^r_{1kd,new} = v^r_{1kd} + G_c(i^r_{1k} - |i_{1k}|) \tag{11.13}$$

where

$$|i_{1k}| = \sqrt{i^2_{1kd} + i^2_{1kq}} \tag{11.14}$$

where G_c represents a PI controller, i_{1kd} and i_{1kq} are the active and reactive current flowing through the WPP-side VSC, respectively, and i_{1k} is the upper threshold limit of the VSC current.

11.3 Feedforward DC Voltage Control based FRT Technique for VSC-HVDC-connected WPP

The decoupling of the WPP collector grid and the in-land transmission system (host power system) makes it impossible for wind turbines to directly respond to voltage changes in the main grid without any external influence. In this paper, a feedforward DC voltage control based FRT technique is proposed to control the AC voltage at the WPP collector network during grid-side faults.

In the proposed control method, WPP-side VSC control actively monitors the DC voltage at the WPP end. As soon as the DC voltage exceeds a pre-defined threshold $v_{\mathrm{dc,th}}$, the WPP-side VSC takes over the DC voltage control role.

A single-line representation of the electrical system at the WPP is shown in Figure 11.3. Because the wind turbines are equipped with a full-converter system, the wind turbines represent a current source, and their response to changes in the collector network voltage is very fast.

The derivation of the applied control is presented below. During the derivation, it is assumed that the converter is lossless. Further, if it is also assumed that the voltages and the currents at the AC side are balanced, the expressions for the active power at AC side and DC side can be written as,

$$P_{wpp} = \frac{3}{2}[v_{1kdp}(t)i_{1kdp}(t) + v_{1kqp}(t)i_{1kqp}(t)] \tag{11.15}$$

$$P_{dc,wpp} = i_{dc,wpp}(t) \cdot v_{dc,wpp}(t) \tag{11.16}$$

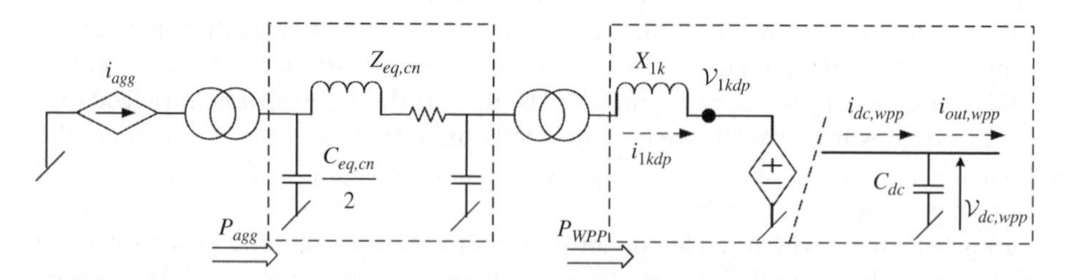

Figure 11.3 Single-line representation of WPP and the WPP-side HVDC-VSC.

The AC-side power and the DC-side power are equal if the system losses are neglected:

$$P_{wpp} = P_{dc,wpp} \tag{11.17}$$

$v^r_{1kqp} = 0$, which implies,

$$i_{dc}(t) = \frac{3v^r_{1kdp}(t)i_{1kdp}(t)}{2v_{dc,wpp}(t)} \tag{11.18}$$

In (11.18), current $i_{1kdp}(t)$ is determined by the wind farm. The collector network voltage is set by the WPP-side VSC, so the AC voltage $v_{1kdp}(t)$ is also the reference value for the HVDC-VSC voltage controller. Under normal operating conditions, $v^r_{1kdp}(t)$ is equal to 1.0 pu. During FRT, the WPP-side VSC is assigned to control the collector network AC voltage and therefore the DC voltage at the HVDC transmission side. Properly implemented DC voltage control maintains the energy balance over the DC capacitor, and any imbalance will lead to an increase in the DC voltage level, as given by (11.19):

$$i_{dc,wpp}(t) - i_{out,wpp}(t) = C_{\mathrm{dc}}\frac{dv_{\mathrm{dc,wpp}}(t)}{dt} \tag{11.19}$$

Integrating the above equation over the sampling period nT_s to $(n+1)T_s$, the following expression can be derived:

$$i_{\mathrm{dc,wpp}}(n) - i_{\mathrm{out,wpp}}(n) = \frac{1}{T_s}C_{\mathrm{dc}}[v_{\mathrm{dc,wpp}}(n+1) - v_{\mathrm{dc,wpp}}(n)] \tag{11.20}$$

Substituting (11.18), the following expression can be derived:

$$v^r_{1kdp}(n) = \frac{2}{3}\frac{v_{dc,wpp}(n)}{i_{1kdp}(n)}\frac{C_{dc}}{T_s}(v^r_{dc,wpp}(n) - v_{dc,wpp}(n)) + \frac{2}{3}\frac{v_{dc,wpp}(n)}{i_{1kdp}(n)}i_{out,wpp}(k) \tag{11.21}$$

where $v^r_{dc,wpp}$ is the DC voltage reference.

The control of the WPP-side VSC during an FRT event can be established from (11.21). The first part of (11.21) consists of the system gain and the error in the DC voltage, which can be further applied to a PI controller. The second part of (11.21) is the feedforward term. Compared to the response time of the HVDC voltage controller, the response time of the current control in wind turbines is relatively fast. Therefore, it can be considered as ideal. The feedforward DC voltage control based FRT technique is illustrated in Figure 11.4. The design of the HVDC voltage control can follow the standard procedure, with the plant transfer function given as . The DC voltage reference during FRT is set to 1.05 pu and the pre-defined threshold to detect a fault is set to 1.1 pu.

11.4 Time-domain Simulation of FRT for VSC-HVDC-connected WPPs

Time-domain simulations have been used to verify the efficacy of the proposed feedforward DC voltage control based FRT technique for VSC-HVDC-connected WPPs. The test system is described in Section 11.4.1. Case study scenarios and their results are presented and discussed in Section 11.4.2.

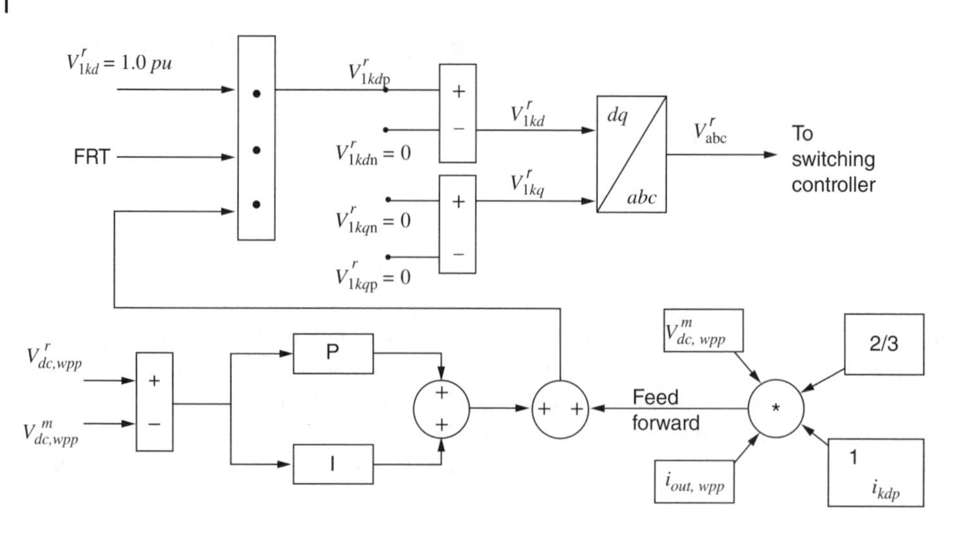

Figure 11.4 FRT control based on feedforward DC voltage control

11.4.1 Test System for Case Studies

The test system used for the case studies consists of an offshore WPP with fully-rated WTGs, connected to the host system by a two-terminal VSC-HVDC transmission line. The test system corresponding to a VSC-HVDC-connected offshore WPP is shown in Figure 11.1. The details of the test system are listed in Table 11.1.

In Table 11.1, N is the number of turbines in the WPP, L_0 is the HVDC transmission distance, subscript wt stands for wind turbine, cn stands for WPP collection network, dc stands for DC, k stands for converter and g stands for grid.

In order to verify the proposed FRT technique for the VSC-HVDC-connected WPPs, a number of time-domain simulations have been performed with faults in the host grid and the WPP collector system. The simulation scenarios are listed in Table 11.2.

11.4.2 Case Study

Case Study 1

The results of Case study 1 are shown in Figure 11.5 and Figure 11.6.

When a three-phase short-circuit fault occurs at the HV side of grid transformer, the active power transfer towards the grid is minimal: $P_g \approx 0$. The only active power sink available in the system is the system losses. Therefore, even though the active power

Table 11.1 Parameters of VSC-HVDC-connected WPP.

Variable	Value	Unit	Variable	Value	Unit
S_{wt}	3.6	MVA	V_{wt}	690	V
V_{cn}	33	kV	S_k	180	MVA
V_{dc}	200	kV	V_k	100	kV
V_g	400	kV	L_0	100	km
N	50				

Table 11.2 Parameters of the VSC-HVDC-connected WPP.

Case no.	Fault location	Fault type	Retain voltage at fault location (pu)	Fault duration time (ms)
1	HV side of the grid transformer	3 Ph	0	250
2	HV side of the grid transformer	1 Ph to ground	0	250
3	LV side of the park transformer	3 Ph	0	250

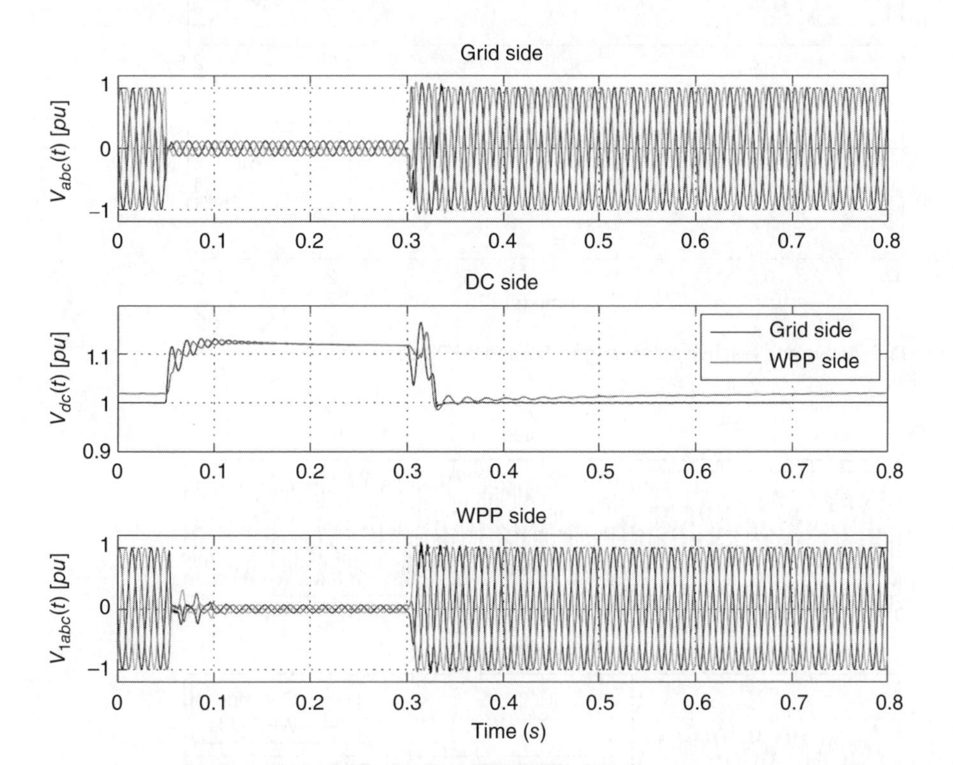

Figure 11.5 Voltages at grid, DC and WPP side for Case 1

produced by the WPP is 0, the HVDC voltage cannot be quickly adjusted to the reference value (1.05 pu) via the control system. Nevertheless, it can be seen from the simulation results that the HVDC voltage remains within a safe limit (slightly higher than 1.1 pu) during a very severe fault (the worst case) with a retained voltage of 0 pu at the HV side of the grid-side transformer.

Case Study 2

The results of Case study 2 are shown in Figures 11.7 and 11.8.

In the event of an unbalanced fault, the grid-side VSC will deliver zero negative sequence current. The active power transfer to the grid is oscillating with a frequency

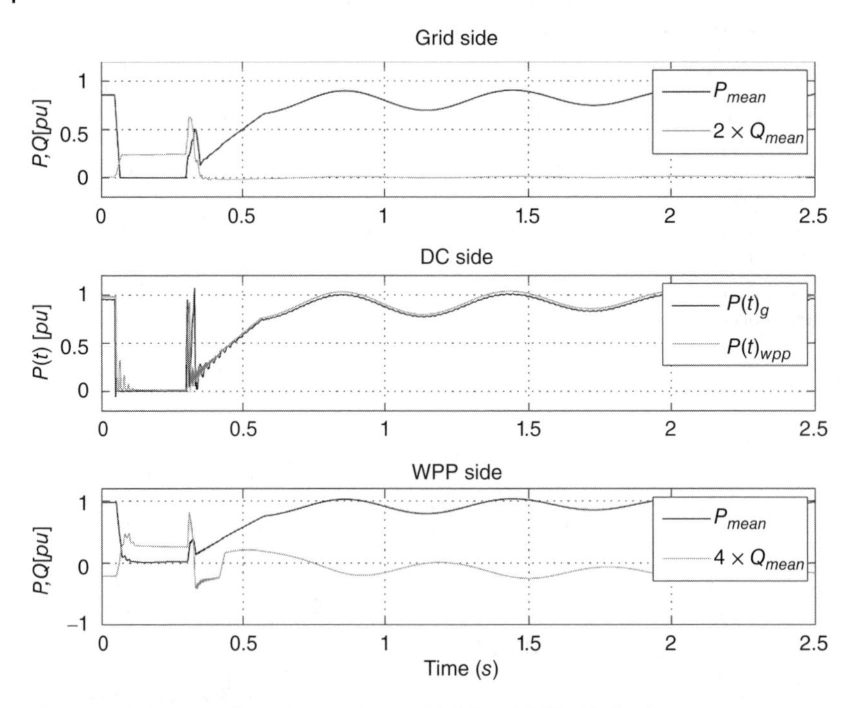

Figure 11.6 Active and reactive power at grid, DC and WPP side for Case 1

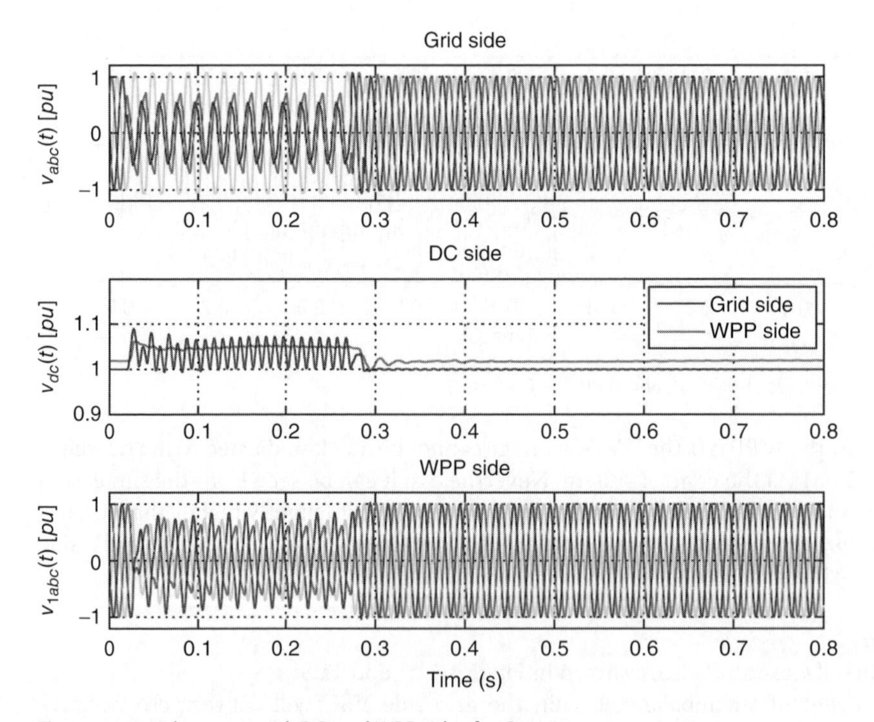

Figure 11.7 Voltages at grid, DC, and WPP sides for Case 2

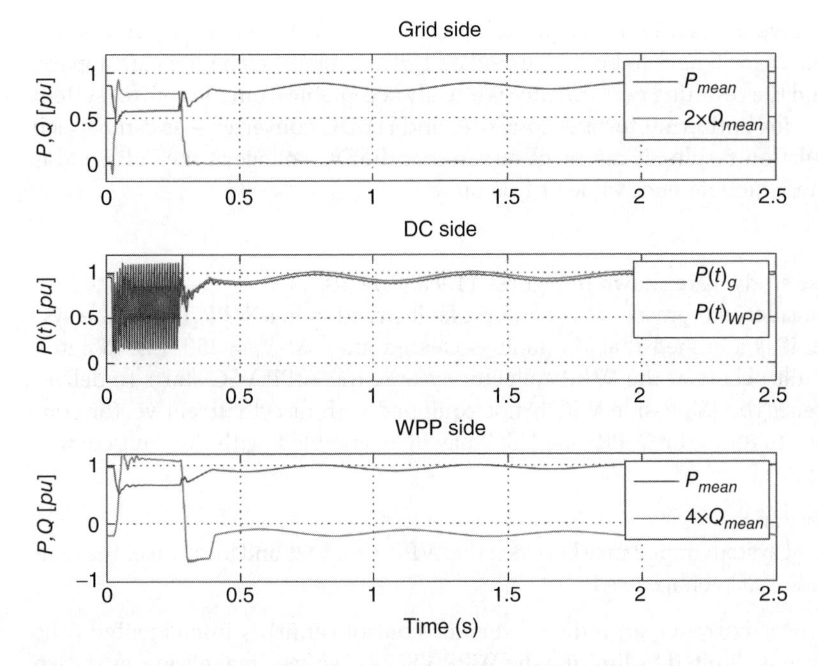

Figure 11.8 Active and reactive power at grid, DC, and WPP sides for Case 2

twice the grid fundamental frequency ($2f_m$). Oscillating power in the DC side means that the DC voltage and the current will also consist of a $2f_m$ ripple. Even though the HVDC transmission cables are long, part of these oscillations in the HVDC voltage and current are reflected at the WPP end. The control of DC voltage at the WPP-side VSC during an FRT mode requires both DC voltage and current as inputs. When such oscillating parameters are applied to the control given by (11.21), the resulting AC voltage peak reference v^r_{1kd} for the WPP-side VSC is also oscillating.

The transformation of this reference voltage into an abc stationary reference frame will generate an unbalanced voltage in the collector grid. The fully rated converter WTG used in the simulation model is already equipped with negative sequence control during unbalanced faults. Therefore, the WPP will act accordingly, injecting balanced three-phase currents, irrespective of the imbalance in the WPP collector network AC voltage. However, it has to be noted that a third-harmonic voltage will be superimposed on the fundamental frequency voltage due to the transformation of the oscillatory reference in the dq rotating reference frame into the abc stationary reference frame. However, the presence of such harmonics is only temporary during the FRT time frame.

In practice, it is possible to get rid of the third-harmonic content at the WPP collector network by adding third-harmonic compensation terms to the modulation signals. Alternatively, a large gain drop at frequency $2f_m$ (such as use of a notch filter) can be directly applied to the collector network AC voltage control reference $v^{\mathrm{r}}_{\mathrm{1kd}}$ or to the feedforward terms $v_{\mathrm{dc,wpp}}$ and $i_{\mathrm{out,wpp}}$. If this condition is satisfied, the three-phase voltages at the collector network can be considerably balanced. An ideal notch filter is not possible to achieve in practice, so a considerable drop in gain for a narrow band of frequency around $2f_m$ will have to be accommodated. However, no such filters are applied in this work and the resulting performance is already acceptable from the control system point of view – for both wind turbine converter and HVDC converter – as can be seen from the simulation results. It can be observed that the DC voltage at the WPP end is adjusted towards the reference value of 1.05 pu.

Case Study 3

Results of Case study 3 are shown in Figures 11.9 and 11.10.

The fault considered represents the harshest fault situation in a WPP. To test the system response, it is assumed that the fault is cleared after $\Delta t_{\text{fault}} = 250$ ms. When a three-phase fault occurs at the WPP collector system, the WPP VSC starts to deliver the fault current. The WPP-side VSC is not equipped with direct current vector control. In such a situation, the WPP-side VSC may be overloaded, with the fault current determined by:

- the fault resistance
- the equivalent system impedance between the WPP-side VSC and location of the fault
- the WPP-side VSC voltage level.

To prevent over-currents, an indirect current control (limit) is implemented. The fault current can be limited by limiting the WPP VSC AC voltage magnitude, as shown in the bottom plot of Figure 11.9. When the current through the WPP VSC exceeds the predefined upper threshold, the AC voltage limit mode is applied. The current feedback is derived from the three-phase currents as measured at the terminals of the WPP-side VSC, which can be transformed into the dq reference frame. The voltage

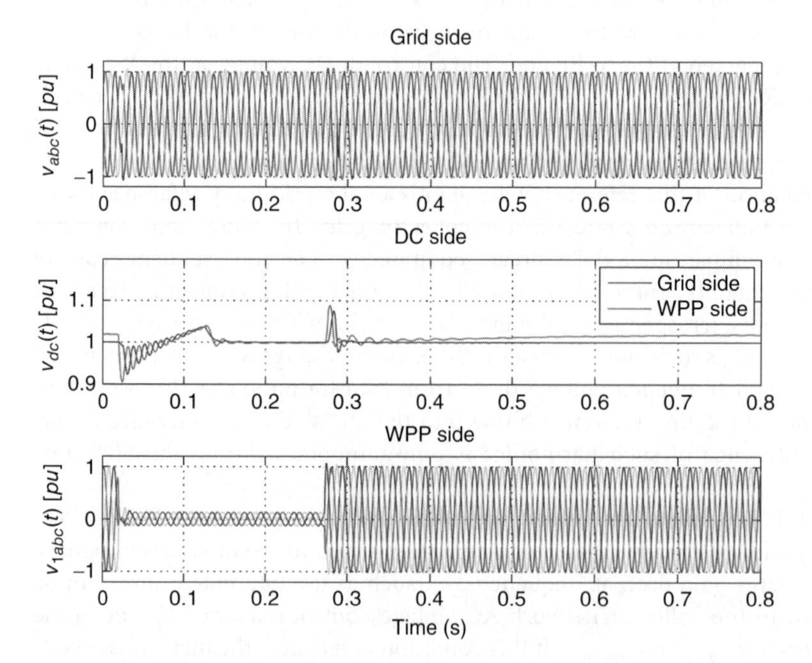

Figure 11.9 Voltages at grid, DC, and WPP sides for Case 3.

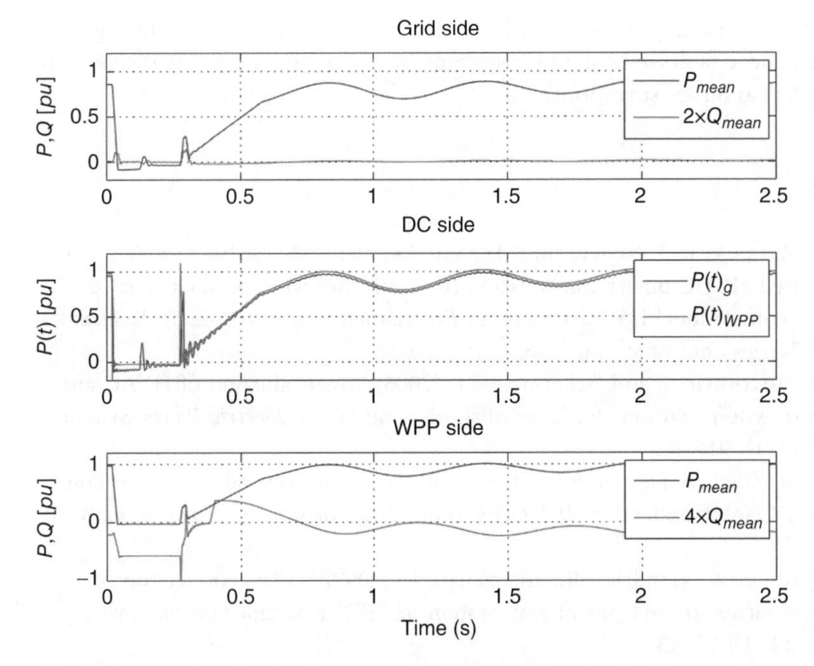

Figure 11.10 Active and reactive power at grid, DC, and WPP sides for Case 3.

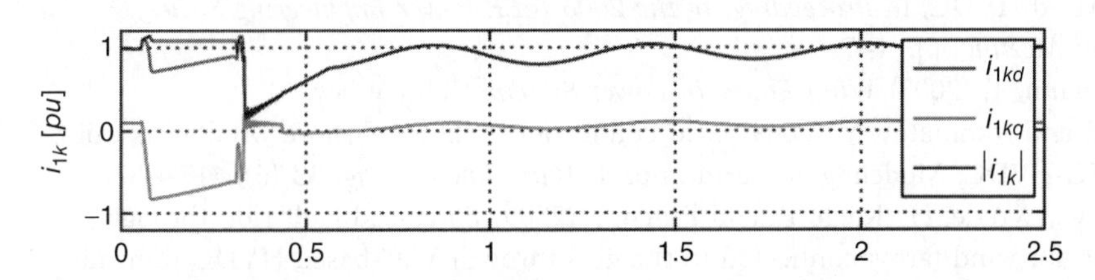

Figure 11.11 WPP-side converter currents in the *dq* reference frame and the total current.

limit mode is set to "on", so as to limit the current within safe limits by lowering the WPP VSC voltage level. The VSC voltage is suddenly reduced to a level that means that the current from the WPP-side VSC is restricted to within its maximum limit, as shown Figure 11.11.

11.5 Conclusions

A feedforward DC voltage control based FRT scheme has been proposed in order to enhance the FRT capability of VSC-HVDC-connected WPPs under both host power system faults and WPP collector system faults. The feedforward DC voltage control based FRT scheme considers both the DC voltage error between reference and measured values and the AC current from the WPP collector system, ensuring fast and robust FRT. Under unbalanced fault conditions in the host power system, the oscillating DC voltage is directly used so that the imbalance in the main grid is also reflected at the WPP collector system voltage. Time-domain simulation results show that the proposed

FRT scheme can successfully enable VSC-HVDC-connected WPPs to ride through balanced and unbalanced faults in host power systems, as well as faults in the WPP collector system, with a fast and robust response.

References

1 Sharma, R., Rasmussen, T.W., Jensen, K.H., and Akhmatov, V. (2010) Modular VSC converter based HVDC power transmission from offshore wind power plant: compared to the conventional HVAC system, in *Proceedings of the 2010 IEEE Electric Power and Energy Conference*, pp. 1–6.
2 Negra, N.B., Todorovic J., and Ackermann, T. (2006) Loss evaluation of HVAC and HVDC transmission solutions for large offshore wind farms, *Electric Power Systems Research*, **76** (11), 916–927.
3 Jovcic, D. and Strachan, N. (2009) Offshore wind farm with centralised power conversion and DC interconnection, *IET Generation, Transmission & Distribution*, **3** (6), 586–595.
4 Bresesti, P., Kling, W., Hendriks, R., and Vailati, R. (2007) HVDC connection of offshore wind farms to the transmission system, *IEEE Transactions on Energy Conversion*, **22** (1), 37–43.
5 Jovcic, D. (2006) Interconnecting offshore wind farms using multiterminal VSC-based HVDC, In *Proceedings of the 2006 IEEE Power Engineering Society General Meeting*, pp. 1–7.
6 Ackermann, T. (2005) *Wind Power in Power Systems*, John Wiley.
7 Xue, Y. and Akhmatov, V. (2009) Grid-connection of large offshore windfarms utilizing VSC-HVDC: Modeling and grid impact, *Wind Engineering*, **33** (5), 417–432.
8 Feltes, C., Wrede, H., Koch, F., and Erlich, I. (2009) Enhanced fault ride-through method for wind farms connected to the grid through VSC-based HVDC transmission, *IEEE Transactions on Power Systems*, **24** (3), 1537–1546.
9 Ramtharang, G., Arulampalam, A., Ekanayake, J.B., Hughes, F.M., and Jenkins, N. (2009) Fault ride through of fully rated converter wind turbines with AC and DC transmission systems, *IET Renewable Power Generation*, **3** (4): 426–438.
10 Jiang-Hafner, Y. and Ottersten, R. (2009) HVDC with voltage source converters-a desirable solution for connecting renewable energies. Paper presented at 2009 conference on *Large-scale Integration of Wind Power into Power System.*
11 Arulampalama, A., Ramtharan, G., Caliao, N., Ekanayake, J.B., and Jenkins, N. (2008) Simulated onshore-fault ride through of offshore wind farms connected through VSC HVDC, *Wind Engineering*, **32** (2), 103–113.
12 Vrionis, T. Koutiva, X. Vovos, N. A., and Giannakopoulos, G.B. (2007) Control of an HVDC link connecting a wind farm to the grid for fault ride-through enhancement, *IEEE Transactions on Power Systems*, **22** (4), 2039–2047.
13 Xu, L., Yao, L., and Sasse, C. (2007) Grid integration of large DFIG-based wind farms using VSC transmission, *IEEE Transactions on Power Systems*, **22** (3), 976–984.
14 Xu, L. and Andersen, B.R (2006) Grid connection of large offshore wind farms using HVDC, *Wind Energy*, **9** (4), 371–382.

15 Xu, L. and Yao L. (2011) DC voltage control and power dispatch of a multi-terminal HVDC system for integrating large offshore wind farms, *IET Renewable Power Generation*, **5** (3), 223–233.

16 Zhou, H., Yang, G., Wang, J., and Geng, H. (2011) Control of a hybrid high-voltage DC connection for large doubly fed induction generator-based wind farms, *IET Renewable Power Generation*, J. (**1**), 36–47.

17 IEC (2015) *Wind Turbines Part 27–1 Electrical simulation models – Wind turbines*, IEC 61400–27–1:2015.

18 Perdama, A. (2008) Dynamic models of wind turbines, PhD thesis, Chalmers University of Technology, Göteborg, Sweden.

19 Conroy, J. and Watson, R. (2009) Aggregate modelling of wind farms containing full-converter wind turbine generators with permanent magnet synchronous machines: transient stability studies, *IET Renewable Power Generation*, R. (**1**), 39–52.

20 Mohseni, M., Islam S., and Masoum, M.A.S. (2011) Fault ride-through capability enhancement of doubly-fed induction wind generators, *IET Renewable Power Generation*, **5** (5), 368–376.

21 Akhmatov, A. (2004) An aggregated model of a large wind farm with variable-speed wind turbines equipped with doubly-fed induction generators, *Wind Engineering*, **28** (4), 479–486.

22 Shafiu, A. Anaya-Lara, O. Bathurst, G. And Jenkins, N. (2006) Aggregated wind turbine models for power system dynamic studies, *Wind Engineering*, **30** (3), 171–186.

23 Song, H. and Nam, K. (1999) Dual current control scheme for PWM converter under unbalanced input voltage conditions, *IEEE Transactions on Industrial Electronics*, **46** (5), 953–959.

24 Saccomando, G. and Svensson, J. (2001) Transient operation of grid-connected voltage source converter under unbalanced voltage conditions, in *Proceedings of 2001 Thirty-Sixth IAS Annual Meeting Industry Applications Conference*, pp. 1–6.

25 Grid code high and extra high voltage 2006 [2013–08–25]. http://www.nerc.com/docs/pc/ivgtf/German_EON_Grid_Code.pdf.

12

Power Oscillation Damping from VSC-HVDC-connected Offshore Wind Power Plants

Lorenzo Zeni

DONG Energy Wind Power A/S, Denmark

This chapter describes the implementation of power oscillation damping (POD) on voltage sourced converter high voltage direct current (VSC-HVDC) connected wind power plants (WPPs). The state of the art of POD from WPPs connected to the grid using power electronic sources and HVDC connections is described and used as a starting point to set out the practical considerations for the implementation of POD on VSC-HVDC-connected WPPs in real systems. Eigenvalue analysis and dynamic simulations of simplified and more complex systems are used to support the description and demonstrate the concepts.

12.1 Introduction

12.1.1 HVDC Connection of Offshore WPPs

The connection of WPPs to shore via HVDC technology becomes advantageous above a certain distance to the onshore connection point. Technically, the possibility of transmitting power over uncompensated AC cables is limited, depending on amount of power, the voltage level and the cable technology [1]. However, different AC solutions have been proposed, which could extend the reachable transmission distance: examples are AC transmission with intermediate compensation [2] and low-frequency transmission inspired by traction applications [3]. It is hence still unclear where the economic break-even distance between AC and DC technology lies.

However, further improvements in HVDC technology are expected to mature the technology to such an extent that is will become attractive at shorter transmission distances. These developments are expected to lead to lower risk, shorter lead times, and optimized electrical designs for VSC-HVDC technology. On the other hand, cost-reducing uncontrolled line commutated converter technology may be used too, provided that the necessary modifications in wind turbine (WT) control are developed [4, 5]. Hence the prospects for more extensive application of HVDC technology are good.

The focus in this chapter is solely on VSC-HVDC technology. Focusing on the rather short term, the configuration depicted in Figure 12.1 represents a standard point-to-point VSC-HVDC-based connection for offshore WPPs, consisting of:

Modeling and Modern Control of Wind Power, First Edition. Edited by Qiuwei Wu and Yuanzhang Sun.

Companion website: www.wiley.com/go/wu/modeling

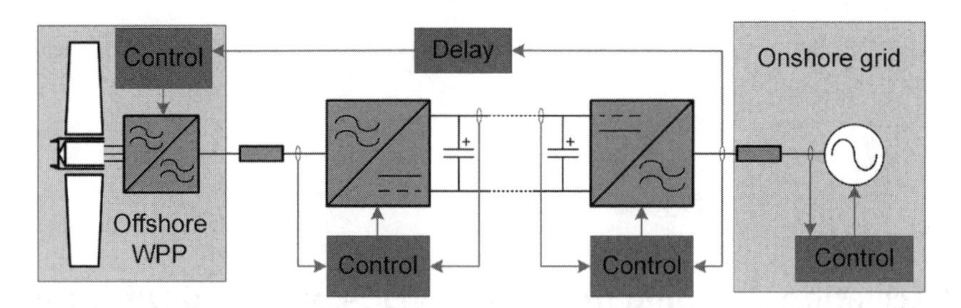

Figure 12.1 High-level diagram of VSC-HVDC connection for WPPs.

- onshore HVDC converter station
- DC cables
- offshore HVDC converter station
- WPP.

The WPP can consist of a cluster of several plants, as has been the case for all installations to date in German North Sea waters.

Besides climbing up the technology learning curve to bring down costs for the standard configuration, the industry can increase the attractiveness of HVDC technology for wind power by capitalising on its significant controllability and readying it to deliver advanced features such as POD. Although this service has not been used by asynchronous generators based on power electronics to date, the latest grid connection requirements do include this functionality for DC power park modules [6, 7]; POD is implemented on power electronic devices such as static compensators. Thus the purpose of this chapter is to illustrate how POD services may be implemented on VSC-HVDC-connected WPPs. The next subsection gives an overview of the state-of-art for POD provision from static power sources, which could be HVDC converters, WPPs, or a combination of both.

12.1.2 Power Oscillation Damping from Power Electronic Sources

Poorly damped low-frequency oscillations related to the electromechanical interaction between synchronous generators (SGs) have historically been a concern for power system engineers [8, 9]. The frequency of such oscillations is usually between 0.2 and 2 Hz, depending on their nature [10], although frequencies as low as 0.1 Hz have been observed in real systems [11]. Lower frequencies are related to governor dynamics, while higher frequencies are usually related to resonances between turbine shafts and electrical networks, or to purely electrical resonances. The POD problem is also known as "small-signal stability" and has been efficiently analysed by application of tools borrowed from linear control theory and mitigated by deploying so-called power system stabilizers (PSSs) on the automatic voltage regulators (AVRs) of selected SGs. Mathematically, the solutions aim at dragging the system eigenvalues further into the left half plane, improving the stability. Physically, this translates into more rapidly decaying power and speed oscillations and reduced mechanical stress.

Attention in recent years has shifted towards tackling the problem by exploiting power electronics controllability. The list of publications which have looked at the topic is long [10, 12–24]. There are many others too. The focus in the cited publications is mostly on:

- assessment of the influence of state-of-art power electronic sources on small-signal stability
- input–output selection for POD control from power electronics, mainly based on system-wide linear mathematical representation
- development of more innovative POD controllers for power electronics.

However, the available literature has the following gaps which this chapter aims at filling:

- Complex mathematics often tend to cause disconnection from the physics of the phenomena. Raising the abstraction level may hinder the necessary dialogue between utilities, transmission system operators (TSOs) and original equipment manufacturers (OEMs) for POD implementation.
- Crucial elements for the small-signal stability are often neglected in the analysis. For instance, excitation systems (ESs) and AVRs are neglected by several references, although they play a dominant role in small-signal stability.
- Limitations imposed by WPPs and collateral effects for WTs have only partly been touched upon in the cited studies, although they could play an important role in a practical realization of POD services.

By considering the cited references as valuable starting points, this chapter contributes to improving knowledge of POD from power electronics in general and in HVDC-connected WPPs in particular by examining the three items above. The scope is restricted to electromechanical power oscillations in the frequency range 0.1–2 Hz. Consequently, the term POD will generally refer to such a frequency range in the remainder of the chapter.

12.2 Modelling for Simulation

As elucidated previously, electromechanical oscillations are related to interaction between SGs rather than between SGs and an electrical network. The fact that frequency is practically constant in modern power systems allows for modeling in a phasor-based (RMS) environment. This section describes each component's modeling requirements.

12.2.1 HVDC System

Electrical System

VSC-HVDC converters can be modeled according to the simplified diagram in Figure 12.2. The capacitance C_f represents the harmonic filter, which may not be needed for modern modular multilevel converters. Z_{ph}, Z_T and $Z_0 = Y_0{}^{-1}$ are the phase reactor and transformer short-circuit and no-load impedances respectively (the transformer can be modeled in detail using power system simulation software). C_{DC} is the equivalent DC side capacitance calculated as in the paper by Glasdam et al. [25]. For RMS studies regarding electromechanical oscillations and the related POD service, the AC voltage source may be substituted by an AC current source.

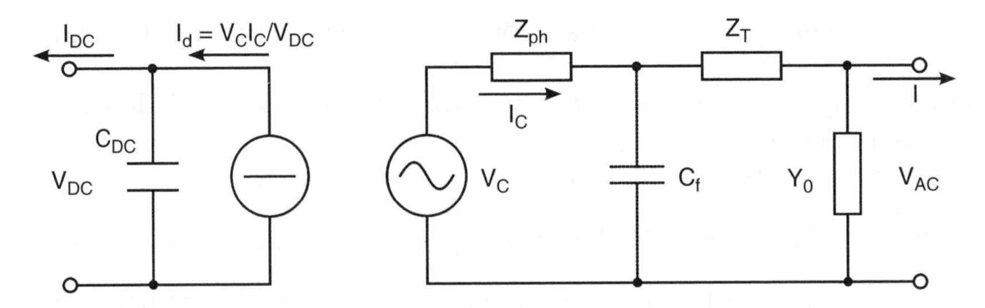

Figure 12.2 VSC-HVDC converter electrical model for POD studies.

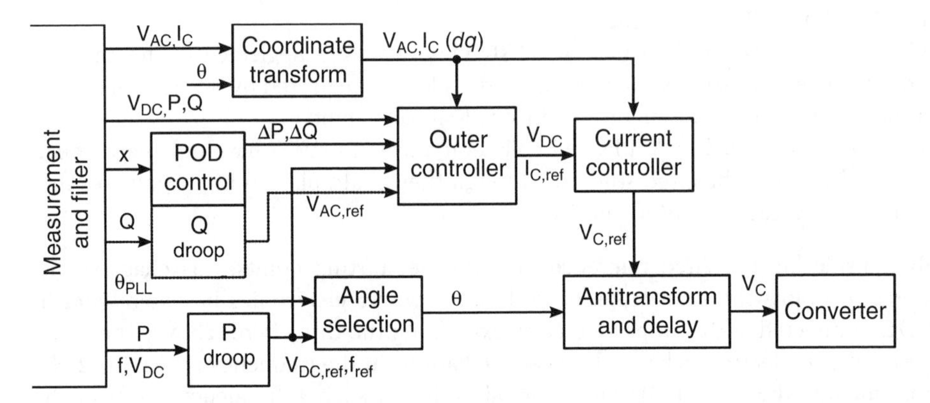

Figure 12.3 Generic block diagram of VSC-HVDC converter control.

Controls

The converter control is usually laid out as illustrated in Figure 12.3. Most of the blocks, and particularly the *Current controller* generating the converter voltage reference, are based on well-known schemes [1, 26]. The difference between onshore and offshore HVDC converters mostly lies in the *Outer controller* block. If, as hinted above, a current source is used to model the converter on its AC side, the blocks *Current controller* and *Antitransform and delay* can be neglected and current references $i_{d,qref}$ are directly provided to the AC current source.

Onshore converter In the onshore converter, the *Angle selection* block selects the measured angle from the PLL ($\theta = \theta_{PLL}$). The *Outer controller* block is based on the diagram in Figure 12.4. Usually, DC voltage (or energy) control is performed: the *d*-axis control makes use of the uppermost PI loop, guaranteeing fixed voltage (energy) at the DC side of the onshore station (i.e. across capacitance C_{DC}). Guidelines for tuning such a controller are found in the book by Yazdani and Iravani [26].

Offshore converter In the offshore converter, the *Angle selection* block selects an angle generated internally and based on integration of a constant frequency ($\theta = \omega_0 t$). Moreover, the *Outer controller* block can rely on different solutions [27]. Many publications use the scheme suggested by Chaudary [28]. However, let us here make use of the scheme shown in Figure 12.5. Combined with a standard current control, the scheme provides a control law similar to what described by Zhang et al. [29]. G_f can be a low-pass filter, G_{HP}, if needed, is a high-pass filter, $X_{ph} + X_T$ is the phase reactor plus

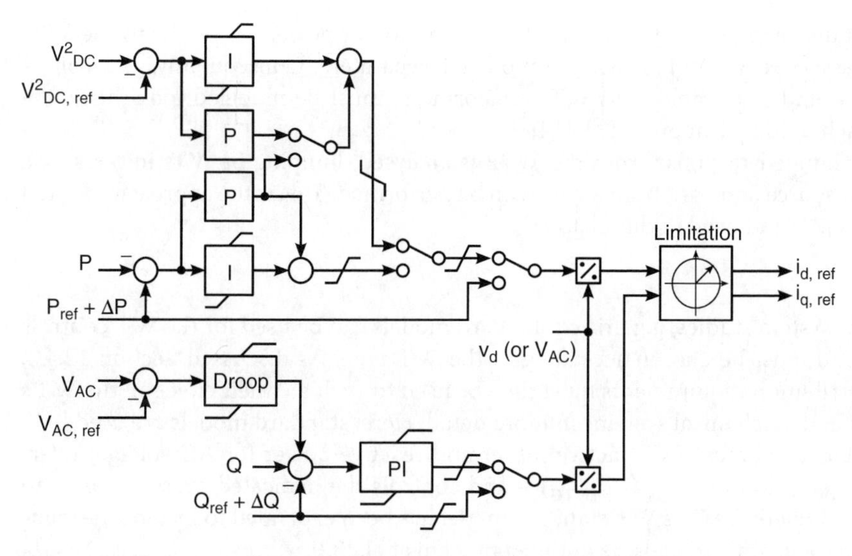

Figure 12.4 Outer controller for onshore HVDC converter control.

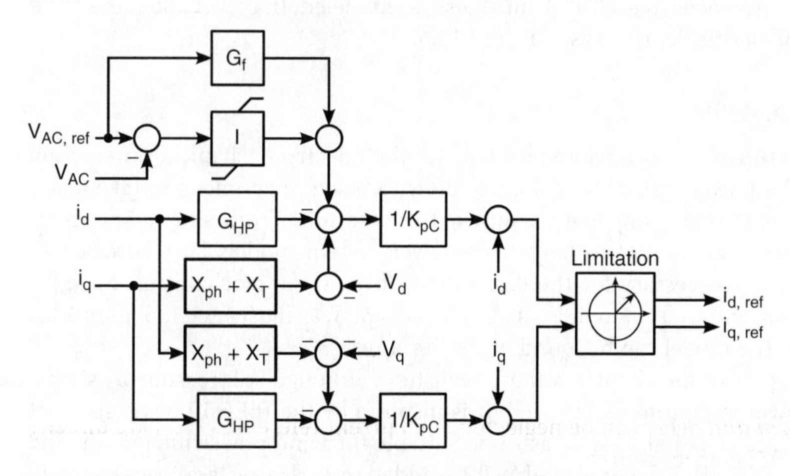

Figure 12.5 Example of outer controller for offshore HVDC converter control.

transformer reactance and K_{pC} is the current control gain – just as in the *Current controller* block. It should be noticed that, for the purpose of POD studies performed in RMS environments, a simple voltage source model with fixed voltage and frequency may be adequate as an offshore converter model.

12.2.2 Wind Power Plant

Electrical System

The WPP's electrical system can be modeled in different levels of detail, depending on the scope of the investigation. Generally, a WT representation through an AC current source is sufficient for RMS stability studies [30]. According to the scope of this chapter:

- If a detailed analysis and design of the distribution of power set-points to the WTs is necessary, each WT needs be modeled separately. Consequently, the collection system (array cables and WT transformers) must be modeled too. This is the approach used by Knuppel et al. [14].
- If only lumped response from the WPP is analysed, lumping of WTs into a single current source and park transformer can be performed. This is the approach adopted by Zeni [31] and used in this chapter.

Controls

For power system studies, generic or standard models can be used for the WT controls [32, 30] and must be chosen according to the WT type. As discussed Section 12.4.3, POD contribution requirements must then be used to evaluate their effect on the WT's electrical and mechanical systems in more detail. Here, standard models are used [30]. Each WT control receives an active power and reactive power (or AC voltage) reference – P_{refWTG} and Q_{refWTG} (V_{refWTG}) – and controls the dedicated current source to track such a reference. The WT standard model has been expanded to include mechanical and aerodynamic models, as outlined by Zeni et al. [32].

Plant controllers acting at the WPP level and dispatching active and reactive power (or AC voltage) references to each WT must also be modeled. In this chapter, the WPP controller is modeled as depicted as in Figure 12.6.

12.2.3 Power System

In order to perform investigations related to the electromechanical interaction between SGs, the power system model must allow for such phenomena to take place. This is done to different complexity levels in the literature, from simple single- or two-machine bus systems [24] to large, real power system models [11]. In order to retain the simplicity necessary for the illustration of the basic mechanisms, a simple single-machine–infinite-bus system is used in this chapter, as illustrated in Figure 12.7. Generic data for the model can be found in the literature [33].

In the second part of the chapter, a more realistic – although still reasonably simple to analyse – power system model is used. It is inspired by the IEEE 12.bus system, as modified by Adamczyk et al. [34] so as to be suitable for wind power integration and POD studies. A VSC-HVDC-connected WPP is added to such a system, as depicted in Figure 12.8. A complete dataset for the model is available [34].

12.3 POD from Power Electronic Sources

12.3.1 Study Case

Generally speaking, it can be initially assumed that the onshore HVDC converter is a controllable static power source, injecting P,Q at its terminals. It is further assumed that the control of P and Q is performed instantaneously: very fast P and Q control is realized according to feedforward or feedback control in Figure 12.4. For the time being, the single-machine–infinite-bus system is utilized to illustrate the main concepts, and the HVDC converter is connected at a variable location along the line between the SG and

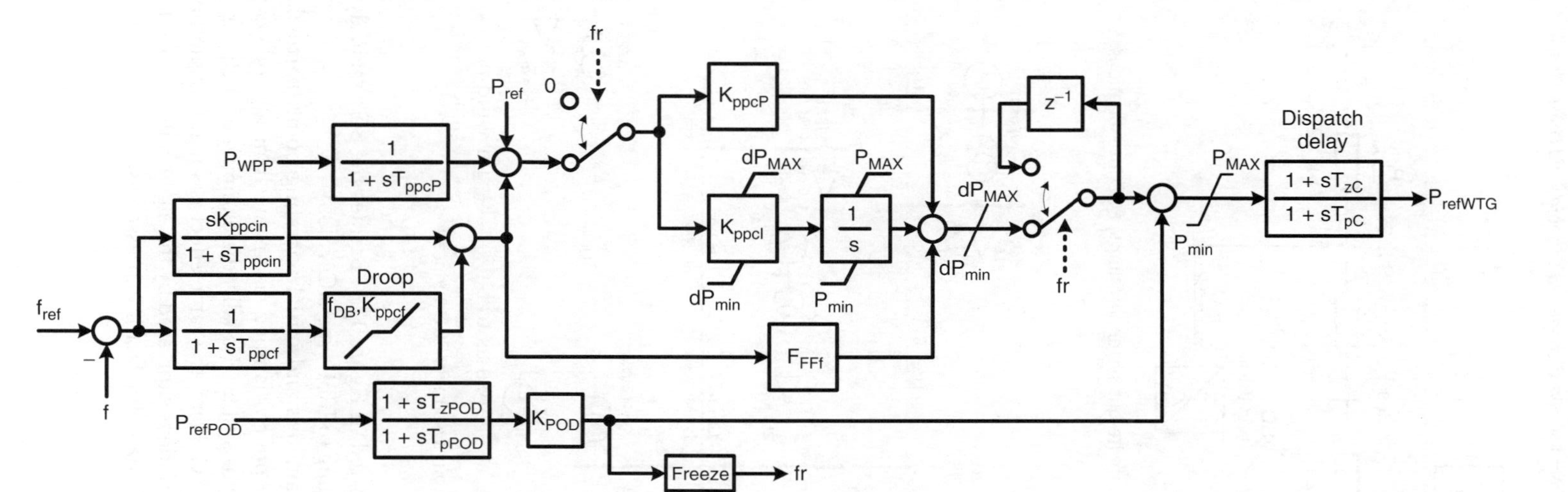

Figure 12.6 Example of WPP controller.

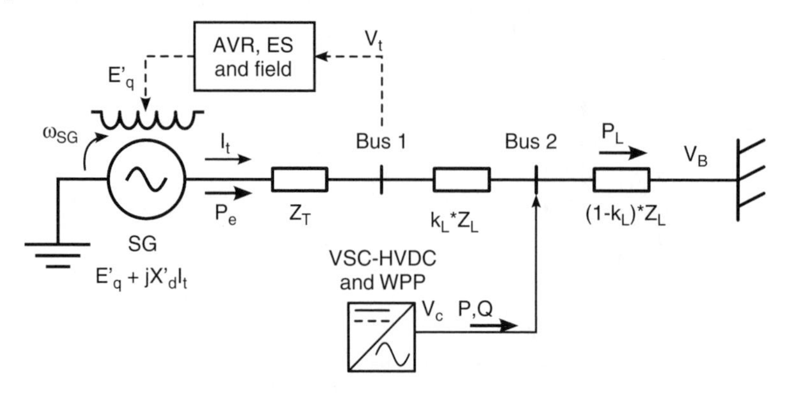

Figure 12.7 Single-machine–infinite-bus system for simple POD investigations.

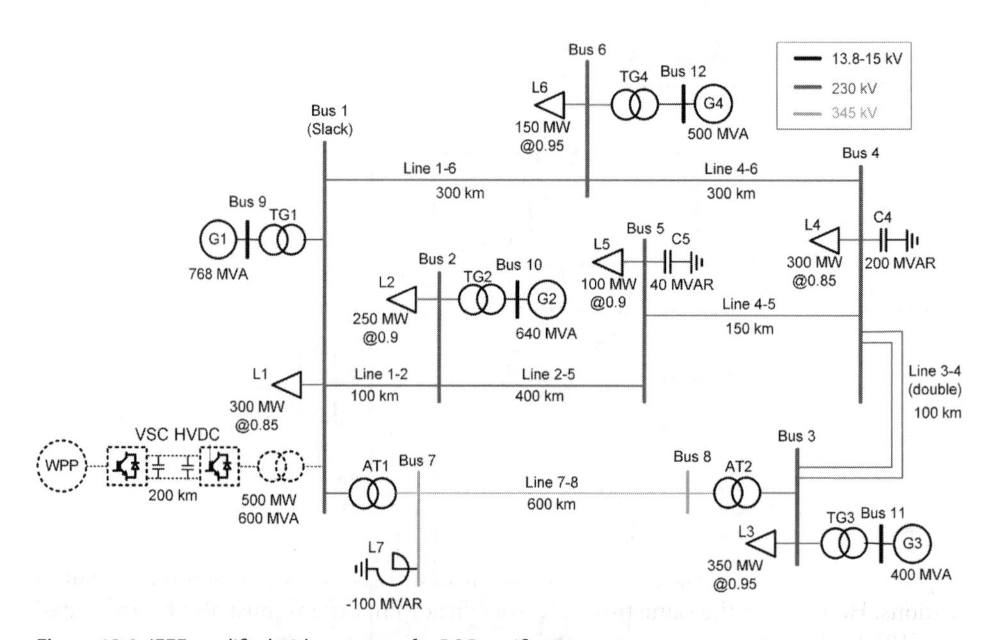

Figure 12.8 IEEE modified 12.bus system for POD verification.

the infinite bus, as illustrated in Figure 12.7. Besides the SG's main dynamics, ES and AVR are modeled, according to IEEE type AC4A [35].

Depending on the parameters and initial operating point (particularly line length, AVR settings, and P_{e0}), a poorly damped electromechanical mode can appear in the single-machine–infinite-bus system. Usually, the damping of such a mode is enhanced by applying a PSS to the SG. Here, the same effect is sought by modifying the HVDC converter control. If the mode is expressed by the eigenvalue $\lambda = \alpha + j\omega_\lambda$, the damping is usually measured by the damping factor expressed as:

$$\zeta = -\frac{\alpha}{\sqrt{\alpha^2 + \omega_\lambda^2}} \tag{12.1}$$

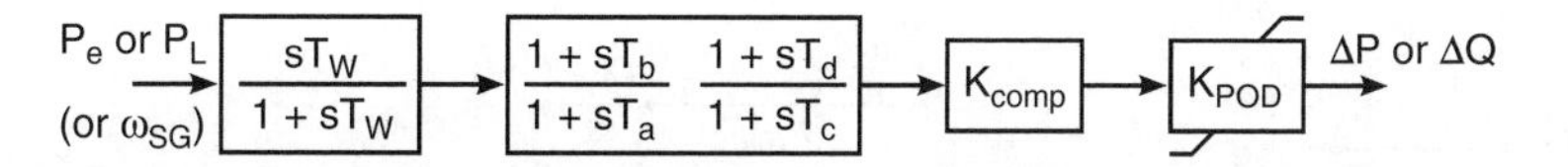

Figure 12.9 PSS-like POD controller.

In order to evaluate the performance of the POD contribution from the static power sources, a control objective is formulated as follows. The target eigenmode λ must have a damping factor $\zeta \geq 0.1$. If this is not possible, ζ must be increased as much as possible as compared to its value without the POD controller. Moreover, the eigenfrequency ω_λ should not be changed by more than 3%. At the same time, other modes must not be negatively affected by the POD control action.

12.3.2 POD Controller

A variety of controllers are available in the literature. While most studies use PSS-like schemes, an interesting new concept to effectively discern relevant modes has been presented by Beza and Bongiorno [36]. This may be well suited to power electronic sources. For the sake of simplicity, here the PSS-like scheme shown in Figure 12.9 is used.

The control block diagram is generic in that it can accept any kind of input. A washout block is followed by a compensation stage and a gain, analogous to classical PSSs [8]. The tuning of the control parameters can be done with techniques developed in the earlier literature, such as the residual method. Here, a more practical approach will be used for parameter tuning.

Input–output Selection

In the realm of POD control from power electronic sources, researchers have mostly focused on the selection of the most appropriate input–output pair. The most natural output for power electronic sources are P and Q at the terminals, or anyhow a linear combination of them. More degrees of freedom are available for selecting the control input. Generally, high controllability and observability are desired from the selected input–output couple; see, for example the thesis of Adamczyk [10] and related publications. However, at the same time, control efficacy limitations must also be accounted for [19]. In the simple single-machine–infinite-bus system, let us assume that the control input will be the machine speed ω_{SG} in the ideal case and the generator active power P_e (or equivalently the line active power P_L) in the more realistic case.

12.3.3 Practical Considerations for Parameter Tuning

In the equation of motion of SGs, "providing damping" physically means to provoke an electrical torque in phase with rotor-speed deviations. The electrical torque acting on and active power generated by the SG can easily be calculated using well-known formulae based on the initial phasor diagram depicted in Figure 12.10. The mismatch between such electrical torque (power) and the input mechanical torque (power) governs the SG's equation of motion [8].

Neglecting sub-transient dynamics and AVR, model linearization and further calculations provide the small-signal expression of the SG's electrical torque and terminal

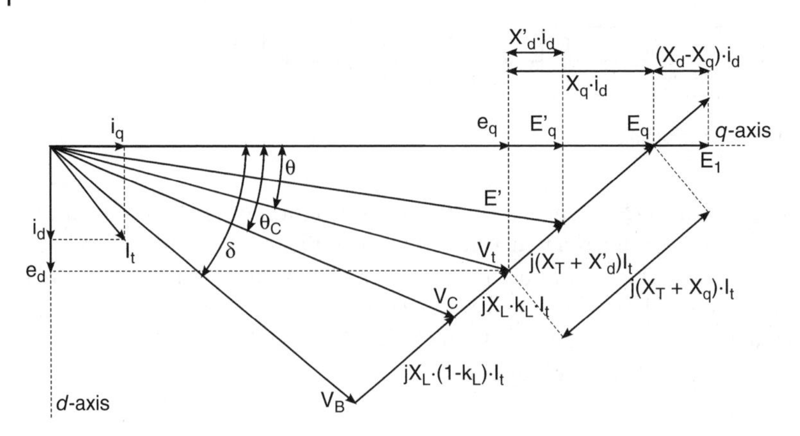

Figure 12.10 Steady-state phasor diagram for single-machine–infinite-bus system.

Table 12.1 Control settings for ideal POD controller.

	P POD	Q POD
K_{POD} [pu]	$K_{POD,P} < 0$	$K_{POD,Q} > 0$
$T_a = T_c$ [s]	0	0
$T_b = T_d$ [s]	0	0
K_{comp} [pu]	1	1

voltage [31], as follows:

$$\Delta T_e = \Delta P_e = k_{P_e\delta}\Delta\delta + k_{P_eE'_q}\Delta E'_q + k_{P_eP}\Delta P + k_{P_eQ}\Delta Q \tag{12.2}$$

$$\Delta V_t = k_{V_t\delta}\Delta\delta + k_{V_tE'_q}\Delta E'_q + k_{V_tP}\Delta P + k_{V_tQ}\Delta Q \tag{12.3}$$

Let us neglect, for now, the terminal voltage expression. It can be concluded that both active and reactive power injections from the power electronic source affect the SG's electrical power proportionally. In a predominantly inductive network, P does so by mainly acting on the electrical angles, while Q provokes voltage magnitude variations that translate into the SG's electrical power variations. These effects are accounted for by the coefficients k_{PeP} and $k_{PeQ,}$ which, in the sign convention used in Figure 12.7, will be negative and positive respectively.

Ideal POD Controller

In the ideal case, the SG's speed ω_{SG} can be sensed and directly used by the HVDC converter to generate damping signals: $u_{POD} = \omega_{SG}$ in Figure 12.9. By examining (12.2), a proportional controller can be selected according to Table 12.1, which should guarantee the desired effect on the damping of speed and power oscillations.

However, if one proceeds in this way and selects $K_{POD,P} = -K_{POD,Q} = -10$ pu, the results shown in Figure 12.11 are obtained. Here, the variations in the real and imaginary parts of the targeted eigenvalue are plotted. This is done for both P and Q POD, two values of initial SG power, two different positions of the HVDC converter along the line

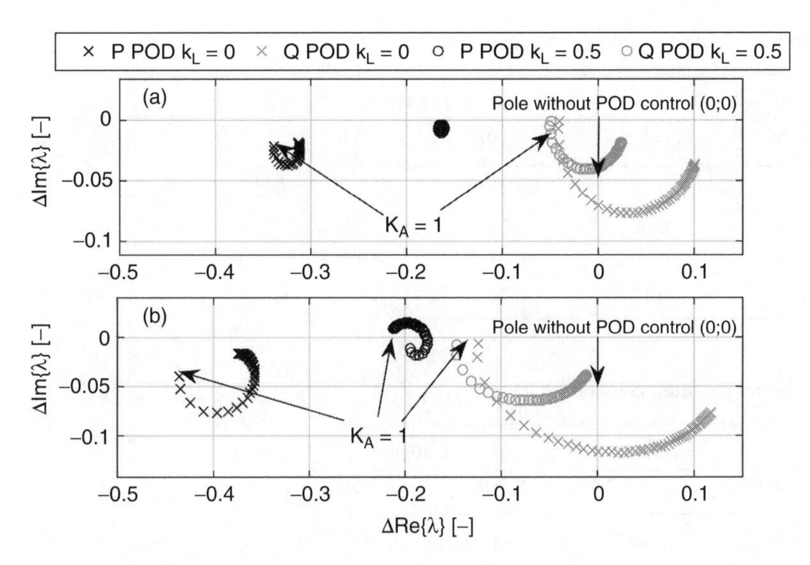

Figure 12.11 Variation in real and imaginary part of target pole: (a) $P_{e0} = 0.417$ pu and (b) $P_{e0} = 0.833$ pu.

(two values of k_L), and varying AVR gain K_A. The superior performance of P POD for $k_L = 0$ has already been pointed out in the literature [24] and is apparent from the plots.

In terms of performance, it is clearly noticeable that the improvement initially provided by the POD controller deteriorates for increasing AVR gains in most of the cases, particularly for Q POD. For three out of four Q POD cases, the effect even becomes negative for high AVR gains. This is because P and Q, besides affecting P_e, also influence the terminal voltage V_t, which is processed by the AVR. A field voltage variation is derived, and this affects the eigenvalue in a counterproductive way. Usually, PSSs are installed on generating units *to counteract* the detrimental effects of AVRs on small-signal stability. On the other hand, POD control from power electronic sources must be tuned *so as not to trigger* further detrimental effects of AVRs. Actually, the AVR's presence can be exploited to make Q POD more effective. This important mechanism cannot be seen if AVR effects are neglected in the analysis.

Real POD Controller

Sensing the SG's rotational speed may not be always possible, since the power electronic asset may be at some distance from the target machines and have different owner. Hence, a different kind of input is usually employed. In order to understand how the POD control must be tuned, the angle relation between the input and the speed oscillations must be known.

As an example, let us assume that P_e is used as an input to the POD controller. The SG's electrical power is, according to the network equations, proportional to the SG's rotor angle. This means that oscillations in P_e will lag oscillations in ω_{SG} by 90°. P and Q affect P_e proportionally, and do so with the sensitivities contained in the first two rows of Table 12.2. The controller must therefore include a phase shift of 90° to provide damping effectively. This is achieved by applying the parameters listed in Table 12.3, where the control gain is swept from 0 to 1 and $K_{comp} < 0$ for P POD, to compensate

Table 12.2 Per-unit SG power and voltage sensitivities to P and Q.

	$P_{e0} = 0.417$ pu		$P_{e0} = 0.833$ pu	
	$k_L = 0.0$	$k_L = 0.5$	$k_L = 0.0$	$k_L = 0.5$
k_{PeP}	−0.54	−0.27	−0.71	−0.32
k_{PeQ}	0.06	0.07	0.19	0.22
k_{VtP}	−0.05	0.00	−0.07	0.02
k_{VtQ}	0.28	0.15	0.30	0.17

Table 12.3 POD control parameters for real POD controller.

	$K_A = 1$ pu		$K_A = 40$ pu	
	P POD	Q POD	P POD	Q POD
K_{POD} [pu]	0–1	0–1	0–1	0–1
$T_a = T_c$ [s]	0.096	0.096	0.088	0.088
$T_b = T_d$ [s]	0.558	0.558	0.512	0.512
K_{comp} [pu]	−0.172	0.172	−0.172	0.172

for the negative sign of k_{PeP}. The parameters are set so as to have flat phase curve at the target eigenfrequency. The resulting pole movement for the varying gain is plotted in Figure 12.12.

As can be seen, while in (a), for low AVR gain, the pole movement is approximately as desired (i.e. horizontal leftwards), in (b) for high AVR gain, the trace is far from the target, particularly for Q POD. The explanation can be found in the last two rows of Table 12.2. High SG terminal voltage sensitivities provoke a stronger interaction between the POD controller and the AVR. The design can therefore not be implemented without considering AVR effects. The design can be corrected, in this simple case, by visualising the phenomena on a phasor-like diagram, such as that depicted in Figure 12.13.

Taking Q POD as an example, the expression for the electrical torques caused by the POD control and by the AVR are the following:

$$\Delta T_{eQ} = k_{P_eQ} \Delta Q \tag{12.4}$$

$$\Delta T_{eAVR,Q} = k_{P_eE'_q} \cdot G_{AVR}(s)|_{s=j\omega_\lambda} \cdot k_{V_tQ} \Delta Q \tag{12.5}$$

where G_{AVR} is the modulus of the AVR and ES transfer function. From trigonometry, it can be seen that the compensation to the initial control settings needed is the angle γ_Q calculated as follows:

$$\gamma_Q = arctg \frac{\Delta T_{eAVR,Q}}{\Delta T_{eQ}} \tag{12.6}$$

The compensations in Table 12.4 were calculated for this simple case and Q POD, and the resulting variations in the eigenvalue's frequency and damping ratio are also listed in the table.

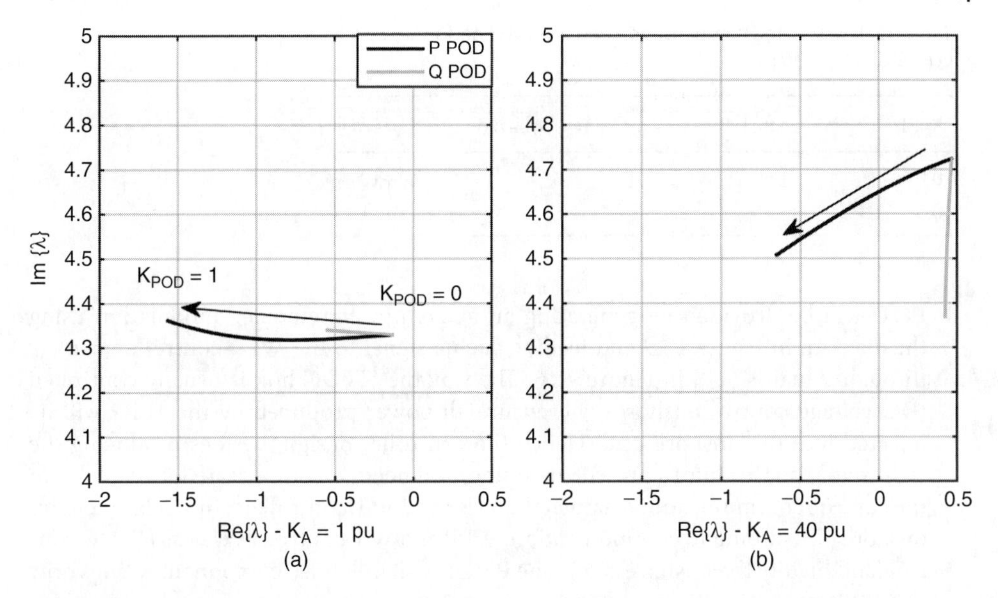

Figure 12.12 Eigenvalue movement for real POD controller and varying gain: (a) $K_A = 1$ pu, (b) $K_A = 40$ pu.

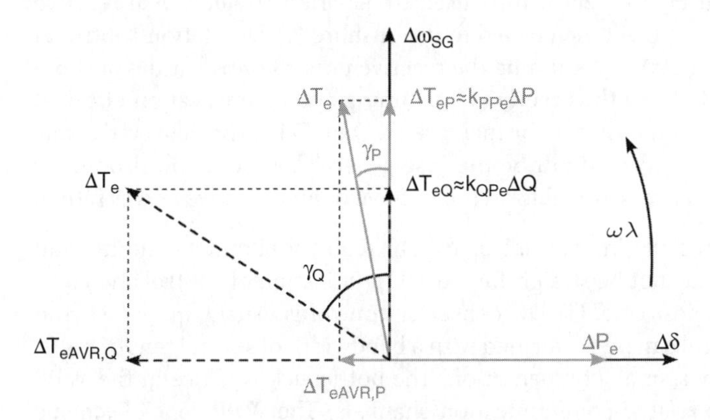

Figure 12.13 Phasor-like diagram of electrical torques acting on SG rotor.

The results illustrate that the calculation can effectively be used in this simple case to correct the control parameters. In more complex scenarios the method cannot directly be applied, but the physical mechanisms governing the phenomena remain the same and can thus be interpreted in a similar way.

12.4 Implementation on VSC-HVDC-connected WPPs

12.4.1 Realization of POD Control

When and HVDC-connected WPP needs to provide the POD service described in this chapter, it will for example be in the following way:

Table 12.4 Calculated compensation and achieved performance for Q POD and $K_A = 40$ pu.

k_L [-]	γ_Q [°]	$\Delta\omega_\lambda$ [%]	$\Delta\zeta$ [%]
0.0	65	−3	+10
0.5	45	<1	+11

P POD will be provided by generating an active power reference modulation using the diagram in Figure 12.9 and feeding such a signal to the WPP controller, namely through signal P_{refPOD} in Figure 12.6. The way the HVDC link is usually controlled (DC voltage control onshore) will ensure all power produced by the WPP will be injected into the onshore grid. Hence the controller design, besides producing the right signal phase shift to provide damping, will need to compensate for the delay in the control/communication chain in the WPP and WT controllers; this delay includes onshore measurement, communication, WPP and WT control, as well as HVDC control (and mainly the onshore side). The P POD will influence component ratings only if WPP over-production is allowed.

Q POD on the other hand, can be directly delivered by the onshore HVDC station. Once again, the controller in Figure 12.9 will be used to generate an additional reactive power control signal, which will then be fed to the onshore HVDC station control in Figure 12.4, through signal ΔQ. Assuming the reactive power control at the onshore HVDC station is fast, Q POD thus requires less tuning and compensation effort. It may require onshore station ratings to be increased if Q POD is to be delivered at any active power level, but this would not be the case generally, as some headroom for grid code compliance is always available on top of the maximum active power rating.

It is apparent that, especially in the case of P POD, control parameter tuning and compensation of delays may not be straightforward. Optimization of control and communication setup is thus desirable. HVDC converter control is usually quite fast, and onshore DC voltage control can be performed with a bandwidth of several tens of rad/s, which would require only a small compensation. The bottleneck is hence in the WPP and WT control and the related communication channels. The WPP control scheme with POD signal feedforward according to Figure 12.6 aims at minimizing WPP delays. However, further effort must be put into reducing communication delays, sample times, and WT control limitations in order to facilitate the delivery of POD.

12.4.2 Demonstration on Study Case

The implementation of POD control on a VSC-HVDC-connected wind farm used the power system model depicted in Figure 12.8. Such a power system has a poorly damped model, with frequency around 0.7 Hz and damping factor $\zeta \approx 1\%$. The mode is an inter-area oscillation, where SGs G1 and G2 swing against G3 and G4, creating large power oscillations visible, for example, in Line 78 [34].

The implementation of POD on the WPP+HVDC system was performed according to the approach set out above. Based on the mode characteristics, the chosen control input is P_{78} for both P and Q POD.

Parameter Tuning and Results

Once the control input is chosen, and knowing the system characteristics, the initial parameter tuning can be performed according to the practical guidelines outlined above, as follows:

- Being the result of an angle displacement between SGs, the input P_{78} is lagging speed deviations by roughly 90°.
- Although the SG with highest participation in the mode is G4, the machine that is most affected by the VSC-HVDC-connected WPP is G1. Approximately 70% of the small-signal active power variations from the WPP flow into G1, while G1's per-unit terminal voltage sensitivity to reactive power from the onshore HVDC station is as high as 10%.
- For P POD, due to the negligible effect P has on the SGs' terminal voltages, one may expect to need a phase compensation of around 90°, with sign opposite to that of $K_{POD}K_{comp}$.
- For Q POD, on the other hand, the AVR action of G1 and G2 is quite significant, as they have high gain, and is most likely predominant over the direct effect of Q modulation from the HVDC. Considering the field circuit, the AVR-related electrical torque will lead voltage variations by 90°. Hence, a near-zero phase shift may be a good choice. However, it should be taken into account that the voltage of both G1 and G2 is affected by the Q modulation from the HVDC converter. G1 leads G2 slightly in the power oscillations, which are to a good approximation in phase with G2's angle. Therefore, a slightly positive value of the compensation may be a better choice for an initial tuning. Let us set it to 10° and then refine it.

After approximate parameter tuning as above, the gain was adjusted so as to achieve $\zeta > 4\%$ in the targeted mode, with the phase shifts refined so as to not have a variation of the eigenfrequency ω_λ larger than 3%. The resulting parameters and achieved performance are shown in Table 12.5.

Two interesting things can be noticed from the table:

- The parameter tuning rules give results quite close to final optima.
- Given that the control input is the same, P and Q POD require essentially a similar control effort (the values of K_{POD} differ only by 30%) to achieve similar performance. This is due to the AVR action; that is, Q POD would not be effective per se, but it becomes so when it can exploit AVRs.[endbl]

In order to verify the fidelity of the linear calculation results, a non-linear simulation was run. A three-phase fault was applied in the middle of Line 3–4a at time t = 1 s

Table 12.5 Control parameters and resulting performance for POD implementation on VSC-HVDC-connected WPP in IEEE 12.bus system.

POD type	Phase comp. [°]	K_{POD} [pu]	ζ [%]	$\Delta\omega_\lambda$ [%]
P	−90	5.5	4.5	<1
Q	+20	7.15	4.2	<2
P&Q	-	-	6.7	<1

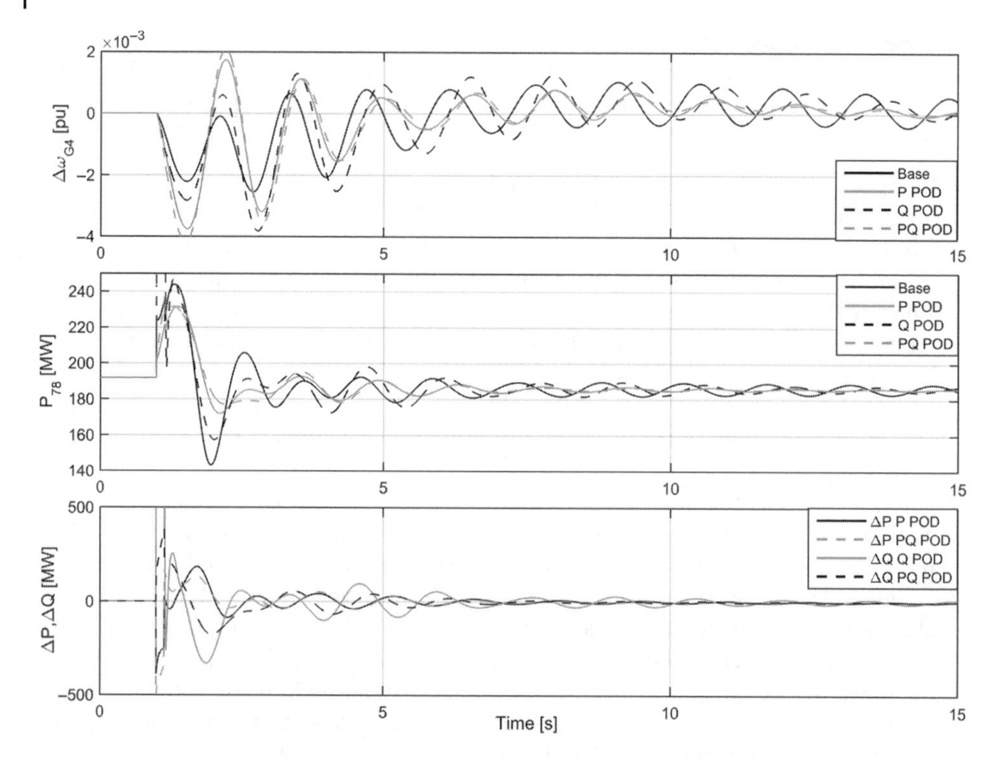

Figure 12.14 Selected results from non-linear simulation of POD control on IEEE 12.bus system.

and subsequently, at t = 1.15 s, the same line went out of service, clearing the fault. A selected set of relevant results are plotted in Figure 12.14 and confirm the linear analysis, by showing the positive damping contribution of the HVDC-connected WPP.

Effect of AVRs

The AVRs have a significant influence on the parameter tuning and performance. Zeni has demonstrated that the effect of the AVRs can be intuitively determined in a similar way to the approach for the simple single-machine–infinite-bus system [31]. The control parameter tuning must take that into account and changing voltage regulation conditions for the network the WPP connects to may represent a challenge from this perspective.

12.4.3 Practical Considerations on Limiting Factors

Stability and Performance Limitations

This section highlights some more interesting aspects related to the implementation of POD from power electronic devices and for VSC-HVDC-connected WPPs in particular.

Let us assume that the RMS representation can be used for the analysis (no electrical resonances in the target frequency range) and that the control outputs (P and/or Q) have an algebraic influence on the control input (this being reasonable if the first assumption is accepted, since both may be network quantities). A SISO control block diagram of the kind illustrated in Figure 12.15 can be derived.

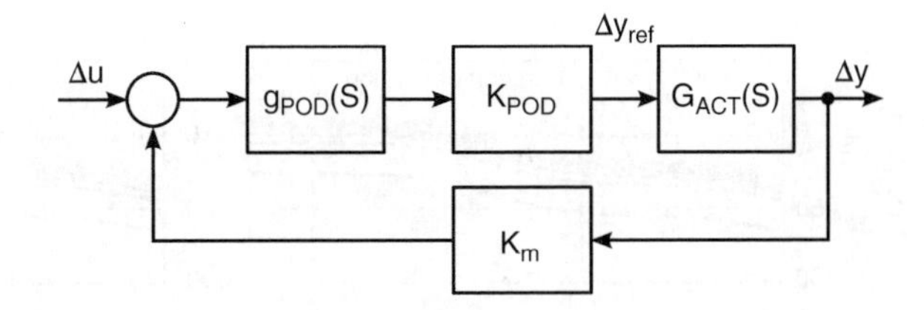

Figure 12.15 Generalized closed-loop control block diagram for POD controller.

In the diagram, $g_{POD}(s)$ is a normalized POD control transfer function, containing all blocks except for the gain K_{POD} in Figure 12.9, which is extracted as a separate block. This is the control transfer function, with unity gain at the target eigenmode frequency. $G_{ACT}(s)$ is the transfer function of the "actuator", meaning all elements in the control chain from the generated POD signal reference Δy_{ref} to its actual value delivered to the grid. In the case of P POD for HVDC-connected WPPs, for example, it would include WPP control, WT control, and communication and control delays imposed by the HVDC system control. Finally, K_m is the measurement feedback: the algebraic influence of the control output Δy on the control input Δu.

Closed-loop stability By inspecting the block diagram in Figure 12.15 and applying linear control theory tools, the SISO stability of the POD controller can be evaluated in a straightforward way. The loop transfer function $K_m g_{POD} K_{POD} G_{ACT}$ must fulfil suitable stability criteria for the controller to be stable. Assuming that $G_{ACT} = 1$ (perfect actuator), varying the modulus of $K_m K_{POD}$ in the range 0.01–1 pu and observing four permutations of the cases given by $K_m K_{POD}$ greater or less than 0 and phase compensation being lead or lag type, the Bode diagrams shown in Figure 12.16 are derived. From the figure it is apparent that the most favorable scenario results from $K_m K_{POD} < 0$ and lead compensation (top right), as stability is guaranteed, even if G_{ACT} deviates substantially from the ideal and introduces significant delays. On the other hand, with $K_m K_{POD} < 0$, as illustrated in the next section, performance limitations may arise. Clearly, $K_m K_{POD} > 0$ and lag compensation must be avoided, as even an ideal actuator can potentially cause instability. This is important to realise, since some freedom is available in the choice of the sign of $K_m K_{POD}$. For example, referring to Figure 12.7, the eigenmode is reflected identically into the two possible inputs P_L and P_e, meaning that the gain K_{POD} will have to be the same for the two inputs. However, K_m will have opposite sign for the two cases, and the most convenient input for stability purposes can therefore be chosen. Finally, as demonstrated by Zeni [31], erasing the assumption of ideal actuator, instability can institute even for the three potentially stable cases in Figure 12.16.

Achievable control performance The closed loop transfer function of the block diagram in Figure 12.15 is the following (Laplace variable dropped for brevity):

$$G_{CL} = \frac{g_{POD} K_{POD} G_{ACT}}{1 - K_m g_{POD} K_{POD} G_{ACT}} \tag{12.7}$$

As seen in the previous section, when the real part of the second denominator's element (loop transfer function) is positive, there exists a possibility for instability. On the other hand, when it is negative, the achievable performance is limited by the following constraint:

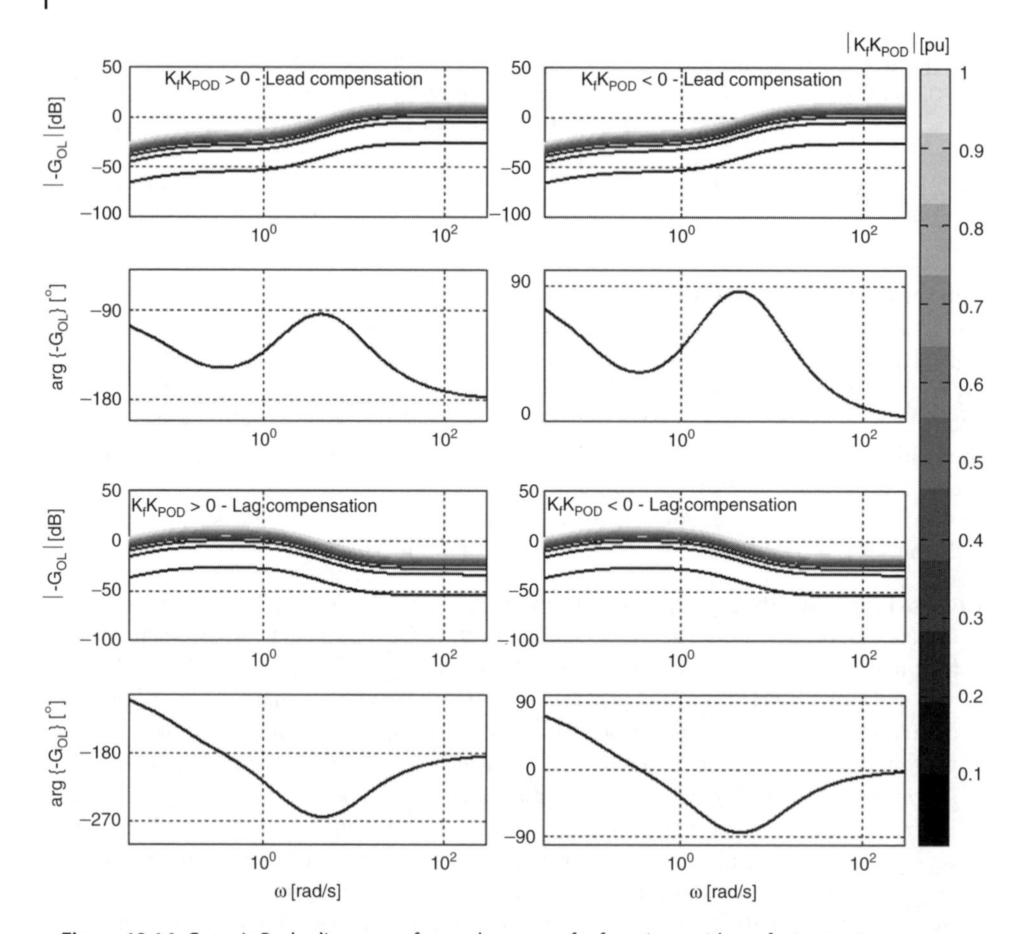

Figure 12.16 Generic Bode diagrams of open-loop transfer functions with perfect actuator.

$$|G_{CL}| \leq \frac{1}{K_m} \tag{12.8}$$

Such performance limitation is important, in that the actual achievable amplification from control input to control output has an upper bound. This means that the desired total damping may not be achieved, even by perfectly tuning the compensation and indefinitely increasing the control gain. In practice, this may be a SISO perspective to view the control limitations, as first explored by Domínguez-García et al. [19].

Recommendations From the above analysis, the following recommendations descend:

- The feedback gain K_m should be minimized. This will make the POD control as much independent of the controlled system as possible, meaning that the control limitation in (12.8) will not be restrictive. In other words, it is desirable that the control output affect the controlled system, but not the control input. Obviously, a compromise is necessary in real applications.
- As already known from the literature, good controllability for the output Δy and good observability for the input Δu should be chosen; this minimizes the necessary

loop gain to achieve the desired effect and thus prevents the occurrence of instability.

- A design with $K_m K_{POD} < 0$ should be pursued wherever possible, to minimize the probability of instability.

Sensitivity to Delays

POD service, compared to frequency control, is very sensitive to control and communication delays. Precise phase shift of the signals is needed in order to deliver the desired performance. Even small phase errors can lead to compromising at least one of the control objectives formulated earlier in this chapter.

Hence the recommendation is to minimize control and communication delays to the extent possible. This is done, for example, by employing POD signal feedforward in the WPP control, as shown in Figure 12.6, or by designing an optimal communication system between the various controllers. Improvement of state-of-art WPP controllers and their communication to WTs may be very beneficial to the implementation of POD services.

Ramp-rate Limiters

WPPs may apply active power ramp-rate limitations. This is done, on the one hand, to protect the mechanical system of the WTs and to optimize their overall design and, on the other hand, to comply with grid connection requirements [37].

However, such ramp-rate limitations may be an obstacle to the effective delivery of power system services such as fast frequency control [32]. Since POD control is usually faster than frequency control, ramp-rate limitations become very relevant in delivery of such a service.

A simple way to derive a constraint in these terms is to consider that the WPP is to deliver a damped perfectly sinusoidal POD signal, with maximum allowed amplitude ΔP_{MAX}, frequency ω_λ, and decay rate α:

$$\Delta P_{POD} = \Delta P_{MAX} \cdot \sin(\omega_\lambda t) \cdot e^{\alpha t} \tag{12.9}$$

The maximum value of the derivative of such signal is $\Delta P_{MAX}\omega_\lambda$. Assuming $\Delta P_{MAX} =$ 0.1 pu and a realistic POD frequency range, the maximum necessary ramp-rate to have undistorted signals would be:

$$0.063\,\frac{pu}{s} \leq \left.\frac{dP}{dt}\right|_{max} \leq 1.26\,\frac{pu}{s} \tag{12.10}$$

The upper end of such a range is significantly higher than the state-of-art limits set out by Zeni et al. [32]. Modern installations may have relaxed these limits. If not, it is probably a pre-condition for successful delivery of P POD in future installations. It should be noted that ramp-rate effects would not only reduce the effectiveness of POD control by quenching its gain, but also potentially cause counterproductive effects by introducing undesired phase shifts. Generally, as POD is a linear control and ramp-rate limitations are a non-linear phenomenon, one should avoid their appearance in POD-related signals.

Due to the perfect decoupling with the mechanical system, ramp-rate limiters are not usually applied to the reactive power. Hence the effectiveness of Q POD should not be endangered by this kind of control block.

Collateral Effects on WPPs

Available energy and recovery period Another aspect that limits WPP performance when they are requested to deliver active power over-production (that is, above available wind power) is the limited available energy and the consequent need to recover the WT's rotor speed. A period of active power over-production may, in certain conditions, extract energy from the rotating masses of WT rotors [38, 39], because more electric power is produced than the available incoming aerodynamic power. In modern variable-speed WTs controlled at MPPT point, the aerodynamic power decreases with the rotor speed; below rated wind speed, pitch control cannot be used to increase the intake of aerodynamic power. This means that a period of active power over-production requires a subsequent period of under-production to speed up the WT rotor and re-equilibrate aerodynamic and electric power. The problem has been encountered and analysed in publications regarding inertial or fast frequency control from WPPs, where the necessary power drop for speed recovery may hinder the beneficial effects of initial power over-production. In the case of P POD, the same problem may occur, with the differences that

- power modulation, depending on the control setup, may be of smaller amplitude
- some kind of power drop allowing for speed recovery is intrinsically present in POD modulation signals, with the oscillating nature of the power modulation and nil net delivery of energy to the network.

In order to understand whether the WT rotor speed recovery may be a problem for P POD provision from WPPs, the available energy can be calculated for the worst case [38]. Let us denote it $\Delta E_{av,min}$. This parameter practically depends on the WT technology and design and the allowed WT rotor-speed variations. Let us further assume ourselves to be in the very worst case of active power over-production, when the POD controller is fully saturated during the first swing and required to deliver perfectly rectangular active power modulation with peak–peak amplitude $2^*\Delta P_{MAX}$ and frequency f_λ. ΔP_{MAX} is a control parameter determined by a compromise between performance requirements, capability, and cost. On the other hand, f_λ is simply the frequency of the target eigenvalue. If it can vary, its minimum value should be used. For the POD service to not pose a problem, the following constraint must be satisfied:

$$\Delta P_{MAX} \cdot \frac{2}{f_\lambda} \leq \Delta E_{av,min} \tag{12.11}$$

The above constraint can easily be verified during design and, in the author's experience, should not be a problem in most cases. Only for extremely low values of f_λ may the constraint be violated. Furthermore, considering the sinusoidal shape of POD signals and the consequent alternative nature of the power modulation, this phenomenon is not expected to be an issue.

Mechanical side oscillations A last aspect which must be accounted for during the design phase of WTs if they are required to modulate their active power production for POD service is the presence of any mechanical-side resonance falling in the same frequency range as POD. If proper control means are not arranged, active power oscillations required for P POD can find their way to the mechanical system of the WTs. Some WT elements, such as shaft, tower, and blades can give rise to resonances in frequency ranges comparable with that of POD. The interaction between POD and such

Table 12.6 Realistic parameters for WT rotor, generator and shaft.

Parameter	Value	Unit
K_{SH}	150	pu/s
H_{ROT}	5.81	s
H_{GEN}	0.93	s

mechanical modes must be prevented from occurring, in order to avoid premature wear of the mechanical system. Taking the shaft oscillation as example, the resonance pulsation can be computed as follows:

$$\omega_{SH} = \sqrt{\frac{K_{SH}}{2}} \cdot \sqrt{\frac{H_{ROT} + H_{GEN}}{H_{ROT} H_{GEN}}} \tag{12.12}$$

where K_{SH} is the shaft's stiffness, H_{ROT} is the WT rotor inertia constant, and H_{GEN} the WT generator inertia constant. Sample values for such parameters are listed in Table 12.6.

Usually, even for significant variations of the parameters above, shaft-related resonances may fall in a frequency range relevant for POD: from a fraction to a couple of hertz. This means that the possible interaction must be accounted for in the design phase, and other resonance sources such as tower and blades must be included in the analysis too. If overlap between the frequency of required POD service and mechanical resonances occurs, the following remedial actions may be considered, selecting the most cost-effective one:

- Reach an agreement with the TSO(s) to avoid POD provision at specific frequencies. This may be particularly relevant if the WPP is already online and upgrading it would be too costly.
- Modify hardware and control (with, say, notch filters) so that dangerous active power oscillations cannot reach the mechanical side. This would require more DC-link energy storage (capacitor) and/or converter voltage over-rating.
- Agree with the TSO(s) that, during POD service, the average power production can be stepped down momentarily so as to be able to show constant torque to the WT generator and burn the excess energy in a resistor. In some WTs, a resistor is already installed for fault-ride-through (FRT) purposes and only a slight over-rating may be required.
- Use an approach similar to the previous item, but installing the resistor at the onshore HVDC station. The resistor would be more expensive, but only one would be needed, instead of several, and it would be conveniently installed onshore. Moreover, no modifications to the WPP and WT controls would be needed in this case. It should also be noticed that a breaking resistor is already installed in state-of-the-art onshore HVDC stations for FRT purposes.
- Use a separate energy storage unit, more conveniently located onshore.

It is clear that early dialogue between WPP developers, TSO(s), and WT OEMs is needed in order to ensure proper design in these terms.

12.5 Conclusion

An overview of POD services from VSC-HVDC-connected WPPs was presented in this chapter. The state-of-the-art of POD service from energy sources connected to the grid by power electronics was taken as a starting point, and the necessary additions for the particular application of HVDC-connected WPPs have been discussed.

Practical guidelines for initial parameter tuning, which relies on power system physics, have been presented, as a complement to tuning techniques based on linear systems mathematics, as found in the literature. The importance of accounting for ESs and AVRs has been highlighted and illustrated by the use of phasor-like diagrams. The effectiveness of the parameter tuning guidelines has been illustrated on simple and more complex power system models.

Finally, an overview of possible limiting factors has been given, focusing on performance and stability limitations as well as limitations brought about by realization of the service on complex and modular units such as remote offshore WPPs.

Acknowledgement

The author would like to thank Poul Ejnar Sørensen and Anca Daniela Hansen from the Technical University of Denmark, Bo Hesselbæk from DONG Energy Wind Power A/S, as well as Philip Carne Kjær from Vestas Wind Systems A/S and Aalborg University for their valuable contribution to the work which forms the background for this chapter.

Müfit Altin, Ömer Göksu, Andrzej Adamczyk, Remus Teodorescu, and Florin Iov are also warmly thanked for permission to use the modified IEEE 12.bus power system model.

References

1 R. Sharma (2011) Electrical Structure of future off-shore wind power plant with a high voltage direct current power transmission, Technical University of Denmark, Lyngby.

2 Lauria, S., Schembari M., and F. Palone (2014) EHV AV interconnection for a GW-size offshore wind-farm cluster: Preliminary sizing, in *IEEE International Energy Conference (ENERGYCON)*, Dubrovnik.

3 Fischer, W. Braun, R., and Erlich, I. (2012) Low frequency high voltage offshore grid for transmission of renewable power, in *IEEE PES Innovative Smart Grid Technologies Europe*, Berlin.

4 Blasco-Gimenez, R., Añó-Villalba, S., Rodríguez-D'Derlée, J., Morant F., and Bernal-Perez, S. (2010) Distributed voltage and frequency control of offshore wind farms connected with a diode-based HVDC link, *IEEE Transactions on Power Electronics*, **25** (**12**), 3095-4005, 2010.

5 Menke, P. (2015) New grid access solutions for offshore wind farms, in *EWEA Offshore Conference*, Copenhagen.

6 ENTSO-E (2014) Draft network code on high voltage direct current connections and DC-connected Power park modules, ENTSO-E, Brussels.

7 National Grid Electricity Transmission (2014) The Grid Code, Issue 5, Revision 7, NGET, Great Britain.
8 Kundur, P. (1994) *Power System Stability and Control*, McGraw-Hill.
9 Kundur, P., Paserba, J., Ajjarapu, V., Andersson, G., Bose, A., Canizares, C., Hatziargyriou, N., Hill, D.A., Stankovic, D., Taylor, C.T., van Custem C., and Vittal, V. (2004) Definition and classification of power system stability, *IEEE Transactions on Power Systems,* **19** (2), 1387–1401, 2004.
10 Adamczyk, A. (2012) Damping of Low frequency power system oscillations with wind power plants, PhD thesis, Aalborg University.
11 MED-EMIP (2010) Medring Update Vol.II – Analysis and proposal of solutions for the closure of the ring and north-south electrical corridors, Euro-Mediterranean Energy Market Integration Project.
12 Adamczyk, A., Teodorescu, R., Iov, F., and Kjær, P. (2012) Evaluation of residue based power oscillation damping control of inter-area oscillations for static power sources, in *IEEE PES General Meeting*, San Diego.
13 Knuppel, T. (2012) Impact of wind power plants with full converter wind turbines on power system small-signal stability, PhD thesis, Technical University of Denmark, Lyngby.
14 Knuppel, T., Nielsen, J.N., Jensen, K.H., Dixon, A., and Østergaard, J. (2012) Power oscillation damping capabilities of wind power plant with full converter wind turbines considering its distributed and modular characteristics, *IET Renewable Power Generation,* 7 (5), 431–442.
15 Knuppel, T., Nielsen, J., Jensen, K., Dixon A., and Østergaard, J. (2011) Small-signal stability of wind power system with full-load converter interfaced wind turbines, *IET Renewable Power Generation,* **6** (**2**), 79–91, 2011.
16 Knuppel, T., Nielsen, J., Jensen, K., Dixon A., and Østergaard, J. (2011) Power oscillation damping controller for wind power plant utilizing wind turbine inertia as energy storage, in *IEEE PES General Meeting.*
17 Tsourakis, G., Nomikos, B.M., and Vournas, C.D. (2009) Contribution of doubly fed wind generators to oscillation damping, *IEEE Transactions on Energy Conversion,* **24** (3), 783–791.
18 Domínguez-García, J.L., Bianchi, F.D., and Gomis-Bellmunt, O. (2012) Analysis of the damping contribution of power system stabilizers driving wind power plants, *Wind Energy,* **17** (2), 267–278.
19 Domínguez-García, J.L., Ugalde-Loo, C.E., Bianchi, F., and Gomis-Bellmunt, O. (2014) Input-output signal selection for damping of power system oscillations using wind power plants, *Electric Power & Energy Systems,* **58**, 75–84.
20 Elkington, K. (2012) The dynamic impact of large wind farms on power system stability, PhD thesis, KTH, Stockholm.
21 Ndreko, M., van der Meer, A.A., Rawn, B.G., Gibescu, M., and van der Meijden, M.A. (2013) Damping power system oscillations by VSC-based HVDC networks: A North Sea grid case study, in *12th Wind Integration Workshop,* London.
22 Pipelzadeh, Y., Chaudhuri, B., and Green, T.C. (2010) Wide-area power oscillation damping control through HVDC: A case study on Australian equivalent system, in *IEEE PES General Meeting.*

23 Harnefors, L., Johansson, N., Zhang, L., and Berggren, B. (2014) Interarea oscillation damping using active-power modulation of multiterminal HVDC transmissions, *IEEE Transactions on Power Systems,* **29** (5), 2529–2538.
24 Bongiorno, M. (2010) *Use of power electronic converters for power systems stability improvements,* Chalmers University of Technology, Gothenburg.
25 Glasdam, J., Zeni, L., Gryning, M., Hjerrild, J., Kocewiak, L., Hesselbæk, B., Andersen, K., Sørensen, T., Blanke, M., Sørensen, P.E., Bak C.L., and Kjær, P.C. (2013) HVDC connected offshore wind power plants: Review and outlook of current research, in *12th Wind Integration Workshop,* London.
26 Yazdani A. and Iravani R. (2010) *Voltage-Sourced Converters in Power Systems: Modeling, control and applications,* John Wiley & Sons.
27 Zeni, L., Hesselbæk, B., Sørensen, P.E., Hansen, A.D., and Kjær, P.C. (2015) Control of VSC-HVDC in offshore AC islands with wind power plants: Comparison of two alternatives, in *PowerTech 2015,* Eindhoven.
28 Chaudary, S.K. (2011) Control and protection of wind power plants with vsc-hvdc connection, Aalborg University, Aalborg, 2011.
29 Zhang, L., Harnefors L., and Nee, H.-P. (2010) Power-synchronization control of grid-connected voltage-source converters, *IEEE Transactions on Power Systems,* **25** (2), 809–821.
30 IEC (2014) Standard 61400-27-1 – Electrical simulation models – wind turbines (FINAL DRAFT), IEC.
31 Zeni, L. (2015) Power system integration of VSC-HVDC connected wind power plants (2015) PhD thesis, Technical University of Denmark, Roskilde.
32 Zeni, L. Margaris, I., Hansen, A., Sørensen, P., and Kjær, P. (2012) Generic models of wind turbine generators for advanced applications in a VSC-based offshore HVDC network, in *10th IET Conference on AC/DC Transmission,* Birmingham.
33 Zeni, L., Goumalatsos, S., Eriksson, R., Altin, M., Sørensen, P., Hansen, A., Kjær, P., and Hesselbæk, B. (2015) Power oscillation damping from VSC-HVDC connected offshore wind power plants, *Accepted for publication in IEEE Transactions on Power Delivery.*
34 Adamczyk, A., Altin, M., Göksu, Ö., Teodorescu, R., and Iov, F. (2012) Generic 12-bus test system for wind power integration studies, in *EPE Joint Wind Energy and T&D Chapters Seminar, Aalborg.*
35 Lee D. and al., IEEE recommended practice for excitation system models for power system stability studies, IEEE, 1992.
36 Beza M. and Bongiorno, M. (2015) An adaptive power oscillation damping controller by STATCOM With energy storage, *IEEE Transactions on Power Systems,* **30** (1), 484–493.
37 Energinet.dk (2010) Technical regulation 3.2.5 for wind power plants with a power output greater than 11 kW, Energinet.dk, Fredericia.
38 Tarnowski, G.C. (2011) Coordinated frequency control of wind turbines in power systems with high wind power penetration, PhD thesis, Technical University of Denmark, Lyngby.
39 Zeni, L., Rudolph, A., Munster-Swendsen, J., Margaris, I., Hansen, A.D., and Sørensen, P. (2012) Virtual inertia for variable speed wind turbines, *Wind Energy,* **16** (8), 1225–1239.

Index

Modeling and Modern Control of Wind Power, First Edition. Edited by Qiuwei Wu and Yuanzhang Sun.

Companion website: www.wiley.com/go/wu/modeling

h

i

l

m

r

s

t